A careful and intensive study of historical sources and a review of the instrumental data of this century have led to this detailed catalogue of earthquakes. Egypt, Arabia, the Red Sea region and the surrounding areas of Libya, Sudan and Ethiopia are studied from the earliest times to the present day. Each earthquake is described as fully as possible from the available data, and is analysed in a geographical and historical context. The completeness of the earthquake catalogue over time is analysed and the range of sources and problems associated with the scrutiny of historical sources are discussed. The information is then placed in a geophysical framework.

The seismicity of Egypt, Arabia
and the Red Sea:
a historical review

The Seismicity of Egypt, Arabia and the Red Sea

a historical review

N. N. AMBRASEYS

Department of Civil Engineering, Imperial College, London

C. P. MELVILLE

Faculty of Oriental Studies and Fellow of Pembroke College, Cambridge

and

R. D. ADAMS

International Seismological Centre, Newbury and PRIS, University of Reading

KING ABDULAZIZ
CITY FOR SCIENCE & TECHNOLOGY

CAMBRIDGE
UNIVERSITY PRESS

CAMBRIDGE UNIVERSITY PRESS
Cambridge, New York, Melbourne, Madrid, Cape Town, Singapore, São Paulo

Cambridge University Press
The Edinburgh Building, Cambridge CB2 2RU, UK

Published in the United States of America by Cambridge University Press, New York

www.cambridge.org
Information on this title: www.cambridge.org/9780521391207

First published 1994
This digitally printed first paperback version 2005

A catalogue record for this publication is available from the British Library

Library of Congress Cataloguing in Publication data

Ambraseys, N. N. (Nicholas Nicholas), 1929–
    The seismicity of Egypt, Arabia, and the Red Sea : a historical
review / N. N. Ambraseys, C. P. Melville, and R. D. Adams.
        p.   cm.
    Includes bibliographical references and index.
    ISBN 0-521-39120-2
    1. Earthquakes – Middle East – History – Chronology.   2. Earthquakes –
Egypt – History – Chronology.   I. Melville, C. P. (Charles Peter),
1951–   . II. Adams, R. D. (Robin Dartrey), 1930–   . III. Title.
QE537.2.M628A43   1995
551.2′2′0956–dc20   94-265 CIP

ISBN-13 978-0-521-39120-7 hardback
ISBN-10 0-521-39120-2 hardback

ISBN-13 978-0-521-02025-1 paperback
ISBN-10 0-521-02025-5 paperback

To Xeni, Alison and Thelma,
for putting up with all this

Contents

Preface

Collecting earthquake records can be like collecting stamps, or butterflies, or rare books – a harmless pastime, or an obsession. Different interests have contributed to producing this book. First, the scientific urge to accumulate, to map and to quantify the reported effects of past earthquakes. In addition, the fascination of the texts themselves, not only for what they record, but why and how they did so. Then there is the true collector's obsession for the most accurate, up to date and comprehensive catalogue of the trophies jealously acquired from years of bibliographical detective work and weeks of numerical calculation.

One problem with the collector's mentality is that the work of acquisition never stops. The greatest earthquake hoarder of the last century, Alexis Perrey (d. 1882), got round this by publishing regional and global catalogues, constantly updated by supplementary lists. This resulted in an unwieldy output spanning four decades. Modern publishing conditions, as well as the objectives of a modern earthquake catalogue, make this solution impracticable, and the present work represents both the culmination and the temporary interruption of a long-term process.

The immediate origins of the book lie in a research project carried out by the Civil Engineering Department at Imperial College, London, for the Saudi Arabian National Centre for Science and Technology (SANCST), at the King Abdulaziz City for Science and Technology in Riyadh (KACST), between 1985 and 1988. The report arising from this research project on 'The Seismicity of Saudi Arabia and Adjacent Areas' was not widely circulated. We are grateful to KACST for permission to publish this revised version, and to Cambridge University Press for agreeing to do so – doubtless encouraged and surprised by the fact that an earlier work, on earthquakes in Persia (Iran), published in 1982, is now out of print.

Since one conclusion of the report was that Saudi Arabia itself (with the exception of the Hejaz) is almost totally aseismic, the concentration of the book tends to focus on the wider context of the 'adjacent areas' and particularly, in view of work already done on Iran, those areas to the west of the Arabian Peninsula.

Although Egypt, too, is an area of relatively low seismicity, it has experienced damaging local shocks, as well as the effects of larger earthquakes in the Hellenic Arc and eastern Mediterranean. It has also been affected by earthquakes in southern Palestine and the northern Red Sea. The Red Sea itself is an active plate boundary and at its southern end, long-term moderate seismicity associated with volcanism has been observed in the Yemen, as well as in Ethiopia. The complexities of the geology of the Red Sea make a good knowledge of regional seismicity useful for an understanding of the tectonic processes at work there. Furthermore, both Egypt and the Yemen have long and well-documented histories, a prerequisite for undertaking macroseismic studies. Finally, the human geography and distribution of population in the area – particularly its concentration in a narrow band along the river Nile – create challenging problems in identifying and assessing the origin and effects of Egyptian earthquakes.

The cut-off dates for the catalogue in the original report were 1983 for macroseimic data and 1987 for instrumental data. These cut-off dates have been essentially retained in the present work, although the instrumental catalogue in particular has been selectively updated to the end of 1992. In contrast with Iran, where several important earthquakes have occurred since our work was published, few major events have affected our area of interest since the Yemen earthquake of December 1982. Of these, the earthquake of 20 May 1990 in southern Sudan was one of the largest known in Africa, but its location in an area of conflict seriously inhibited the gathering of macroseismic information. For the sake of completeness, this earthquake and its largest associated events have been included in the instrumental catalogue. The Egyptian earthquake of

October 1992 occurred when this book was in the final stages of preparation, and has also been included in the catalogue. Occurring within 10 km of Old Cairo, it caused substantial damage to ageing buildings and vulnerable historical monuments, and provides a useful yardstick for the analysis and interpretation of earlier, less well-known events.

Although our actual cut-off date is thus largely determined by the publishing schedule, we consider this to be an appropriate moment to present the results of our work so far. The likely return from pursuing our researches further does not at the moment justify a longer delay in publishing the catalogue. A similar expectation in the case of Iran turned out not to be accurate, for quite a few additional data have since been retrieved from sources not used by us at that time; but this is largely a function of the higher level of seismicity of Iran. Although there are substantial gaps in the macroseismic information presented here, there is little likelihood of filling them in the foreseeable future, without disproportionately laborious and systematic researches that we are not now in a position to carry out.

A second problem with the collector's instinct is that there is a tendency to hoard but not to discard. Most previous earthquake catalogues are accumulative, in the sense that data from one are absorbed by the next, errors and all, and 'new' events are added. However, the new events often turn out to be the old events appearing under a new date. The single most common failing in several generations of earthquake catalogues, for the Middle East as elsewhere, is the problem of multiple entries for the same event. The only catalogues to avoid these problems are those that refer back to the original sources of information, rather than relying on secondary evidence and a slavish repetition of previous lists.

A considerable amount could be said on this subject, and has been said elsewhere; but despite recent advances in the field of historical seismicity, to the extent that a European working group now meets regularly, it is worth reiterating the basic concept that underlies our work and the way it has been presented. Above all, we emphasise the need for critical analysis of historical sources. This involves identifying primary historical and instrumental evidence, distinguishing false reports and resolving conflicting information, particularly over dates and locations. This in turn has determined the layout and presentation of the main catalogue. Unlike our book on Iran, in which the notes were kept at the end, we have thought it preferable to present all the textual evidence and critical analyses together, to demonstrate the intimate connection between source criticism and the

interpretations reached about individual earthquakes.

In the course of this, we aim to demonstrate why certain events should be removed from existing catalogues. It is not enough merely to remove (or ignore) them without comment, since later compilers simply reinstate them. It has to be shown conclusively why false earthquakes are false. The problem is usually chronological, but may be locational (Tripoli in Syria mistaken for Tripoli in Libya, or degrees West taken as degrees East, for example); or the 'earthquake' might actually have been a meteorite impact, a destructive flood, or a landslide. As a general principle, no earthquake that has not been confirmed or verified in a primary source of information has been included in our list.

This does not mean that we believe our catalogue to be perfect, with all sources of error or uncertainty removed. Nor would we wish it to be copied straight into computer databanks, any more than those of our predecessors. On the contrary, we emphasise that different interpretations of the evidence are possible, and that ours are not definitive. Rather, we aim to present the data available to us and the reasons for our own conclusions as transparently as possible, so that the reader or user can form his own opinion. Conclusions are often reached on very flimsy or uncertain evidence; indeed, sometimes it might be better if no conclusion were made, as this gives a misleading impression of exactitude.

These points are examined more fully in Chapter 1, which discusses the historical geography of the region, the sources of information available, and the methods of analysis of macroseismic data. In Chapter 2, a descriptive catalogue of all the earthquakes retrieved is followed by a table summarising our conclusions. We also present a section on earthquakes that have been reported but which it has not been possible to identify, and a summary list of false events.

A similarly critical approach has been adopted with regards to instrumental data, which are covered in Chapter 3. We have collected and reassessed all available reports, and where possible redetermined magnitudes from intensity reports and instrumental readings. One important aspect of this reevaluation of twentieth-century seismicity is the reconciliation between instrumental data and the macroseismic information presented in Chapter 2. Not only do reliable descriptions of the effects of an earthquake help determine its general location and size, thus minimising the chances for gross error, but they also permit correlations to be defined between various parameters of twentieth-century earthquakes, such as felt area, magnitude and depth. These in turn can be used to quantify our assessments of events that occurred in the pre-instrumental period.

The work ends with a discussion of the completeness of the earthquake catalogue and the long-term seismicity of the region.

Comparisons with our book on Iran are perhaps inevitable, but our objectives here are less ambitious and this is not conceived as a comparable work. In the first place, the level of seismicity in the region studied is much lower, which imposes itself on the character of the data at our disposal. In particular, in the absence of major earthquakes and because of other constraints, we undertook almost no field trips to the scene of recent and early twentieth-century events.[1] Such visits could doubt-

less have thrown light on historical events, such as the Yemen earthquake of 11 September 1154. The research out of which this book grew was undertaken with the aim solely to provide a *catalogue raisonné* of earthquakes in the region, and this work was done in a relatively short time. The speed of the work was, however, facilitated in practice by experience gained in Iran and the random accumulation of information over a much longer period. Many points already addressed in our earlier book continue to be applicable and are not repeated. In addition, no attempt has been made to investigate regional tectonic processes or to evaluate seismic risk. We aim rather to produce a detailed source book of earthquakes affecting the area, for others to use for their own purposes, in the hope that they will be fully aware of the completeness and reliability of the information presented.

N.N. Ambraseys
C.P. Melville
R.D. Adams

[1] In January 1982, one of us (R.D.A.) undertook a UNESCO-sponsored mission to investigate the effects of the Aswan earthquake of 14 November 1981. Instrumental analysis of the main shock and aftershocks was evaluated, and damaged buildings were inspected in the Aswan region, where the intensity reached VI (MSK). A second visit to the area was made in March 1985; ground cracking near the Kalabsha fault was still evident.

A note on transliteration

Much of the research undertaken for this book is based on material written in Arabic or in Arabic script. The spelling of proper names, book titles, place names and other terms has been based on a consistent transliteration from Arabic, according to the system used in the *Cambridge History of Islam* (Cambridge, 1970), though for ease of reading, rigorous transliteration is only found in the bibliography.

Only the Arabic vowels a, i, and u are utilised; long and short vowels are not distinguished: thus u for al-Mukha (or Mokha, Mocha) and Qus. The diphthongs are written *au* and *ai*, though some may be more familiar with *aw* and *ay*: thus Hadramaut (Hadramawt). The signs ' and ' represent the *'ain* and the *hamza* respectively; the latter is used only in its medial position and ignored in the final position: thus Taima for Taima' (Tayma'), San'a for San'a'. The *ta marbuta* is spelled -*a* (not -*ah*), or -*at* in the construct state: thus Jidda (Jeddah). The Arabic definite article is retained in personal names, thus al-Maqrizi, and generally in geographical names, as in al-Hudaida (Hudaydah, Hodeidah, etc.), though the *al-* is normally omitted from the maps. In all cases, well-known places are spelled in the current usage: Alexandria, Cairo, Medina, Mecca, the Yemen (Arabic: al-Yaman), the Hejaz (al-Hijaz).

This is fine for the earliest periods, but in the accounts of more recent events, this strict system can seem unnecessarily pedantic. Places in Cairo, for example, may appear as they do on modern maps. Where the same place is spelled in different ways in the book, alternatives are indicated either in brackets or in the index. For Libya and the Sudan, places are spelled as found on modern maps, or on the Geographical Section of General Staff (GSGS) editions of the War Office and Air Ministry maps of 1960. For Ethiopia we generally follow the forms given by Gouin (1979).

Modern Arab authors writing in English spell their names according to a variety of systems and naturally their names appear in the references and the bibliography as they appear on their work.

No reader will be surprised to find both inconsistencies and eccentricities in our spelling. Those who mind will certainly notice; we hope those who notice do not mind, and that all will excuse any shortcomings in this notoriously tedious exercise.

A note on chronology

Perhaps the simplest way of distinguishing one earthquake from another is its date, and yet the single greatest source of confusion and inaccuracy in existing earthquake catalogues is in the realm of chronology.

Arabic chronicles are the main source of information for the present catalogue, and they generally date events according to the Muslim calendar of twelve lunar months (354 days). The Muslim era originated in AD 622, when the Prophet Muhammad left Mecca for Medina on his *hijra* (Migration). This is sometimes called the *hijri* calendar; its use is indicated in the present work by the suffix H. In cases where a conversion is provided, the Muslim year comes first, followed by the Christian equivalent that forms the larger proportion of the Muslim year, e.g. 758/1357. The suffix H is not used when the month of the hijri year is given, since this identifies the calendar concerned.

Three points need to be borne in mind. First, the Muslim day begins at sunset (generally around 6.00 p.m.), so that day follows night. The Muslim 'Monday evening' is equivalent to the preceding Sunday evening according to our reckoning.

A second potential source of dislocation between the two calendars is that, in the past at least, the month began with the sighting of the new moon, particularly Ramadan, the month of fasting. This may account for some local discrepancies over dates. Furthermore, this source of discrepancy is enshrined by the formal existence of two separate hijri calendars, namely the civil or popular reckoning, starting from Friday 16 July 622, and the astronomical reckoning based on the true conjunction of the new moon on Thursday 15 July. It may not be clear which is being used in a given text. Here, for the sake of consistency, all conversions are performed from the tables in Freeman-Grenville (1963), which takes 16 July 622 as the start of the Muslim era.

Thirdly, the Muslim day is divided by the five times for prayer: between dawn and daybreak (*fajr*); shortly after midday (*zuhr*); afternoon (*'asr*); between sunset and dusk (*maghrib*) and night-time (*'isha*). These times are often used as reference points in the dating of earthquake shocks and other events.

The term *daraja* ('degree') is used for the measurement of the passage of time. Since 24 hours is equivalent to 360°, 1° is equivalent to 4 minutes. Although this may be an accurate definition of *daraja*, it sometimes gives rise to a very long duration of shaking, which cannot always be due to the exaggeration of the observer. In the account of the earthquake of 936/12 November 1529, the shock is said to have occurred 10 *daraja* before dawn, while the muezzin was preparing to give the call to the dawn prayers. Forty minutes seems an inordinately long time for an experienced muezzin to be getting ready. Quatremère (1845, II/2, pp. 216–7), using several examples, says it means a short time, or a minute; at the other extreme, Nejjar (1974, p. 85), says it means five minutes. In view of this uncertainty, we generally leave the term *daraja* untranslated.

Another way of measuring the passage of time is by how long it takes to recite certain verses of the *Qur'an* (Koran). We have not attempted an empirical equivalent to such estimates, which are left as they are found. The whole interesting question of the perception and reporting of time in the mediaeval Islamic world would benefit from the type of analysis undertaken, in a parallel context, by Ferrari and Marmo (1985). It is worth considering the vagueness of time measurement, as well as the large areas sometimes involved, when one reads statements such as 'the shock occurred in all these places at the same time' – which may simply be a convenient judgement by the chronicler. In the case of the large North Arabian earthquake of 18 March 1068, for example, it is unlikely to be true, and may serve to disguise the occurrence of more than one shock.

The other calendar frequently referred to is the Christian Coptic calendar in Egypt and its Ethiopian equivalent. Its use in records of Ethiopian earthquakes is explained by Gouin (1979, pp. 19–20). A detailed discussion and concordance of this calendar is given by Chaine (1925); Pellat (1986) provides a recent and convenient publication of some mediaeval Egyptian examples.

Various other calendars in use in the Middle East, such as the Syriac, which may be referred to occasionally in our sources, are treated by Grumel (1958). The latter notes that historical events can be dated by reference to earthquakes, comets and eclipses, and gives lists of each (pp. 458–81). After 1500, references to eclipses have been checked in the tables of Th. Oppolzer (English translation, 1962).

The Christian calendar was reformed by Pope Gregory XIII in 1582, although the unreformed Julian calendar remained in use in England until 1752. In the present catalogue, earthquakes are dated according to the Julian (Old Style) calendar up to 1582, and thereafter according to the Gregorian calendar. Dates between 1582 and 1752 are given the suffix NS (New Style).

Acknowledgements

Research for this book has been carried on intermittently for many years, during which time we have received assistance and information from a large number of people, too many to name. We would particularly like to acknowledge the help of the following: Professor Bob Serjeant (who died shortly before the book went to press) and Professor Rex Smith for their ready response to questions about the Yemen and its historical topography; Abd al-Malik Eagle for providing copies of manuscripts of Yemeni chronicles and many useful references; Dr Robert Wilson for valuable personal communications on Yemen earthquakes; Bill Tucker, of Arkansas, for unselfishly sharing his historical earthquake data; and Dr P. Pattenden (Peterhouse, Cambridge) and Professor Michael Reeve (Pembroke, Cambridge) for some translations. We have benefited in many various ways from discussions with James Jackson (Queen's, Cambridge). Caroline Finkel (Imperial College) and Boyd Johnson (University Library, Chicago) assisted with data acquisition, and we also thank Abdullatif al-Maneefi (Imperial College) for his researches on Yemeni earthquakes; Jean Vogt (Strasbourg) for his tantalising sorties into European archives; Mr P. Pantelopoulos (Ministry of Public Works, Athens) and Dr I. Karcz (Geological Survey of Israel). We would also like to acknowledge the assistance of D.M. McGregor, Mrs D.E. Robertson, Mrs A.-R. Surguy, and Mrs C.A. Tubby of the International Seismological Centre, with data preparation and presentation. Professor H. Kopp's prompt and positive response to our request to use one of his photos for the cover illustration is very much appreciated.

Grateful thanks are due to the Saudi Arabian National Centre for Science and Technology (SANCST), for commissioning the report on which this work is based, and to the King Abdulaziz City for Science and Technology (KACST) for their permission to publish it in its revised form. We would also like to acknowledge Dr Adnan Niazy's assistance and advice.

It is a shame, in a sense, that there are not more earthquakes to record in Saudi Arabia itself, and that the focus of our attention has thus strayed inevitably to the surrounding regions. Nevertheless, one benefit from our research has been to purge the catalogue of some earthquakes previously located in Arabia – it would be a fitting repayment if we could have a similar effect on the earthquakes of the future.

CHAPTER 1

Introduction

1.1   Area of study

Our investigation is concerned with a large, irregular area defined at its greatest extent by the co-ordinates 0°–34° N latitude and 10°–60° E longitude (see Figure 1.1). This area is centred on the Red Sea. To the east, it incorporates the whole of the Arabian Peninsula, bounded to the northeast by southern Iraq and the Persian Gulf and to the south by the top of the Horn of Africa. We have excluded the parts of Ethiopia and Somalia south of 10° N and east of 40° E. To the west, the area includes the whole of Egypt, northern Ethiopia, and most of Sudan and Libya. Central Africa south of 20° N and west of 25° E is excluded. To the north, the land boundary is fixed at the top of the Dead Sea (32° N).

The main focus of our attention is on the regions bordering the Red Sea, namely the Hejaz and Yemen to the east, Egypt and Ethiopia to the west. This is a crucial area in the history of the Middle East, and also one of contrasts in terms of population, topography and seismotectonic activity. We have extended our survey beyond this central region to ensure that its seismicity is located in a wider context. It is useful to know the full extent of areas of both high and relatively low seismicity, and to try to include their boundaries within an artificially delineated study area. Similarly, earthquakes do not respect the political borders shown on Figure 1.1; besides, such lines are of recent origin, and sometimes equally unnatural. They are also subject to change. In terms of seismic hazard, it is obvious that large events at a distance can be important. Finally, the wider regional context allows us to identify earthquakes that should be moved into (or taken out of) our area of interest, on the grounds that they have been mislocated either in previous historical catalogues or in instrumental listings.

One disadvantage of attempting to cover such an area (larger than the United States) is the difficulty of achieving a homogeneous catalogue. Fortunately, parts of the region have already been the subject of modern investigations using primary sources.[1] Our own earlier work has been expanded where possible, although further research has been confined to the area of central interest. Our reliance on the work of Gouin (1979) is very apparent for Ethiopian earthquakes. There is no difficulty in verifying and adopting the information he presents, but the fact that we have undertaken no work of our own in this region means that we are not in a position to judge its completeness. To a lesser extent, the same applies to the other regions peripheral to Egypt, for which we do not claim any specialist expertise: the search for historical data for Libya and the Sudan has been neither intensive nor systematic. For the twentieth-century instrumental record, however, homogeneity has been achieved by a reevaluation of all the instrumental data available, up to the end of 1987.

Even boundaries as widely drawn as those on Figure 1.1 are artificial, and we have been flexible over events occurring at the margins of our area. In view of our previous work on Iran, macroseismic data on events in that region are omitted, with the exception of a few that have since come to light, which are included for the

---

[1] Namely, Gouin (1979) for Ethiopia; Ambraseys (1984) for Libya; Ambraseys and Adams (1986a) for the Sudan; Ambraseys and Melville (1982, 1983) for Iran and the Yemen.

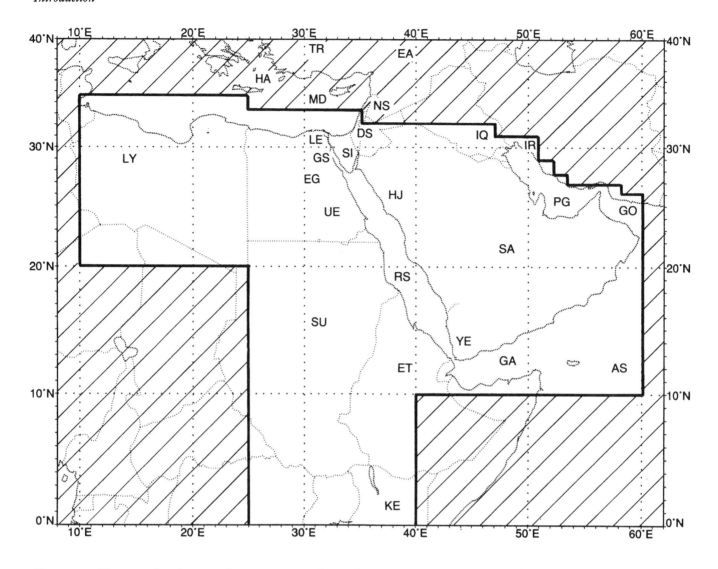

Figure 1.1. The area of study area at its greatest extent. In practice, some events occurring outside this area (particularly to the north) are included if they were reported felt inside these boundaries. It should also be noted that the coverage of events in the twentieth century is not uniform throughout the whole of this area (compare Figure 4.5). The letters denote the location codes assigned to the earthquakes in the catalogues (see Tables 2.1 and 3.1).

record (e.g. 1426 November, Qishm; 1786 July 28, Basra; 1858 June 13, Bushire). For the twentieth-century catalogue, only Iranian earthquakes greater than magnitude 5 occurring in the immediate vicinity of the Persian Gulf are retained. In addition, since many of the earthquakes that have affected Egypt occurred in the Hellenic Arc or elsewhere in the eastern Mediterranean, it has been necessary to extend our research some way to the north, though again without claiming to be comprehensive.

In studying this region from a long-term, historical perspective, it is usually not appropriate to refer to modern nation states. Rather loose terms are used to describe broad geopolitical entities. The concept and definition of the 'Middle East' is itself a favourite bone

for geographers to chew on. In deference to the dominant culture in the region, we use the terms Islamic, or Muslim, or Arab world more or less interchangeably, while recognising that it has always contained large Christian and Jewish populations, and many races. The territory lying within Israel and Jordan is generally referred to as Palestine or the Holy Land, while further north, the terms Syria and Anatolia are used in a broad geographical sense.[2] Our definition of Libya also includes a small area within the border of modern Tunisia. Ethiopia, though physically central to the region, is in many ways a distinctive element.

[2] Useful surveys of the historical and physical geography of the whole region are provided by Fisher (1978) and Wagstaff (1985).

Regional seismicity and seismotectonics

Our area is one of only moderate seismicity compared with the more active plate boundaries to the north, where the Aegean (Hellenic) Arc, the Anatolian fault systems and the collision structure along the Zagros mountains mark the boundary of the Eurasian plate with the African and Arabian plates.[3] Figure 1.2 depicts all events in standard catalogues with magnitude 5 or greater in this wider context, for the period 1964–92. As might be expected, it is clear that the most active parts of our area are the rift systems along the Red Sea

and the Gulf of Aden. Seismicity also occurs where these systems continue northwards along the Gulf of Suez and the Jordanian–Dead Sea Rift, and southwards into the African Rift system at the Afar triple junction near Djibouti.

The Arabian shield is one of the most seismically stable areas in the world, with its only activity in Yemen in the southwest corner. This activity is associated with volcanism which extends up the western side of Arabia. As recently as 1256, an extensive eruption occurred close to the city of Medina, the lava from which still abuts the city's airport.[4] We have noted this, and the reports of other eruptions, as we found them in passing,

[3] See for example, Barazangi (1983), Adams and Barazangi (1984), Jackson and McKenzie (1984).

[4] Camp *et al.* (1987).

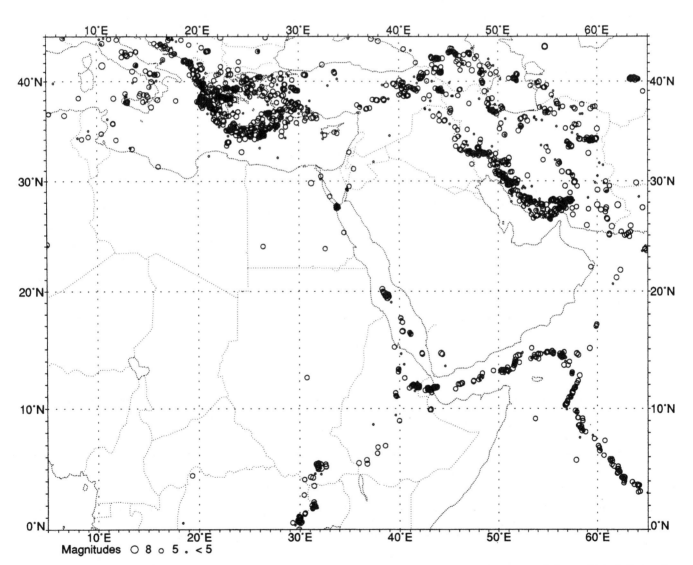

Figure 1.2. Background regional seismicity. This plot shows all the events in the files of the International Seismological Centre for the period 1964–92 of magnitude greater than about 5. Note the large number of earthquakes to the north of our area.

though we have made no deliberate effort to search for such information.[5] Although most of the Arabian Peninsula is shown as aseismic on Figure 1.2, we are aware that local microearthquake surveys have detected small events, in particular near Jidda and elsewhere on the coastal plain.[6] The main scarp, where the western edge of the Arabian plateau rises to about 2000 m, does not appear to be fault controlled, nor associated with any particular seismicity.

In contrast, the stable inland areas of Libya, Egypt and Sudan are not totally aseismic, but are subject to infrequent earthquakes which can be of significant size. An example is the well-recorded event of magnitude 5.0, which occurred on 9 December 1978 about 650 km west of the Nile, in a region previously considered to be stable. The lower part of the Nile Valley is subject to numerous small earthquakes, but the larger events, of magnitude 5 and above, are concentrated along the Gulf of Suez. The central part of Sudan, although generally stable, has also experienced isolated large events, such as the earthquake of 9 October 1966, southwest of Khartoum. The southern province of Sudan, Equatoria, is traversed by a band of earthquakes connecting the main Ethiopian Rift to the West African Rift system. It was here that one of the largest earthquakes known in Africa, of magnitude 7.2, occurred on 20 May 1990. The southeastern part of Ethiopia and Somalia are excluded from our study area, but appear to be relatively stable.

HISTORICAL SEISMICITY   Figure 1.2 and other maps of instrumental epicentres since the 1950s not only reveal very large areas of low seismicity in Arabia and the lands west of the Red Sea; more surprisingly, southern Palestine and the Dead Sea region also appear blank. Earthquake maps based on pre-instrumental data (Figure 1.3) suggest that the recent instrumental record is not entirely representative of the longer-term seismicity of the area, even though the macroseismic maps are based on felt effects rather than the location of epicentres.

It is the question of long-term seismicity that our catalogue addresses. We have attempted to discover references to earthquakes from earliest times, though sources of information are scarce before the eighth century. Macroseismic (that is, descriptive or non-instrumental) data are often interchangeably referred to as historical

data, in as much as they relate to earthquakes that have occurred in the past and are reported in historical sources. However, macroseismic data can also assist the analysis of modern events, and supplement the use of instruments to measure and define an earthquake. In particular, the presence or absence of felt reports can confirm, or cast doubts upon, instrumental locations, especially in regions where recording stations are few or unevenly distributed. Instrumental seismology, too, has a historical period; pre-1963 seismograms have recently been described as 'historical', since after that date they are more uniform as well as more accessible.[7]

The documentary sources available for the reevaluation of the seismicity of the area are reviewed in the next section, followed by a discussion of the problems of assessing the data that have been retrieved. Such topics cannot be discussed without reference to some of the geographical and historical factors that have influenced the reporting of earthquakes in the area.

1.2   Sources of macroseismic information

The objective of the present study is to provide a uniform account of the seismicity of the region, based on the retrieval and assessment of original sources of information. Our area, thanks to its position at the intersection of ancient and important continental and maritime routes, has a long and relatively well-documented history preserved in a variety of sources. For the period before around 1900, these chiefly comprise mediaeval Arabic chronicles and, later, European travel literature and technical studies. These are supplemented by Byzantine, Syriac, Amharic and other non-Arabic sources of Middle Eastern history, and later by official diplomatic correspondence and newspaper reports. For the period after 1900, such sources of macroseismic data are supplemented by the study of original seismographic station bulletins and instrumental records (see Chapter 3).[8]

The reporting of earthquakes is determined partly by the nature of the events themselves, and partly by the context in which they occur. Throughout history, earthquakes have been recorded by the societies that suffered them, whether in legends, inscriptions, chronicles or

---

[5] These data can be used to supplement the lists found in standard catalogues, such as Simkin *et al.* (1981).
[6] See e.g. Merghelani and Gallanthine (1980).

[7] Lee, Meyers and Shimazaki (1988). Sources of instrumental data are discussed in Chapter 3.
[8] A blueprint for the work of retrieval and evaluation of macroseismic data for the whole Arab region can be found in our contribution to the UNESCO volume edited by Cidlinský and Rouhban (1983), chapter 2, on the sources of information available.

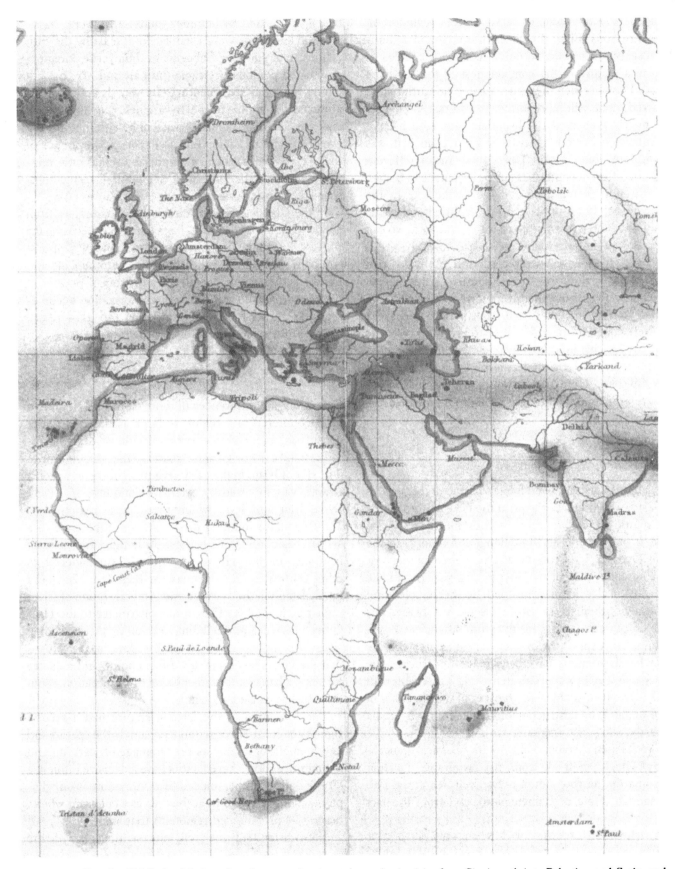

Figure 1.3. Detail of Mallet's global earthquake map, showing enhanced seismicity from Sinai north into Palestine and Syria, and fainter activity along the Red Sea.

newspapers. Knowledge of these sources is important for two reasons. First, it enables thorough and systematic research to be carried out. Secondly, it enables us to form a picture of the completeness of the surviving record. However rich the sources are, if no earthquakes occurred, none will be reported. In contrast, no matter how poorly preserved the historical record, if a large earthquake occurred, there is a good chance it was reported in some form. 'Large' may mean different things in different contexts: a large earthquake in Britain is a different matter from a large earthquake in the Middle East. If we understand the cultural and social milieu in which written sources were produced, we can estimate how large an earthquake had to be before it could not fail to have been recorded. It is obvious that many small earthquakes are missing from the record, particularly the further back one goes in time, but we should be able to judge from what period we have a relatively complete record of the larger events.

We will return in a later chapter (see Section 4.1) to the question of the completeness of the catalogue in time and space. Here, we review briefly the main sources that are available for different periods and areas in the region of our study.

The pre-Islamic period

The Middle East has been home to some of the world's earliest civilisations. In Egypt, at the crossroads of three continents, dynastic powers based on the river Nile go back, if one accepts the traditional chronology, to the First Dynasty (*c.* 3100 BC). The chief sources for the Pharaonic period are inscriptions, papyri and the archaeological evidence provided by the temples and monuments themselves. So far as we are aware, this large corpus of material for the study of ancient Egypt contains no explicit reports of earthquakes.[9] On the other hand, attempts have been made to associate the evidence of geological events in the Old Testament with particular dates, such as an eruption of Mount Sinai in 1606 BC, and the destruction of various tribes in Arabia by earthquakes, not long afterwards.[10]

The Yemen (Arabia Felix to the classical geographers), has benefited from its favourable location guarding the narrow mouth of the Red Sea at the Bab al-Mandab (Gate of Lamentation), and from the fact that it is touched by summer monsoon rainfall. Monarchical kingdoms were established at least by the fifth century BC, and the Sabaean or Himyarite kingdom dominated peninsular Arabia until around AD 300, not falling until the sixth century. Its wealth and position attracted the rivalry of the Byzantines, the Abyssinians and the Sasanian Persians, who only withdrew on the eve of the rise of Islam. The archaeological evidence of these early civilisations in southern Arabia, only relatively recently the object of scientific investigations, seems also to include no details about earthquakes.[11]

In the Classical period, Greek historiographers from about 500 BC and Latin from 200 BC, occasionally provide information about events in Libya, Egypt and southern Palestine, reflecting the commercial and imperial interests of the Greek and Roman powers in the eastern Mediterranean. More information becomes available during the Byzantine period (AD 450–1453), mainly from ecclesiastical histories.

Islamic period (from AD 622)

Despite all the differences of topography and climate to be encountered in a region so large as the area under consideration, it enjoys a particular cultural unity, forming as it does the original heartland of the Muslim world. The rise of Islam in the early seventh century, in western Arabia, was followed by extensive conquests in Egypt, Syria, Iraq and Iran, which fused these regions into a great empire founded on a common adherence to a set of basic religious beliefs. Although the political centrality of the Hejaz was quickly surrendered to other centres, such as Damascus, Baghdad, Cairo and finally Istanbul, the spiritual and symbolic home of Islam remained in Arabia. The annual pilgrimage to the Holy Cities of Mecca and Medina, as well as the individual visits of countless travellers and traders, have ensured that the Hejaz has remained a focal point, and a centre for the exchange and transmission of information about the Muslim world as a whole.

The formation of the Islamic empire had a second significant cultural consequence, namely the spread and development of Arabic as the language of learning and of literature. Both these factors have a decisive influence on the nature of the sources of information at our disposal, and allow us to treat the area as a coherent whole, despite its pronounced regional variations.

Notwithstanding the religious significance of the

---

[9] Generalised literary references do, however, occur, notably in the Pyramid Texts, see e.g. Faulkner (1969), pp. 184, 187, 309. For the sources of ancient history, see Drioton and Vandier (1962).

[10] See e.g. Mallet (1853), p. 1; some similar episodes are reported in the Qur'an (Koran). See also Bentor (1989).

[11] For the early history of South Arabia, see Shahid (1970); art. 'Marib', *E.I.*², **VI**, 559–67.

Hejaz and its relatively well-documented history throughout the period,[12] and a relatively rich tradition in Hadramaut,[13] it is Egypt and the Yemen that provide the richest historical sources to draw upon. These two countries have been the most populous, culturally flourishing and politically important regions within the area of interest – as well as the most seismically active – and they furnish the bulk of the macroseismic data we have retrieved. In both cases, the nature and type of the documentary sources in which their history is preserved are essentially the same.

ARABIC SOURCES From an early date, the Muslims took a serious interest in history, not least because of the need to document the origins and experiences of the early Islamic community, which became the basis for later developments. To a great extent, Muslim historiography arose out of biographical literature. In common with the Byzantine writers who served in this respect as models, Muslim annalists generally preserved a strict chronological framework in their histories, and paid considerable attention to 'natural phenomena' – earthquakes, plagues, floods, hailstorms, comets, eclipses, meteorite falls, monstrous births, famines and epizootics.[14] A similarly morbid or nervous preoccupation with celestial and terrestrial 'signs' is common to the annalists of mediaeval Europe.[15] Devastating earthquakes were one of the portents of the Day of Resurrection,[16] and allusions to the terrors of judgement day are common in the accounts of earthquake scares and false predictions.[17] This religious and superstitious dimension to the perception of earthquakes in mediaeval Islamic society encourages us to believe that even minor shocks will be reported quite systematically, when circumstances are favourable. Indeed, the wealth of data in the Arabic chronicles has led a recent researcher to the rather over-enthusiastic conclusion that 'Arab and Islamic historians have taken care of the earthquake problem'.[18]

Arabic chronicles are the primary source for the history of the Middle East, and certainly for its earthquake history, from the ninth century at least until the end of the seventeenth century.[19] They continue to provide useful information even in the nineteenth century, especially in the Yemen, long after this genre of historical writing ceased in Europe.

There is a distinction to be made between universal and local or dynastic histories, and it is necessary to know not only when, but also where, chronicles were being written. In the first centuries of Islam, up to around AD 1000, universal chronicles, generally written in the capital, Baghdad, cover events throughout the empire, though inevitably concentrating on Mesopotamia and Iraq. The political break-up of the Abbasid caliphate in Iraq by the end of the tenth century was accompanied by a gradual resurgence of the Red Sea trade at the expense of the Persian Gulf, and a readjustment of the relations between the component parts of the empire. Regional aspirations were reflected in the production of more local chronicles. Many of those known to have been produced in Egypt in the Fatimid period (969–1171) are now lost and the same is true of the Yemen, where there is the additional problem that numerous works remain unedited. However, the repetitive character of Muslim historiography, and the extent to which chroniclers borrowed verbatim text from earlier authorities, to some degree ensures the survival of information from early sources that have since been lost. Indeed, for the Yemen, many valuable and clearly authentic reports of early earthquakes are preserved in the annals written by Yasin al-'Umari of Mosul (d. 1811).[20]

This in turn makes a clear distinction between primary and secondary sources, on the basis of their date of composition, rather difficult. The same chronicle can be a primary source for the period contemporary with the author, a secondary source for earlier material derived from previous works, but also a primary source when these earlier works have not survived. Even when we are dealing with the account of a contemporary or near-contemporary author, we cannot always judge his

---

[12] See e.g. the collection of Meccan histories edited by Wüstenfeld (1857–61).

[13] Serjeant (1950, 1962); extant chronicles, preserving some earlier material, date back to the fifteenth century.

[14] Tucker (1981) has drawn attention to the wealth and importance of these data; his own collection remains unpublished.

[15] See e.g. 904/1498, 919/1513, 1077/1667 and 1085/1674 in the catalogue.

[16] Qur'an, *sura* 99, 'the earthquake'.

[17] In our catalogue, the earthquakes of 702/1303, 836/1433 and 886/1481 are associated with the Resurrection. Earthquake scares and false predictions are reported on 8 Rabi' II, 928/7 March 1522, by Ibn Iyas, **V**, 440–1/trans. Wiet (1960), pp. 423–4, who refers to a similar episode in the reign of al-Ashraf Qayitbey. He probably intends al-Ashraf Barsbay, and the incident reported on 8 Shawwal 841/4 April 1438, when it is was predicted that the Resurrection was at hand, see Ibn Taghribirdi, ed. Cairo, **XV**, 98–9/trans. Popper (1958), pp. 149–50. A similar incident occurred as late as Jumada I, 1205/February 1791, al-Jabarti, **IV**, 132/trans. **V** (1890), p. 89.

[18] Al Hakeem (1988), p. 19.

[19] For a useful discussion of the characteristics and scope of these narrative sources, see Cahen (1965), esp. pp. 29–35.

[20] Cf. Ambraseys and Melville (1982), and al-Chalabi (1974). The seventeenth-century Yemeni author, Yahya b. al-Husain, also contains much material not identified in earlier sources.

reliability, nor do we usually know what sources he himself had access to. In the case of conflicting accounts of the same earthquake, criticism of the sources has to rely on the merits of the case in question.

In both Egypt and the Yemen, the greatest flourishing of historical literature occurred in the period between the sack of Baghdad by the Mongols (1258) and the Ottoman conquest of the Middle East (1517 onwards). During this period, Egypt and Syria were ruled from Cairo by the Mamluks (1260–1517). In the Yemen, the Rasulid sultans (1229–1454) were based in Zabid and Taʻizz, and the Tahirids (1454–1526) in Aden. Thereafter, though local chronicles continue to be useful, a quantitative and qualitative change can be seen, particularly in the historiography of those areas that became merely far-flung provinces of the Ottoman empire. From the sixteenth century onwards other, non-Arabic, sources of information become increasingly important.[21]

Apart from chronicles, we should mention some other categories of Arabic source material. The connection between Islamic history and biography has already been noted, and historical and anecdotal information may be contained in biographical dictionaries or obituaries. Geography was also a flourishing form of scholarship, particularly between the tenth and thirteenth centuries. The descriptions of the various regions and cities of the Muslim world, and of the routes between them, contain incidental evidence of the occurrence of earthquakes, and at the same time give a picture of the cultural homogeneity of Islamic civilisation and the ease with which news and ideas circulated. Topographical works, such as al-Maqrizi's systematic description of Egypt and Cairo, contain evidence of the effects of earthquakes on particular buildings, which may also be recorded in inscriptions dating repair work.[22] In addition, it goes without saying that any significant and damaging earthquake to affect a society as culturally diverse as mediaeval Egypt could be expected to leave its trace in a variety of other documentary materials outside the main fields of narrative history.[23]

ETHIOPIA Although parts of Ethiopia are central to the area of study, we have undertaken no search for data of our own. Despite its proximity to Arabia, the ties between the Ethiopian church and the Coptic church of Egypt, and its commercial contacts with both, the country is usually discussed in the context of African, rather than Middle Eastern history, and it is frequently excluded from general surveys of the Muslim world. Nevertheless, Islam made considerable progress in Ethiopia at different periods, and thanks to the control that Muslim merchants exercised over Ethiopian trade, particularly in its outlets to the Red Sea, information about the internal affairs of Ethiopia can be preserved in Yemeni as well as indigenous Arabic sources. Indeed, the earliest reference to volcanic activity in Ethiopia, in 1203, is reported in this way. From the thirteenth century onwards, Arabic chronicles can therefore supplement the considerable body of native Amharic historical documents.[24] In the early sixteenth century, Ethiopia became involved in the struggle for control of the trade in the Indian Ocean and the Red Sea between the Ottomans and the Portuguese, one result of which was the arrival of the Jesuits. Until their expulsion in the 1630s, they played an active part in the affairs of Ethiopia.[25] It could be expected that Portuguese archival sources might provide some evidence of earthquake activity at this time. For later periods, European travel narratives provide important information on the social and political history of Ethiopia, as of other countries in the area.

The same applies to the Sudan, for which very few Arabic sources exist before the nineteenth century.[26] The earliest earthquakes in the Sudan identified in existing studies occurred in 1850, and we have not pursued a search for data into earlier periods.

EUROPEAN TRAVEL LITERATURE By the seventeenth century, European sources begin to contribute in a significant way to the information at our disposal about the Middle East, particularly Egypt and to a much lesser extent, the Hejaz and Yemen. The main category of this source material is travel literature, of which a vast corpus exists.[27] We cannot pretend to have read systematically all the travel narratives available, which would be a lifetime's work, but we have found some justification for this neglect in the relatively disappointing results obtained from the material that we have studied. Travel-

---

[21] For Yemeni sources, see al-Hibshi (1972) and Sayyid (1974). Early Mamluk historiography is discussed by Little (1970); for the historical sources of Ottoman Egypt to 1798, see Holt (1968).

[22] Al-Maqrizi (d.1441), *Khitat.* For inscriptions, see *Corpus Inscriptionem Arabicarum*, especially parts I (Egypt) and IV (Arabia), and the *Répertoire chronologique d'épigraphie arabe* (currently to 783/1381).

[23] Cahen (1965) conveniently surveys the range available.

[24] For which, see particularly Huntingford (1989).

[25] See the interesting work of Abir (1980).

[26] See e.g. Holt and Daly (1988), pp. 26–8.

[27] For bibliographies, see Carré (1956) and Volkoff (1973) for Egypt; many accounts of French travellers to Cairo in the seventeenth century have been published by the Institut Français d'Archéologie Orientale. See also Bevis (1973), for the Middle East as a whole.

lers to Egypt, particularly in the wake of the Napoleonic expedition at the end of the eighteenth century, followed a regular itinerary and were mainly concerned to visit Biblical sites and the ancient monuments of the Pharaohs. There is relatively little concern with contemporary events or recent history; furthermore, one has to presume that the generally low seismicity of the Nile Valley militates against earthquakes being a subject of perennial interest, in the way that they were, for instance, in Persia.

As for Arabia, inland penetration by European travellers was relatively recent as well as infrequent, though the sea route through the Bab al-Mandab Strait was more commonly taken following the arrival of the Portuguese on the scene in the early sixteenth century. Again, more sources of this nature exist than have actually been studied, though European travel accounts can be useful for identifying and locating places mentioned in Arabic texts.[28]

DIPLOMATIC CORRESPONDENCE Official political and commercial reports, preserved in archives in Europe and, to a far lesser degree, in the Middle East, can also provide evidence of earthquakes.[29] Reports and documents from these sources have not been used systematically, but rather to look for further information about specific earthquakes, such as the Egyptian earthquakes of the early nineteenth century. Foreign Office files for Tripoli (Libya) for the period 1850–95 were read at the Public Record Office. Yemen files have been read, though not systematically, for the period since 1938. Ottoman archives, in particular, in both Istanbul and Cairo, as well as Venetian collections, remain a largely unexploited resource, and it is likely that the greatest opportunity for future progress in the retrieval of macroseismic data lies in this area.[30]

NEWSPAPERS Newspaper reports of earthquakes provide an important source of information from the nineteenth century onwards. Of the British press, *The Times* is indexed from its origin in 1785, while the French *Gazette de France* was extensively used by the nineteenth-century compiler, Alexis Perrey (see below). Very few and incomplete collections of Arabic and Turkish newspapers exist outside the region, but some press reports have been recovered for specific earthquakes, and a systematic search has been made in Chicago through certain Cairo newspapers: *al-Ahram* (for 1958–74); the *Egyptian Gazette* (1895–1927, and then unsystematically up to 1955); *al-Mu'ayyad* (1900–6); *al-Muktataf* (1887–1927) and *al-I'lam* (1886–8). The Saudi Arabian papers *Saut al-Hijaz* (April 1932–July 1941) and *Umm al-Qura* (Mecca, January 1925–June 1960), have also been read, but they contain no reports of earthquakes. Ottoman newspapers in collections in Istanbul have also provided some information for specific events, and a systematic search would doubtless yield further data.[31]

Previous earthquake catalogues

Most early catalogues of earthquakes, such as Batman (1581), Beuther (1601) and Coronelli (1693), list a few of the most celebrated earthquakes in the Middle East, generally on the authority of Classical or Renaissance authors. There is little intrinsically wrong with these early works except their incompleteness. Landslides and other natural phenomena, however, are sometimes misleadingly listed as earthquakes, and they do not always specify their sources.

The second generation of catalogues appeared in the late eighteenth century, in the wake of the Lisbon disaster of 1755. Authors such as Seyfart (1756) and Berryat (1761) contain details of a few Middle Eastern earthquakes, but are not systematic. Berryat, in particular, who cites no authorities for his list, is not reliable. The first thorough catalogue of global seismicity is that of Hoff (1840), who incorporates most of the data found in earlier works, while adding new material. He gives references for his entries, including a small number of Arabic sources available in translation. It is worth noting briefly what these Arabic sources were, as it illustrates a general point. The first Arabic chronicle to be edited and translated was the work of al-Makin (by Erpenius in 1625); Pococke's edition and translation of Eutychius

---

[28] Pirenne (1958); there is a useful list of early maps in Tibbetts (1978) and a bibliography in United Arab Republic (1963). On the problem of identifying place names, see the note by P.J.M. Geelan in Grimwood-Jones, Hopwood and Pearson (1977), pp. 129–33.

[29] For the distribution of British consulates in the area of interest in the mid-nineteenth century, see the map in Melville (1984a), p. 107.

[30] Cf. Ambraseys (1983), for references to collections of diplomatic correspondence, and Cahen (1965), pp. 193–6 for the Ottoman archives. We are grateful to Boyd Johnson for his archival searches, and for various data supplied by Jean Vogt. Netton (1983) lists the main library holdings of Middle Eastern material in the UK.

[31] We are grateful for the assistance of Caroline Finkel and Boyd Johnson in their searches for data in the Turkish and Arabic press respectively. Ottoman newspapers were not searched systematically for earthquake reports from our area of study, only for shocks in Anatolia and the Bosphorus. For references to the Arabic press, see the chapter by Derek Hopwood in Grimwood-Jones *et al.* (1977), pp. 101–22. A catalogue of Arabic newspapers available in British libraries is given by Auchterlonie and Safadi (1977).

(Sa'id b. Bitriq) and Bar Hebraeus (Abu'l-Faraj) followed in 1658 and 1663 respectively. All three were Christian Arab authors, and their works are primarily compilations relying on earlier authorities. It is precisely in this that their popularity lay, but they have acquired totally unmerited importance as primary sources, simply because of their early availability in translation. The first Muslim author to gain this sort of renown was Abu'l-Fida; John Greaves translated selections from his geographical work in 1650, and a partial edition and translation of his chronicle was published in 1723.[32] We may also note that a translation of a short chronology by Hajji Halife (d. 1657) was published in 1697. The work of translation gathered pace in the 19th century, following Napoleon's Egyptian expedition.

Hoff's use of these and other materials made his catalogue an important source for later ones, particularly Mallet (1853–5). Mallet's great catalogue is again a global work and not specifically concerned with the Middle East. It is, however, distinguished by a genuine scientific purpose, and seeks to remove earthquakes 'from the thrall of superstition and tyranny of the Middle Ages'.[33] Despite a critical approach to its sources, it preserves numerous errors which, because of the authority Mallet's catalogue acquired, were passed on to subsequent generations.

Mallet was almost an exact contemporary of Alexis Perrey (1807–82), who together with Hoff provided Mallet with a great number of references. Perrey's work, though very different in aim and character from Mallet's, marked a departure in that in addition to his general lists he also produced regional catalogues. His paper on earthquakes felt in Turkey, Greece and Syria (1850a) is the first European catalogue devoted to parts of the Middle East. Perrey was not a critical compiler, but he was an extremely assiduous one, and his regular publication of supplements provides a wealth of undigested information, with the authority cited.

Other uncritical nineteenth-century compilers of Middle Eastern earthquakes include Tholozan (1879), who mentions 27 earthquakes in Egypt, many of them inaccurate, and does not name his sources. Regional lists are also given by Schmidt (1879), who refers to

almost all the shocks in the Hellenic Arc that were felt in Egypt up to 1878, Fuchs (1886, pp. 529–35 for Africa, including Egypt, 1865–85) and Montessus de Ballore (1906, pp. 160–3 for Egypt).

The first catalogue devoted to earthquakes in Egypt is Lyons's 'preliminary list' (1907). It includes 29 events between 27 BC and 1906, some of them duplicated, and generally gives its sources of information. Lyons's catalogue has a large gap between 1303 and 1698, into which Sieberg (1932b), is able to insert only two events, both erroneous.

During the twentieth century, numerous catalogues of Middle Eastern earthquakes have appeared, most of them fatally flawed by a remarkable error by Willis (1928), which was noticed by the author himself (1933) but not before erroneous entries from his list had found their way into the catalogues of Sieberg (1932a, b). Willis's mistake lay in failing to notice that dates in al-Suyuti's catalogue, as translated by Sprenger in 1843, were given in the Muslim *hijri* calender, with the result that many earthquakes in Willis's list are dated about 600 years too early.[34] While Willis is therefore rightly treated with caution, the same misgiving is not applied to Sieberg's work, which remains almost universally regarded as authoritative. Sieberg, however, never gives his sources (which explains the survival of false information taken from Willis), and his lists (including those on Egypt and Arabia) contain a large number of errors, due partly to his obvious reliance on secondary works, and partly to his own inaccuracy. Sieberg's work is prominent among those used in the most recent 'authoritative' catalogue of Middle Eastern earthquakes, by Ben-Menahem (1979).[35]

In view of the problems associated with European earthquake catalogues, and their underlying cause, it is perhaps not surprising that one of the best Middle Eastern catalogues of all is not only the earliest, but also an

---

[32] See the collected papers by Holt (1973), part 1. Eutychius (*c.* 940) does, in fact, have some value as a primary source, and was used extensively by al-Makin. Erpenius did not publish the useful portion of al-Makin, which was undertaken by Cahen in 1957. Bar Hebraeus (d.1286) relies chiefly on Ibn al-Athir (d.1234). The useful portion of Abu'l-Fida (d.1331) has now been translated by Holt (1983).

[33] For a detailed appraisal of Mallet's catalogue, see Melville and Muir Wood (1987), pp. 25–32.

[34] Ambraseys (1962). For al-Suyuti, see below.

[35] Sieberg's work was also adopted by Rothé (1969), from whom it resurfaces in Maamoun (1979). A fuller list, still heavily dependent on Sieberg, is given by Maamoun and Ibrahim (1978). Here we pass over several catalogues whose main focus is on Syria and Palestine, outside the area of our study. Being more seismically active, and for other reasons, the Holy Land has been the subject of more studies than other regions, but the greater number of earthquakes and of sources of information has served to magnify the scope for duplications and inaccuracies in works dependent on secondary materials. Comments on these lists are made where relevant in the notes to our catalogue. Kallner-Amiran (1951) remains the most widely used. More recent lists either do not add anything, failing to name their sources (e.g. Ghawanmeh, 1989), or reintroduce false events that are beginning to be filtered out by more critical work (e.g. Al Hakeem, 1988). For further comments, see Melville (1989).

indigenous one. Al-Suyuti (d.1505), an Egyptian polymath, produced a catalogue of about 130 earthquakes in the Islamic world, from Spain to Transoxania. Al-Suyuti very frequently names his authorities, and cites them accurately. As a work of compilation, it could hardly be improved upon; and thanks to the devotion of two of al-Suyuti's disciples, an invaluable continuation extends the list of events affecting Egypt down to 1588.[36]

By far the most valuable compilation of material on the seismicity of the region has been undertaken by Taher (1979), who presents a full corpus of texts from Arabic sources, and a summary French translation. Although this work is not without blemishes and (in common with all other lists) is particularly deficient for the Yemen, its value lies in the unexpurgated data that it provides. Taher's work is the starting point for the present retrieval and reassessment of historical information. Summary results of Taher's research were presented by Poirier and Taher (1980), whose catalogue unfortunately passes on, and even adds to, the inaccuracies in the original work. It nevertheless remains the most authoritative and reliable list of events in the region up to 1800, thanks to its reliance on primary sources.[37]

Finally, we note recent critical catalogues of earthquakes in and around the area of interest, namely Gouin (1979) for Ethiopia, Ambraseys and Melville (1983) for the Yemen, Melville (1984a) for the northern Red Sea, Ambraseys (1984) for Libya, and Ambraseys and Adams (1986a) for the Sudan. These works are incorporated and extended where possible in the present work.

As already mentioned, experience has shown that it is not sufficient merely to exclude false earthquakes without showing why they have been omitted. The tendency remains for them to be reinstated in later uncritical lists, presumably on the grounds that they were omitted by oversight. We have therefore gone to some lengths to point out the errors in earlier works, in the footnotes

to our catalogue, and to provide a handlist of false events.[38] The chief victims of this operation are Sieberg and Ben-Menahem. The intention towards them is not vindictive; precisely because they are authoritative, as well as relatively recent works, the problems they contain are the most extensive and pervasive, and therefore most need to be exposed.

In conclusion, we may note that macroseismic information is less readily available for earthquakes occurring since approximately the end of World War II. Seismological bulletins of the first half of the century contained a fair amount of macroseismic information even for weak shocks, since it formed part of the seismologist's work to assess and evaluate such data. Macroseismic information was gradually displaced by the larger volume of instrumental data becoming available from more sensitive networks and seismologists lost their interest in small events. Felt information of low intensity is seldom reported in any national or regional bulletin, unless as part of the description of a large event; even then, there are hardly any far field reports from outside the national territory of the investigator (even wave amplitude and period data are no longer published). Press reports and field visits can supply macroseismic data for modern earthquakes, but it is generally hardly worth the effort to search for information about all the small events that are recorded, unless for the purpose of confirming the accuracy of the instrumental record, or for seismic hazard studies.

1.3 Assessment of macroseismic data

The macroseismic data retrieved from the sources discussed above have been interpreted following the methods described by Ambraseys and Melville (1982). This interpretation includes epicentral location, assessment of the maximum observed intensities, and a calculation of the magnitude of the earthquakes as defined by their reported felt area (or radii of perceptibility). Some attention is given to the role of population distribution, communications and the availability of historical documents, in evaluating the evidence for individual events and the completeness of the dataset as a whole (see also Chapter 4). The significance of negative evidence and lack of data also needs to be considered.

Some of the problems encountered in reaching these

[36] For al-Suyuti's life and works, see Saartain (1975); for a discussion of his earthquake catalogue, see Clément (1984). The confusions introduced by Willis's misuse of Sprenger's 1843 translation have been noted above. An annotated French translation was published by Nejjar (1974).

[37] Poirier and Taher's list is used almost exclusively by Guidoboni (1989), pp. 622–723, for the Arabic component of her catalogue of earthquakes in the eastern Mediterranean before the year AD 1000. Despite her critical approach to the sources with which she is familiar, she is obliged to take Poirier and Taher on trust. This highlights the interesting problem of how to achieve consistency and homogeneity in a catalogue relying on very diverse materials, which ideally need a specialist in each relevant field. The present authors are prone to the same difficulty, in accepting the work of Gouin (1979) for an area in which they have undertaken no original research themselves.

[38] See pp. 107–8. This has already been proved a necessary and useful exercise for North European historical earthquakes, see Alexandre (1990).

interpretations for historical earthquakes in our study area are discussed below. Assessment of the data in terms of magnitude is performed on the basis of correlations between intensity, felt area and magnitude derived from the study of the twentieth-century record (see Chapter 3, where assessment techniques adopted for the instrumental record are also discussed).

Location

The minimum we would expect to achieve by reassessing historical evidence is to locate different earthquakes in time and space. This simple classification underlies most existing catalogues: when and where an earthquake happened are its two most fundamental distinguishing features.

While problems of dating can be resolved relatively simply by reference to original sources, it is often much harder to determine a sufficiently accurate location for historical earthquakes.[39] Mediaeval chroniclers were generally most interested in the mere *fact* that an earthquake had occurred, and what it might signify; the significance of its location is a modern concept. Nevertheless, Arab chroniclers almost invariably give some idea of the location: an earthquake in Egypt, or in the Yemen, and specifying one or more well-known towns. This basic level of reporting reveals first, the well-developed communications between and within different regions of the Middle East and, secondly, a widespread interest in what was going on in the component parts of the Islamic empire. Partly, as suggested above, this reflects the intrinsic fascination of natural events; partly a concern for matters affecting commercial and economic activity.

The importance of communications, and of the distribution of population for the reporting and location of historical earthquakes, has been fully illustrated in our study of Persia. The larger region under consideration here poses particular and contrasting problems in this respect. In the first place, during the early centuries of Islamic history, when Baghdad was the main reporting centre, Egypt was sufficiently central and important a province for events there to attract attention, whereas the Yemen, despite its proximity to the Hejaz, was far more remote. With the fragmentation of the empire and increased regional independence, Egypt remained of interest not only to local chroniclers but to those in, for example, Syria and Iraq. Arabia continued to be something of a backwater, except for its coastal regions on the Persian Gulf to the northeast, and in the Hejaz to the west: a region that tended to be regarded, sometimes rather optimistically, as directly within the Egyptian sphere of influence, particularly after 1258. Libya, the Sudan and the Yemen were peripheral regions, an isolation that explains the almost total absence of Yemeni earthquakes from most lists, including that of al-Suyuti (writing in Cairo). As already noted, however, the Yemen has a long and well-documented history, though this to be found almost exclusively in works of local provenance. Only while the Yemen, along with the rest of the Middle East, was incorporated within the Ottoman Empire (after *c.* 1540), was news again current in an imperial capital, this time Istanbul; continuing Ottoman interests in the Hejaz kept the connection alive, even after their expulsion from the Yemen in 1635.

These considerations apply to the position of the study area in the general context of the Arab world. When we focus on the specific regions involved, internal conditions can be seen to influence the reporting of earthquakes and the possibilities for locating them. There are two key factors: population distribution and political geography. Both are influenced by the physical environment (see Figure 1.4).

The desert terrain and the aridity of the climate in both central Arabia and Libya deter large settlements. Long-term macroseismic data from such regions will always be scarce and the chances of accurate location small: the low density of population militates against any precision in mapping the few earthquakes that may chance to be reported. Throughout most of the period studied, there was no centralised political authority in either Libya or central Arabia and no metropolitan city to act as a literary centre. Circumstances for reporting earthquakes were slightly more favourable in coastal areas, but such reports raise their own problems of location, with the possibility of offshore events. Earthquakes in Suhar (Sohar) in 879, for example, doubtless originated in the Gulf of Oman, but an accurate location is impossible from the evidence available.

EGYPT Egypt, by contrast, has invariably been a highly centralised political and geographical entity, thanks to the unifying role of the Nile. Population is distributed almost exclusively in the Delta, and along the Nile Valley in a narrow band seldom more than ten miles across, bounded by desert. Political centralisation has concentrated the densest population, with its contingent cultural monopoly, in Cairo (with a secondary long-term centre in Alexandria), to the virtual exclusion of

---

[39] See Melville (1984b), pp. 110–12, and (1985), for a general discussion, particularly with respect to Britain.

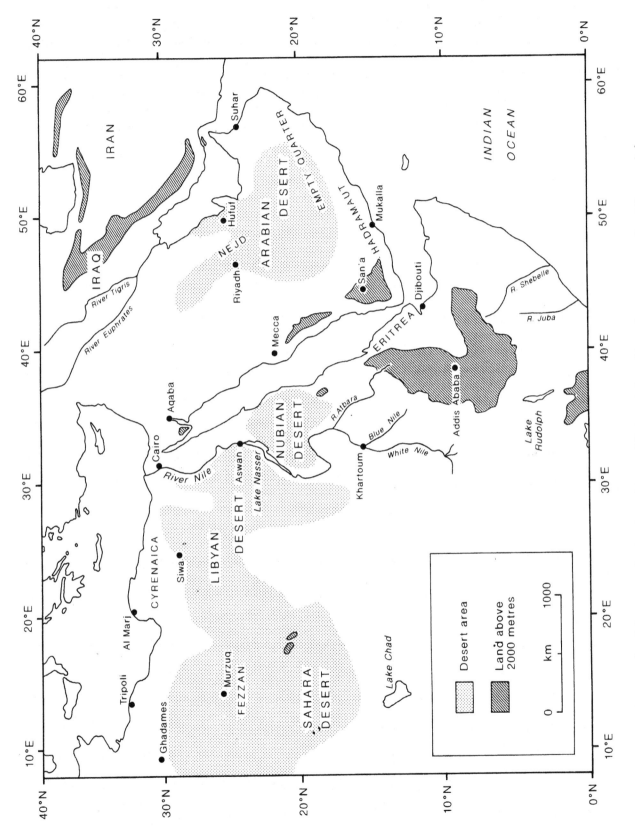

Figure 1.4. Physical topography of the region, showing the main natural features influencing settlement patterns. Apart from capital cities, a few places and areas that do not appear on the maps in Chapter 2 are shown here.

13

local centres with significant literary output.[40]

These geographical facts have various implications for the location of historical earthquakes. In the first place, until the late seventeenth century, the vast majority of Egyptian earthquakes are reported only in Cairo. It can generally be assumed that the shock was felt there, even when this is not specifically stated. It is rare for other places to be mentioned, even when it is clear that the shock was widely felt (a rare example is the large earthquake of 8 August 1303). A related problem is that the Arabic word for Egypt, *misr* (literally, 'city') is also the word used for the capital, al-Fustat, founded by the Muslim conquerors in the seventh century. An earthquake in *misr* could therefore signify a shock in Egypt generally or in the capital. Even after the foundation of Cairo (al-Qahira 'the Victorious') in 969 the problem is not removed, because an earthquake in *misr wa'l-qahira* could mean in Old and New Cairo, or in Egypt and Cairo. Even if all these tremors were experienced in greater Cairo, it is hardly plausible to put the epicentre directly under the city, though when there are no indications that other places were affected, in practice there is no choice.[41] The recent Egyptian earthquake does at least demonstrate that shocks can originate relatively close to Cairo, and that such interpretations need not necessarily be too inaccurate.

Even when information does exist from other Egyptian localities, the distribution of population, and therefore of intensity data-points, is effectively one-dimensional, making it difficult to draw meaningful isoseismal maps. While it may be possible to determine the highest intensities experienced along the north–south axis of the Nile Valley, the absence of control to east or west makes it difficult to determine the true epicentral region. An example of this difficulty is presented even by such a relatively well-documented event as the earthquake of 7 August 1847. The October 1992 earthquake again provides a plausible interpretation of the location of this event.

More than half the earthquakes that affected the Nile Valley and the Delta originated from epicentres outside Egypt. These are generally large-magnitude earthquakes in the Hellenic Arc, or in the Dead Sea system. This is demonstrated by the date of the shock coinciding with the date of a known event in these regions.[42] These events, especially those offshore, present their own problems of location, but they are not our immediate concern. An additional distant source of earthquakes affecting Egypt is the northern Red Sea. A recent example is the earthquake of 31 March 1969, felt throughout the Nile Valley with isolated damage at epicentral distances greater than 300 km. The 1969 event can to some extent serve as a model for earlier, less well-documented earthquakes, and at least three such events have been identified.[43] The northern Red Sea is, however, a barren and inhospitable region, yielding few macroseismic data – were it not for the fact that the pilgrimage routes from Egypt and Syria traverse the area, and for the records of St Catherine's monastery in Sinai, there would probably be no information at all before the twentieth century. There are obvious problems in locating earthquakes in such a region, especially purely on the basis of felt reports from Cairo, but occasionally this seems plausible when long-period effects there cannot be associated with any known event in the Hellenic Arc or southern Palestine. In general, we would expect earthquakes of this size to leave some stronger trace in the sources available, even for a region such as the northern Red Sea.

A good example of these problems is provided by the earthquakes of 25 July 950 and 15 September 951.[44] Contemporary authors apparently distinguish two shocks in Egypt in successive years, both of which caused damage. Both could be local earthquakes. On the other hand, an earthquake affecting Sicily and another affecting southeast Anatolia are also reported at this period, without precise chronological details. Both these events could have been felt in Egypt, though not so as to cause much damage. Furthermore, accounts of the effects in Egypt of both earthquakes allow the possibility of long-period shaking: in the 950 event, part of the mosque in al-Fustat fell down and in the 951 event, the lighthouse at Alexandria was damaged. In the absence of strong chronological evidence for the timing of these various earthquakes, it is very difficult to know which, if either, of the Egyptian shocks should be associated with which of the distant earthquakes. Our own assessment is just one of the possible interpretations.

---

[40] Qus, in Upper Egypt, may be a partial exception, for which see Garcin (1976).

[41] In Iraq, scene of a riverine civilisation comparable to that of Egypt, there is a similar concentration of felt reports for Baghdad and Mosul, but this does not mean that these two cities lie in the most highly seismic regions; see nevertheless, Alsinawi and Ghalib (1975), p. 541 and their fig.1. A distinction has to be made between hazard and risk.

[42] Various catalogues for Greece, Turkey and the Levant, such as Gutenberg and Richter (1948), Papazachos and Comninakis (1982), Ayhan and Alsan (1988) and Arieh *et al.* (1983) are usually sufficient to make such connections, but these events have also taken our researches outside the main area of study.

[43] Melville (1984a), Ambraseys and Melville (1989).

[44] See below, pp. 28–9.

The effect of Egypt's peculiar topography and population distribution is clearly seen in Sieberg's earthquake map of Egypt (Figure 1.5), which depicts seismic activity in terms of intensity zoning. Clearly, this is more correctly a map of seismic risk rather than seismic hazard; but the epicentres assigned to pre-instrumental Egyptian earthquakes inevitably tend to follow the course of the Nile, artificially distorted by cultural factors.

THE YEMEN   In contrast with Egypt, the mountainous terrain of the Yemen has not facilitated the growth of a predominant cultural centre, but different centres have flourished at different periods. The chronicles nevertheless concentrate on events in the different regional capitals and information about earthquake effects on communities in isolated or inaccessible valleys is almost non-existent: a rare example being the case of the earthquake of 11 September 1154. Here, the difficulty is identifying the places mentioned. Generally, however, earthquake reports are associated with one place only (Zabid, or San'a, or Aden), without an indication of the area worst affected. It can also be the case that the city mentioned is evidently not in the epicentral area but no idea is given of the other districts involved. Sometimes, two or three such centres are mentioned, so there is a reasonable chance of locating the earthquake in the area between them, even though no details from that area survive. Furthermore, although the urban population is distributed more evenly through the country than it is in Egypt, the major towns themselves are aligned along the north–south axis of the central uplands, and are relatively far apart. Locational precision for historical earthquakes in the Yemen is therefore poor, and similarly influenced by cultural factors, though seemingly in a less obvious way than in Egypt.[45]

The great majority of historical earthquakes reported in the Yemen have been located on land, where their effects were documented. Modern maps of instrumentally recorded earthquakes (e.g. Figure 1.2) demonstrate that the greatest rate of activity occurs along the zone of seafloor spreading in the Red Sea. Evidence for the occurrence of earthquakes on land seems to be reasonably clear and we do not believe that these are simply poorly located earthquakes that in fact occurred offshore. On the other hand, it is doubtful whether any but the largest offshore earthquakes would be reported in Yemeni historical sources: cases when reports are available from coastal regions are rare (as in the series of earthquakes in 1788–9), and depend also on similar reports from the Ethiopian side of the Red Sea (as in the earthquake of 30 August 1504). Unfortunately, the Ethiopian littoral was even less densely populated and less frequented than the Arabian side, and so there are few opportunities for identifying and locating offshore events.

In Ethiopia itself, Gouin (1979, pp. 246–7) notes that 'the regions exhibiting the highest number of earthquakes unfortunately coincide with the areas of highest population density.' He does not discuss to what extent the absence of information from other regions may be due to cultural factors, but it is safe to assume that similar considerations apply to the location of historical events in Ethiopia as to the other regions of our area of interest.

In general, it should be borne in mind that the epicentres of historical earthquakes are macroseismic epicentres (sometimes referred to as macrocentres). They are assigned on the basis of maximum observed intensity, or otherwise as the geographical centre of the reported felt area. The most confident locations are those at which these positions coincide. Generally, small earthquakes can be located more accurately than large ones. All macroseismic epicentres reported in the catalogue are given a quality grade which reflects the level of information available (see Chapter 2, Table 2.1).

Intensity assessment

Intensities have been assessed in accordance with the principles discussed in detail elsewhere with respect to Persia.[46] As noted above, it is a feature of early accounts of earthquakes that very few places are mentioned. Details of the effects of the shock in individual places are lacking for all but the most destructive events. When details do become fuller, even for smaller earthquakes, particularly during the nineteenth century, it is still generally only for the largest places.

The collapse of a single house, or damage to a few, resulting in a handful of casualties in large population centres such as Alexandria or Cairo, should be given very little weight in assigning intensities.[47] Sporadic collapse of houses and buildings in such cities with ageing

---

[45]   The Yemen resembles Syria and Palestine in its regional particularism, but in view of the greater density of population and concentration of large towns, more data points and correspondingly more precision can be expected in reports of earthquakes in the Holy Land.

[46]   Ambraseys and Melville (1982), pp. 25–32. See also the pertinent observations in Poirier and Taher (1980), p. 2188 ff.
[47]   Sometimes the chroniclers show themselves aware of this, see e.g. under 872/1467.

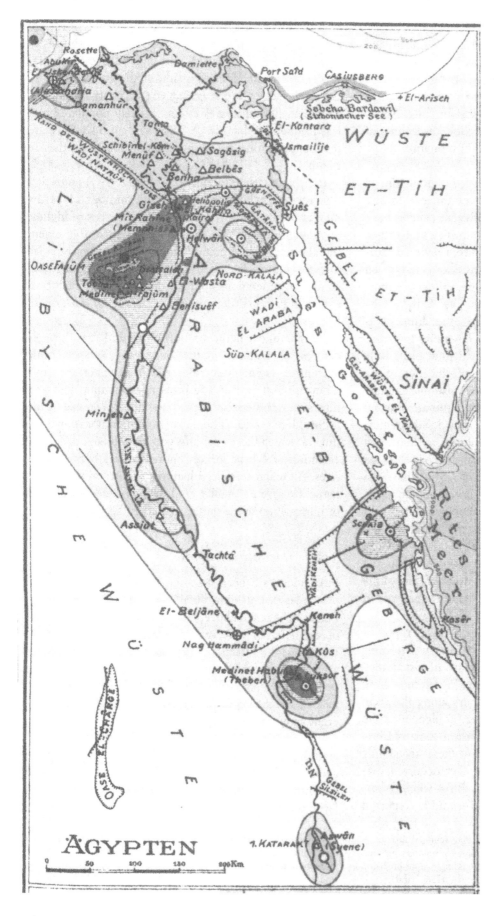

Figure 1.5. Earthquake map of Egypt, from Sieberg (1932b).

building stock does occur regularly without the help of earthquakes and is not an indication of the intensity of shaking.[48] In smaller places too, accounts of the occasional collapse of a house, particularly the multistorey dwellings in the Yemen (see Figure 1.6) without an earthquake, are often found in local chronicles.[49] In the accounts of many nineteenth- and twentieth-century earthquakes, damage to buildings of traditional construction and materials can be contrasted with the total absence of damage to modern engineered structures. The Cairo earthquake of 1992 illustrates this to some extent: with the exception of a high-rise apartment block in Heliopolis, most of the damage was concentrated in the poorer areas of the city and in historical monuments, the vulnerability of which has increased with age.

To assess damage reports from regions of vastly different population density, without taking this factor into account, may lead to grossly exaggerated intensity rating and a serious distortion of the hazard potential. Unfortunately, reliable population statistics are unavailable in most countries of the study area until the present century; even estimates for the major cities of the region have to be treated with great caution.[50] Nevertheless, it should be possible to gauge the severity of an earthquake's effect in the context concerned.

Damage or collapse of minarets and tall structures, when not accompanied by damage to ordinary houses, has also been given an almost zero weight. These effects are taken to indicate a relatively large, distant shock, particularly if there is some other evidence of a long duration of shaking or other long-period effects, such

Figure 1.6. Traditional architecture in the Yemen, precariously perched.

as the sloshing of water in pools. An example of this is provided by accounts of the 23 June 1425 earthquake, and more recently by the large intermediate-depth earthquake in the Hellenic Arc on 26 June 1926.

Another factor to be borne in mind when assessing intensity is the way in which earthquakes were perceived and reported. The systematic use of terms such as 'slight', 'strong', 'frightening', or 'great' by Arabic authors, such as al-Suyuti, give a subjective but graded evaluation of the severity of shaking, and have been regarded as equivalent to intensity grades by some writers.[51] However, one man's light is another man's

---

[48] For example, four houses collapsed in Cairo with casualties in 1133/1721, Ibn 'Abd al-Ghani, p. 332. Similar collapses are reported in 1190/1776 and again in 1219/1804, al-Jabarti, III, 126, VI, 189/ trans. III, 232–3, VII, 344–5. The minaret of the mosque of Qausun collapsed in 1215/1801; *ibid.*, V, 203/trans. VI, 271 – he blames this on the French. The same author lists many occasions when houses were destroyed by rains and storms. The minaret of the unfinished madrasa of al-Hasan fell in 762/1361, see catalogue. For a more recent example, of the collapse of a twelve-storey building under construction in Jidda, see *Arab News*, 15 Ramadan 1401/16 July 1981.
[49] For example, a house collapsed in San'a in 682/1283 and another in Aden in 746/1346, both causing casualties, see al-Khazraji, trans. Redhouse, I (1906), pp. 202, 211; II (1907), p. 64. For the traditional architecture of the Yemen, see Serjeant and Lewcock (1983), Bel (1988), Leslie (1986) and Hughes (1988).
[50] A trenchant view of the current state of knowledge is given by Ayalon (1985). Cairo is thought to have had a peak of around half a million in the fourteenth century, and around 250 000 in 1800; Alexandria in the twelfth century, about 65 000. In the late nineteenth century, Cairo was around 590 000 and Alexandria about 230 000. For Cairo, see the various estimates cited by Abu-Lughod (1971); for Alexandria, art. 'al-Iskandariyya', *E.I.*², **IV**, 133–4.

[51] See e.g. Nejjar (1974), p. 101, and a similar tendency in Quittmeyer and Jacob (1979), pp. 775–6. The readiness of the latter to convert published intensities from one scale to another without reference to the original data is not recommended; see Ambraseys *et al.* (1983).

heavy, and al-Suyuti is merely repeating what his various sources wrote, rather than applying a uniform reassessment of his own. There are numerous examples of the same shock being characterised in conflicting ways in the chronicles.[52] The earthquake of 21 December 1694 in Egypt, reported in very different terms by a French resident and an Arabic chronicler, also shows how potentially misleading subjective accounts can be and illustrates the problems of intensity assessment from such data. In view of this, intensity assessments are given symbolically on Table 2.1 (p. **100**).

The 1694 earthquake also illustrates the problem of intensity mapping: though we are told houses were destroyed, we do not know where. In general, macroseismic data have proved inadequate to allow the construction of isoseismal maps, though individual intensity points are shown on the figures accompanying the catalogue, where data permit, by reference to the MSK (1981) scale. In some cases, it is possible to use the absence of information to help limit the size of a shock. Negative evidence can take the form of an explicit statement that the shock was not felt in a certain locality; alternatively, and slightly less usefully, inferences can be made from the accounts themselves (see e.g. 259/873, 668/1269, 816/1413 in the catalogue). Also, the silence of sources known to be available for the period and place in question can be taken to indicate that the earthquake was not felt there, or was felt at such low intensities as not to have merited attention.

Felt area

The problem of intensity mapping is bound up with the question of location, discussed above, since in cases where the epicentral region is uncertain, an acceptable

[52] See e.g. under 775/1373, 828/1425, 881/1476, 891/1486 and 905/1500.

approximation can at least be proposed at the centre of the reported felt area.

Similarly, our inability to define the felt areas of many historical earthquakes means that we cannot estimate their size from their radii of perceptibility. In many cases, it has only been possible to determine a few epicentral distances, rather than a genuine isoseismal radius. Even when the epicentral distance is known to a place mentioned as affected by the shock, the problem often remains of identifying the correct intensity grade at that point. In general, unless there is sufficient evidence to suggest a higher or lower grade, felt reports are taken to be equivalent to intensity V, below which it seems unlikely that an isolated shock would attract sufficient attention to justify being reported in the chronicles. At certain periods in Cairo, however, the threshold of perception seems to have been lower and to have included very weak shocks (e.g. late fifteenth and early sixteenth centuries).

In view of these difficulties, magnitudes have only been calculated for a small number of the better-documented events before 1900. Table 2.1 gives the radii of perceptibility or epicentral distances on which the calculation of magnitude is based, using the equations presented and discussed in Section 3.2.

Depth

Isoseismal mapping, if sufficiently detailed, can also provide evidence of the focal depth of the shock. In practice, it has not been possible to construct enough isoseismals for any particular shock to allow depths to be calculated from macroseismic data. Tectonic evidence suggests that all earthquakes in the area are shallow events concentrated in the crust (down to about 30 km), and there is nothing in the macroseismic data to indicate the contrary. However, intermediate-depth earthquakes outside our area are known to have produced felt effects extending into the study region.

CHAPTER 2

Macroseismic information

In this chapter we present the macroseismic information retrieved and analysed according to the methods outlined earlier. In the first section (2.1), each earthquake is described as fully as possible from the available evidence. For the most part, commentary is confined to the footnotes, which attempt to resolve any conflicting information in the authorities cited, and illustrate how well each earthquake is documented. This itself can be an important consideration in reaching an acceptable assessment of the event in question. In general, if no reference is given to a primary source of information that appears in the Bibliography, this indicates that the work does not mention that particular earthquake. On the other hand, previous catalogues are not necessarily cited systematically, unless their entries require correction (see below).

For the period up to 1899, the catalogue includes *all* earthquakes in the area of interest for which felt reports have been retrieved, with the exception of those occurring in southern Iran, on the edge of the study region, which have already been listed in Ambraseys and Melville (1982, esp. pp. 161–2). As a rule, all the information available is presented for the early periods, largely translated directly from the Arabic sources. Less reliance is placed on sources written in Arabic in later periods, and there is a greater tendency to summarise the more voluminous information found in European sources, particularly for the nineteenth century. In the case of Ethiopia, our data are taken almost exclusively from Gouin's catalogue covering his regions A (Western Plateau), C (the Main Rift, Afar and Southern Red Sea) and D (Aden Western Sector). Some of these events are simply noted briefly in the footnotes to the more significant shocks. The aim is to present as clearly as possible all the information on which our analysis is based. The interpretation of each earthquake is summarised in Section 2.2 (Table 2.1).

From 1900 onwards, almost all earthquakes for which macroseismic data were found have been described.

Exceptions include several events in Ethiopia and Iran that have been studied elsewhere, and some large events originating in the Hellenic Arc or the Mediterranean that were reported to have been felt within the study region. These and a few small Egyptian events, for which some felt information is available but not presented here, are all listed in the catalogue of twentieth-century earthquakes (Table 3.1). With the exception of a few small earthquakes reported in the Yemen, for which no instrumental data have been identified, all the twentieth-century events described in Section 2.1 are listed in Table 3.1. Instrumental information is included in the descriptions of twentieth-century events, and generally becomes increasingly important after around 1950. For reasons noted above (p. 11), macroseismic data are readily available only for the most significant earthquakes after that date. In general, references and comments are kept to a minimum in the description of events occurring after 1900.

Each event discussed in Section 2.1 is identified by its date (including hijri date when given), and a generalised location, such as Hejaz, Sudan (see Figure 1.1). Not all the events discussed in this chapter are genuine earthquakes, and some remain dubious. When only the date or location of an earthquake is uncertain, this is indicated by a question mark. When the event is confidently believed not to be a genuine earthquake, its heading is put in square brackets; such events are excluded from Table 2.1. Genuine earthquakes incorrectly alleged to have been experienced in our area are also distinguished in the same way. Other dubious events – such as earth tremors perhaps triggered by meteorite impacts, or rockfalls perhaps triggered by small earthquakes – are retained in Table 2.1, but the uncertainty is indicated symbolically (see p. 100). Earthquakes that have not been fully identified, of uncertain date or location and probably spurious, are discussed in Section 2.3, and do not appear in Table 2.1.

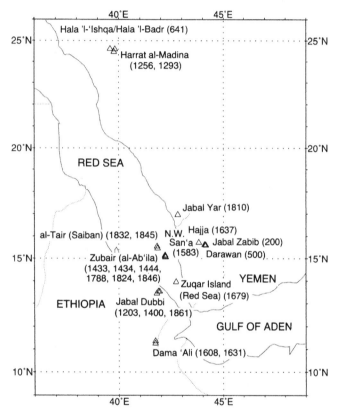

Figure 2.1. Sites of volcanic eruptions in the period up to 1899 (as reported in Table 2.1).

False earthquakes listed in previous catalogues are generally mentioned in passing in the footnotes to the macroseismic catalogue. A summary list of false events before 1900 is given in Section 2.4, which also presents a few case studies to show why these earthquakes have been deleted from the catalogue. False earthquakes in instrumental catalogues are discussed in Section 3.4.

Volcanic eruptions and earthquakes associated with volcanism have been discussed in the macroseismic catalogue, although no attempt has been made to search systematically for such data, nor to analyse the charac-

teristics of the eruptions themselves. These events are included in Table 2.1, with their volcanic origin indicated symbolically. Figure 2.1 plots the sites of known historic eruptions, together with the dates of events discussed in the catalogue.

Most of the larger or better-documented earthquakes are accompanied by a location map, showing intensity assessments where possible (see p. 18). Places where the shock is reported not to have been felt are plotted with an open circle. For many of the earliest events, only one or two places are mentioned. The same places tend to be mentioned throughout the period covered, normally the main towns or cities. Figure 2.2 shows the location of the places most commonly mentioned in the central area of interest, together with a few sites reported to have experienced earthquakes that have not been mapped.

2.1 Descriptive catalogue of earthquakes (184 BC–AD 1992)

184 BC   **Lower Egypt**
A Greek papyrus written in Egypt mentions an earthquake this year.[1] The context in which this event is mentioned is not certain.

1 Preisigke (1915), pp. 615–16.

c. 95 BC   **Gulf of Suez**
A Greek inscription found in Magdolum (near Bir Magdal) refers to the effects of earthquakes on the Temple of Heron, between 97 and 94 BC (See Figure 2.2).[1] The nature of these effects is not mentioned.

1 Preisigke (1926), pp. 262, 330.

AD c. 112   **Dead Sea**
Archaeological evidence suggests an early second-century destruction at Petra, Masada, Avdat and several other sites along the Petra–Gaza road.[1] There is no corroborating documentary evidence for this event (see Figure 2.3).

1 Russell (1985), pp. 40–1.

AD c. 200   **Yemen**
An eruption of Jabal Zabib, north of San'a, occurred sometime during the third century AD (see Figure 2.1). The lava flowed southeast of a small adventive crater on the east side of the volcano, and stopped after a flow of 5.5 km near the village of al-Huqqa, where it overlies

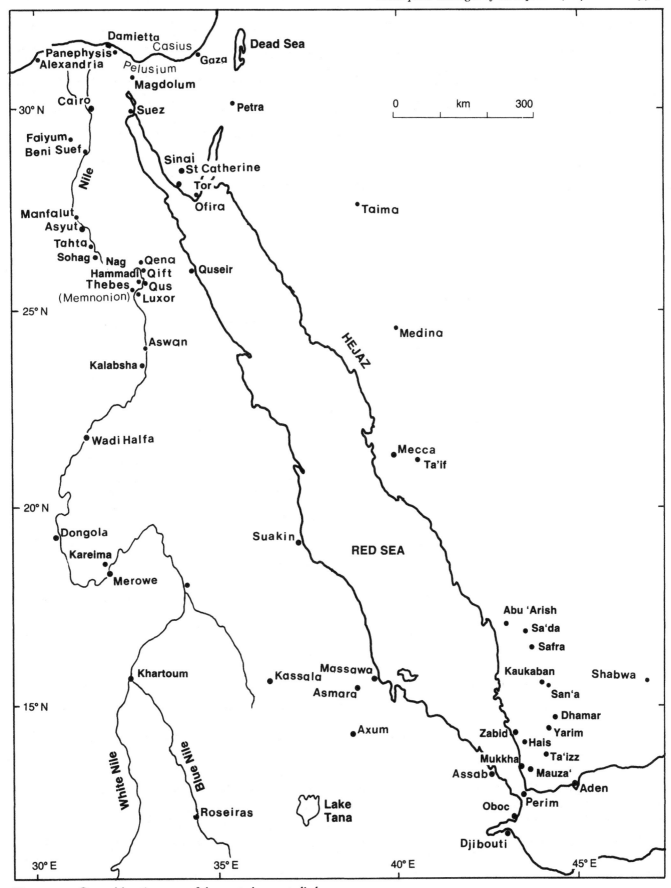

Figure 2.2. General location map of the central area studied.

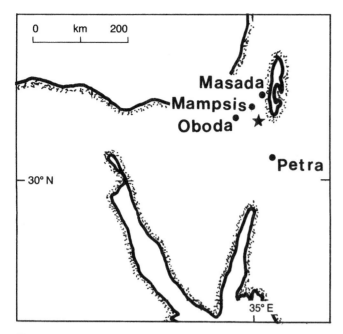

Figure 2.3. Earthquake of *c.* 112, Dead Sea. The symbols used on this and all subsequent maps in this chapter are as follows: ★, macroseismic epicentre; ●, place where earthquake was experienced (with intensity rating when assessed); ○, other places in the area, for which no information is available (including 'not felt' reports); ⇉, tsunamis; +₊+, denotes archaeological site.

loess and recent loams. Inscriptions found here indicate that there was a large temple of the Moon God Ta'lab Ri'am, of which, however, no ruins were found, possibly buried under the lava flow. Ruins of the temple of the Sun Goddess Dhat Ba'dan suggest that the site was burnt down, probably by this eruption. This temple is located a few hundred metres southeast of al-Huqqa (Bait al-Haqr) which, according to an inscription, was still in use in AD 200. Thus, if the temple was burnt by this eruption, the activity must have taken place after that year.[1]

1 Lamare (1930); Neumann van Padang (1963), pp. 16–18. See also below, under AD *c.* 500.

AD **262**   **Libya**

Strong earthquakes caused considerable damage in Libya in the fifth consulship of Gallienus, with Faustinus (i.e. AD 262).[1] It is possible, but not certain, that the shock primarily affected Cyrenaica, where there is archaeological evidence of destruction in the city of Cyrene, which was rebuilt with the new name Claudiopolis, in 268.[2]

Sieberg, without quoting an authority, suggests that this earthquake destroyed the oasis of Siwa in Egypt, 280 km away.[3]

1 Trebellius Pollio, v. 2. He amalgamates this event with others in Rome and particularly Asia in the period from 262 to the siege of Thessalonika by the Eruli in 270.
2 Stucchi (1965, 1975); Goodchild (1968), p. 43. Stucchi is rather too confident in using his source to date the destruction of buildings in Cyrene. He also refers (1975, pp. 228, 234, 333, 351) to earthquakes in Cyrene in the period of Augustus and Tiberias, and again in AD 251 and 306. No confirmation of any of these events has been found in historical sources.
3 Sieberg (1932b), p. 188.

AD **320**   **Mediterranean**

Theophanes (d.818) records an earthquake which damaged many houses in Alexandria in anno mundi 5812 (AD 320–1) and many people were injured.[1] He does not mention whether this earthquake affected other regions, but an earlier chronicle of the sixth century says the earthquake extended to many places in anno Abraham 2335 (AD 319–20).[2] This suggests a relatively large shock of unknown origin. The chronological data indicate the year as being 320.

1 Theophanes, p. 13/24.
2 *Chronicon Pseudo-Dion.*, p. 159.

365 July 21   **Hellenic Arc**

A major earthquake in the Hellenic Arc, over 900 km northwest of Lower Egypt, was responsible for a catastrophic sea-wave that played havoc with coastal settlements on the Nile Delta. Contemporary historians Ammianus (d.395), John Cassian (d.435) and Sozomenes (d.*c.*450) say that in Alexandria '... the sea passed beyond its boundaries and flooded a great deal of land, so that on the retreat of the waters the sea-skiffs were found lodged on the roofs of the houses'. In the region of the lagoon of al-Manzala, east of the Delta between Damietta and Port Said, the land previously rich became a desert: '... the sea rose suddenly due to the earthquake, rushing over its limits, ruining all the villages, covering with salt the land which before was fertile ... only villages on high ground survived'. The anniversary of this inundation, which was called the birthday of an earthquake, was commemorated at Alexandria by a yearly festival.[1]

Later writers add that in Alexandria alone 50 000 houses were flooded and 5000 people were drowned; ships were carried by the waves over the city walls and boats in the Nile were deposited on dry land about three and half kilometres from the river. The region of Tinnis and the town itself were totally destroyed.[2]

The sea-wave had equally destructive effects in other

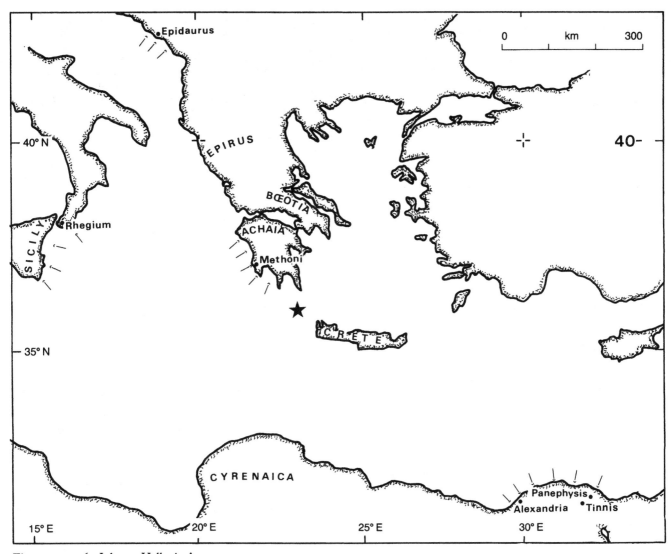

Figure 2.4. 365 July 21, Hellenic Arc.

parts of the eastern Mediterranean region (see Figure 2.4). In Crete, the Peloponnese, the Adriatic and Sicily it was far more serious than the shock itself.

Near-contemporary and later writers grossly exaggerate the effects of this event, describing it as a global catastrophe. This has permitted modern authors to link the earthquake to archaeological evidence of local damage in sites throughout the southern Mediterranean, such as Djemila in Algeria.[3] Although some doubt must remain as to the correct interpretation of the sources, however, it is more likely that the damaging effects of the earthquake were confined between Messenia and Laconia in the Peloponnese, the west part of Crete and Cyrenaica, but not further west. Such a large epicentral area would imply a great event of magnitude perhaps over 8 in size, which would certainly have been felt across the eastern Mediterranean region, with effects in many ways similar to those of more recent

large intermediate-depth shocks originating in the Hellenic Arc.[4]

It is likely that the 365 earthquake, like the AD 551 seismic sequence, consisted of more than one large earthquake in the eastern Mediterranean region. Indirect evidence, in need of authentification, suggests that the period 350–550 was perhaps one of the most seismically active periods in the region during the last two millennia.[5]

1 Ammianus Marcellinus, xxvi.10; John Cassian, xi.3; Sozomenes, vi.2. See also Daressy (1934), pp. 45–9 for the effects of this earthquake on the regime of the Nile.
2 George the Monk, p. 196; Theophanes, p. 47/87; Cedrenus, p. 310/543; Glycas, p. 255/473.
3 For a recent reassessment of this event, and a detailed analysis of the sources, see Jacques and Bousquet (1984) and for Libya and Algeria see respectively Lepelley (1984) and Blanchard-Lemée (1984). The earthquake and the methodological problems it poses are also discussed in Guidoboni (1989), esp. pp. 449–51, 607, 678–80; cf. Russell (1980), p. 53, and Russell (1981).

4 The shock is analysed in these terms in Guidoboni (1989), pp. 552–73. See Ambraseys (1965) for the effects in Cyprus (under AD 370).
5 Pirazzoli (1986).

AD *c.* 500   Yemen

Arab geographers refer to a legend that the 'Nar al-Yaman' or Fire of Yemen was located in the Darawan valley northwest of San'a; fire and lava eruptions became objects of veneration and also sources of judgement between litigants. 'The Fire' continued for about 300 years, i.e. eruptions were possibly fairly continuous in the area since the events noted above under AD *c.*200.[1]

The source of the eruption was the southern foot of the cone of Kaulat Hattab (see Figure 2.1); al-Hamdani says that the origin of the fire was the extreme end of mount Darawan, a statement confirmed by field studies, which showed that lava flowed in a southern direction to the edge of the village of Darawan. The date of this phase of eruption is unknown, but Neumann places it between the fifth and seventh centuries AD.[2]

1 Al-Bakri, p. 621, has a detailed account of the extent of the lava-field; for the places mentioned, Wilson (1989), p. 219 etc. See also Yaqut, III, 470.
2 Al-Hamdani, *al-Ikil*, VIII, 67/trans. Faris, pp. 46–81; Neumann van Padang (1963), pp. 14–16.

520 October 14   *17 Teqempt*   Lower Egypt?

Either this year, or about quarter of a century later, a strong earthquake was felt in Egypt, shaking buildings violently for a long time. A later report mentions that many cities and villages were swallowed up. Shocks lasted for a year, and the event was commemorated in an annual festival.[1]

The correct identification of this earthquake is problematic. The details on the whole suggest a damaging local event, in which case the exact year is uncertain. The shocks could, however, have been associated either with the large earthquakes in Antioch in May 526 and November 528, or with that of August 554.[2] None of these earthquakes, however, occurred on the right day, nor did the 551 event, also felt in Egypt (see below).

Modern writers falsely associate Nikiu's report with the beginning of the formation of the lagoon al-Manzala, which according to al-Mas'udi (tenth century) began in 535 and was completed about five years later.[3] This statement seems to be substantially incorrect and anyway, al-Mas'udi does not mention an earthquake, nor does he associate the formation of Lake Tinnis with seismic activity.

1 Severus of Antioch, trans. Brooks, II/2, p. 340. Severus (d.538) was banished from Antioch in 518 and spent most of the rest of his life in Egypt. According to the editor, this letter was written around AD 520. Severus gives the date as 14 October. The Coptic Bishop, John Nikiu (fl. *c.* 700), I, 143, also reports an unidentified earthquake in Egypt in the reign of Justinian (527–65). The event was commemorated on 17 Teqempt (= 14 October). The coincidence of these dates leads us to suppose that both authors refer to the same event. This would imply that Nikiu has put the earthquake in the reign of the wrong emperor, Justinian instead of Justin (518–27), though the rest of his chronology seems to be correct. Nikiu blames the earthquake on the changes in the orthodox faith introduced by Justinian.
2 For these events, see most recently Guidoboni (1989), pp. 691–5, 699–701, for the problems in the sources: Cedrenus, for example, puts the first Antioch earthquake on 4 October 525, cf. Grumel (1958), p. 478. Guidoboni says the earthquake around midnight on 15–16 August 554 also affected Alexandria, cf. Sieberg (1932b), p. 188 under 553; but see below under 551.
3 Clédat (1923), p. 66, followed by Daressy (1934), pp. 49–50, citing an additional account by al-Maqrizi of the formation of Lake Tinnis. But see al-Mas'udi, *Muruj*, II, 374–7 and al-Maqrizi in Maspero and Wiet (1919), p. 35. Nikiu, furthermore, says nothing of a marine inundation.

551 July 9   Dead Sea

A large earthquake in Palestine was strongly felt in Alexandria, where it caused panic but no damage. For the Nile Delta, which was considered to be almost free of earthquakes, this was unprecedented.[1] The shock was felt throughout Palestine, Arabia, Phoenice and Lebanon as far as Antioch (see Figure 2.5), and was associated with a seismic sea-wave on the coast of Phoenice.[2]

Modern writers place the epicentral region of this event offshore from Lebanon.[3] This is due to the bias of information from the more populous coastal region. The

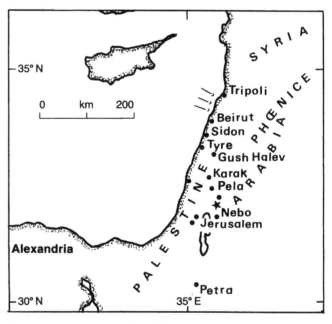

Figure 2.5. 551 July 9, Dead Sea.

data available to us, however, suggest an epicentre in the Jordan rift valley.

1 Agathias, ii.15/ pp. 95–8. Guidoboni associates this with the August 554 event in Constantinople.
2 Malalas, p. 229/485. In the sixth century the Byzantine province of Arabia corresponded approximately to the old Nabatean kingdom, stretching as far as Aila. The capital was Bostra (Busra), in Hauran. For a detailed analysis of the sources, see Russell (1985), pp. 44–6.
3 See e.g. Ben-Menahem (1979), p. 286; see also Sieberg (1932b), p. 199.

c. 626   5 H   Hejaz

Al-Suyuti narrates a tradition that Muhammad and the first three caliphs felt an earthquake on Mt Uhud outside Medina (or Mt Hira outside Mecca). Depending on which mountain it was, the date varies – Muhammad was at Mecca before the hijra (AD 622) and primarily at Medina from then till his death in 632. The later date seems more likely, because al-Suyuti lists the shock as occurring in the Islamic period, i.e. after the hijra, and also after a shock dated approximately 7–11/628–32.[1]

Al-Biruni and others say that the year 5/626 was called the year of earthquakes, and there was an earthquake in Medina.[2] There may be some connection between both the events of uncertain date, reported by the traditionists, and the earthquake in year 5 H, which seems more securely dated.

1 Al-Suyuti, *Kashf*, p. 22, citing various traditionists; trans. Nejjar (1974), pp. 6–7 and notes on pp. 51–2 (hereafter cited as al-Suyuti, p. 22/6–7). Ibn Jubair, p. 113, refers to the shaking of Hira.
2 Al-Biruni, p. 31; Ibn al-Athir, '*Usd al-ghaba*, I, 22. The earthquake is not mentioned by other earlier historians, such as Ibn Hisham and al-Ya'qubi. The seventeenth-century author, Yahya b. al-Husain, *Ghayat*, I, 87, has 6 H. See also Sieberg (1932a), p. 795, under AD 631; Ambraseys (1961), p. 21; Taher (1979), p. 12/146.

641   20 H   Hejaz

Al-Suyuti cites various traditions that report an earthquake in the Hejaz in the reign of 'Umar (d. 644).[1] This is identified as the earthquake of 20/641, which affected Medina,[2] where houses are said to have been destroyed.[3]

The earthquake may be connected with the volcanic activity reported the previous year, 19/640, in the Harrat al-Nar region north of Medina.[4] It is not certain whether the lava flow reported was from Hala'l-'Ishqa (27.58° N – 36.80° E) or from Hala'l-Badr (27.25° N – 37.20° E), or from both locations (see Figure 2.1).[5]

1 Al-Suyuti, p. 22/7, citing indirectly Ibn al-Jauzi, cf. Taher (1979), p. 14/148.
2 Al-Ya'qubi, II, 179; see also Ibn al-Dawadari, III, 231.
3 Al-Suyuti, *l.c.*; Nejjar has 'abandoned', but the Arabic has 'destroyed'. According to al-Suyuti's source, this was the first earthquake in the Muslim era (*sic*); al-Ya'qubi merely says that nothing like it had been experienced before.

4 Al-Tabari, I/5, 2579; Yaqut, II, 252. Harrat al-Nar extended between the Wadi al-Qura (modern al-'Ula) and Taima, where Yaqut states there was a borax mine. For other accounts, see Taher (1979), pp. 13–14/146–8.
5 Neumann van Padang (1963), pp. 2–5. See also Sieberg (1932a), p. 795, under 640.

[644   23 H   Yemen]

Great earthquakes 'in most towns (or countries)' are reported by a seventeenth-century Yemeni annal for this year.[1] There is no particular reason to suppose that these events affected Yemen, although there is little evidence to identify them with earthquakes elsewhere in the Muslim world at this date.[2] The shocks are probably to be associated with events in the Hejaz a few years earlier (see previous entry).

1 Yahya b. al-Husain, I, 87.
2 Taher (1979), p. 15/25 also records an earthquake in Damascus on this date, but Poirier and Taher (1980), p. 2189 locate the shock in the Hejaz and Yemen (*sic*).

742   Yemen

Byzantine authors of the ninth and tenth centuries describe an earthquake in the Spring of 742 in the 'desert of Sava' or 'Sava'.[1] The latter may correspond to the monastery of Savva, south of Jerusalem; but Michael the Syrian (twelfth century) says the earthquake occurred in the 'desert of the Taiyeye' (Arabs), which can be equated with Sava or Saba. This is probably somewhere round the edges of the desert between Shabwa in Hadramaut and Ma'rib. The earthquake caused large landslides and many villages were overwhelmed by collapsing mountain sides.[2]

The date and details of this event have been greatly misrepresented in some earthquake catalogues, where it is located in Libya, 4000 km away.[3] However, a location in the Yemen is supported by the fact that this notice is followed by the account of another event in the Yemen during that year.[4]

With so few details, it is clearly difficult to suggest either an epicentral location or area of perceptibility for this event, which should have been relatively large, considering that its occurrence is recorded by chroniclers writing in Constantinople and Upper Mesopotamia.

1 Theophanes, p. 349/641; Agapius, p. 510/250; Cedrenus, p. 460/II, 5; see also Ambraseys and Melville (1983). 'Byzantine' authors do not again show an interest in events in Arabia till the extension of the Ottoman Empire into that region at the turn of the sixteenth century.
2 Michael the Syrian, XI, 22/II, 507.
3 Sieberg (1932a), p. 872, puts this event near Murzuq in Libya in 704 and again (1932b), p. 188, in 742, saying much damage was caused in Egypt by an earthquake located in Libya. Sieberg misquotes the sources used by earlier cataloguers, such as Hoff (1840), pp. 195–6, and Mallet (1853), p. 11. The account of 600 towns being

destroyed in fact follows Sawirus's account of the earthquake of 747, see next entry. Sieberg's mislocation is followed by later writers, e.g. Campbell (1968) and Kebeasy (1980).

4 Michael the Syrian mentions that monkeys attacked and ate some people in the Yemen!

747 January 18   Dead Sea

A large earthquake centring in the Dead Sea region was felt in Egypt (see Figure 2.6). Some damage was caused in Damietta; in Fustat the shock was strongly felt and caused fear but no damage.[1]

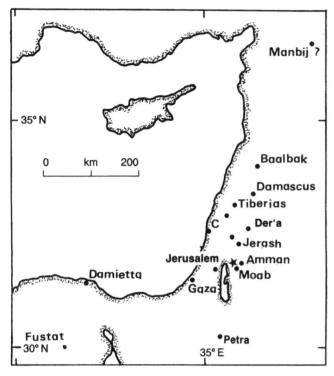

Figure 2.6. 747 January 18, Dead Sea.

There is considerable confusion over the dating of this event, which the Arabic sources put in 130 H (began 11 September 747),[2] and January 748 has recently been proposed as the correct date.[3] The effects of the earthquake are frequently confused with those of another event that affected parts of Syria two years later.[4]

1 Sawirus b. al-Muqaffaʿ (ed. Evetts), p. 139–40. He gives the date 21 Tuba/16 January; Agapius, p. 521 (who does not mention Egypt) has Kanun II/January. Neither specify the year.

2 Caetani, V, 1649 (re. 129/747?), 1664–5 (re. 130/748). Sibt b. al-Jauzi, fol. 235vo, and al-Dhahabi, V, 39–40 both have 130 H. Though late sources, they are generally reliable. Al-ʿUlaimi (ed. Najaf), I, 237–8, has Ramadan 130/May 748. See also al-Suyuti, pp. 23–4/9–10; Taher (1979), p. 18/28–30.

3 See Russell (1985), pp. 48–9, with a detailed discussion of the non-Muslim accounts, of which that by Theophanes (late eighth century) is the most important. The date 748 is also adopted by Gil (1992), pp. 89–90. Sieberg (1932b), p. 193, and Ben-Menahem (1979), p. 261, have 746 January 18, disregarding the systematic error in Theophanes.

4 For the second earthquake, see e.g. al-Khwarazmi (fl. *c.* 847) in Baethgen (1884), p. 126, under 131 H (began 31 August 748). Recently, Tsafrir and Foerster (1992), who discovered a coin dated 131 H buried under earthquake destruction at Bet Shean, assign the year 749 to one event amalgamated from all sources.

796 April   *180 H*   Hellenic Arc

A large earthquake in the Hellenic Arc with an epicentre near the island of Crete was strongly felt in Lower Egypt. In Alexandria the shock caused the collapse of the upper part of the lighthouse and great panic among the people, who left their homes. The earthquake was felt as far as Sicily and Constantinople.[1]

The Bulaq edition of al-Maqrizi's *Khitat* erroneously reads 777 H, as has been demonstrated by Wiet;[2] entries in some catalogues for an earthquake in 777/1375 are therefore spurious.[3]

1 Theophanes, p. 396/645; al-Tabari, III, 645; al-Suyuti, *Kashf*, p. 24, and *Husn*, II, 275; al-ʿUmari, fol. 26ro. See also Baratta (1901), p. 13 (under April 797), and Guidoboni (1989), p. 711, who lists separate earthquakes in Crete, Alexandria and Constantinople.

2 Al-Maqrizi, *Khitat*, ed. Bulaq, I, 156, ed. Wiet, III/2 (1922), p. 117. The date actually given by al-Maqrizi is 179 H.

3 See e.g. Taher (1979), p. 197/246; Poirier and Taher (1980), p. 2193.

827   *212 H*   Yemen

A great earthquake occurred in Sanʿa that was strongest in Aden; houses collapsed, villages were destroyed and many people perished.[1]

These details permit only the most approximate location for the event in the region between Sanʿa and Aden, rather nearer the latter.

1 Ibn al-Athir, VI, 138, followed by Ibn ʿAbd al-Majid, Ms. Arabe 5977, fol. 13ro; Yahya b. al-Husain, I, 152; *Anba al-zaman*, I, 60, cited in Taher (1979), p. 23/176; Ambraseys and Melville (1983). No contemporary source has yet been identified.

857 April   *Dhu'l-Hijja 242*   Egypt

A damaging earthquake occurred in Egypt. The shock was widely felt, shaking walls of mosques and destroying mosques and houses.[1]

There are no further details of this event, although al-Suyuti refers to a meteorite fall in Egypt at this time.[2] These events in Egypt are clearly to be distinguished from the large earthquake sequence in Damghan, north Iran, earlier in 242/856,[3] and from the landslides reported in the Yemen.[4]

1 Al-Yaʿqubi, II, 600.

2 Al-Suyuti, p. 26/12–13; there is some dispute over the location of the village concerned (*Sauda*), which al-Suyuti specifically places in Egypt; a location near Homs in Syria is given by Yaqut, III, 183. However, the same story of a meteorite fall is told by Ibn al-Dawadari, but with reference to Qairawan, which undermines al-Suyuti's identification; Taher (1979), pp. 27–8.

3 Ambraseys and Melville (1982), p. 37.
4 See next entry.

859   *245 H*   **Yemen**

Widespread earthquakes were felt this year, during the reign of al-Mutawakkil (d. 861). Among the events reported is that the spring al-Mushash outside Mecca dried up for some time, leading to a steep rise in the price of water.[1]

In the Yemen, where an earthquake is also alleged to have occurred this year,[2] the diminution of the supply of water below the Rai'an dam at the head of Wadi Dahr (or Zahr) due to undated earthquakes (*sic*) is noted by al-Hamdani (d. *c*.960).[3] This may be associated with the events of this year, or with a collapse of the dam due to other causes.

It should also be noted that the Yemen is reported to have been one of the areas affected by the widespread earthquakes of 242/856–7. A mountain in the Yemen, covered with cultivation, was transported and the landslide buried other cultivated properties.[4]

The seismic origin of all these events in western Arabia at this period is doubtful. It is not certain whether they occurred before or after the main earthquake of 245 H (see next entry), but it is convenient to separate them chronologically.

1 Al-Tabari, III/3, 1439; Ibn al-Athir, VII, 27; al-Suyuti, p. 27/13. For the well of Mushash, see al-Azraqi, p. 444; the drying up of the well is mentioned (but not the earthquake) by al-Nahrawali, p. 129, but not in earlier Meccan histories.
2 *Anba al-zaman*, I, 64, *cit*. Taher (1979), p. 31/149; Yahya b. al-Husain, I, 158–9.
3 Al-Hamdani, *al-Iklil*, p. 62/trans. pp. 42–4: the flow of water is said to have halved since the pre-Islamic period. This statement is followed by al-Razi, pp. 236, 243. See also Wilson (1989), pp. 181–2, 221, and Taher (1979), p. 31/176–7.
4 Al-Tabari, III/3, 1433; al-Suyuti, p. 27/13.

860 January   *Shawwal 245*   **Eastern Anatolia**

A large earthquake in Eastern Anatolia and North Syria, particularly destructive in Antioch, Jabala and Lattakia, was felt in Egypt.[1] The inhabitants of Tinnis heard a loud and prolonged roar. It is implied that the noise was due to the earthquake, but only al-Suyuti specifies that a shock occurred in Egypt.[2] Similarly, the statement that casualties were numerous appears to refer to Tinnis (or Egypt), but it may refer generally to the destruction in Antioch and elsewhere.

The year 245 H began on 8 April 859 and the month of Shawwal began on 30 December. The various events of this year are usually catalogued under 859 and often under 8 April.[3]

1 Al-Tabari, III/3, 1439–40; Ibn al-Athir, VII, 27; al-Suyuti, p. 27/13; Taher (1979), p. 30/32.

2 Al-Tabari and al-Suyuti have Tinnis in Egypt where the sound was heard, while Ibn al-Athir has Sis; this is probably an error arising from a manuscript without diacritical points. Some versions of al-Suyuti read Bilbais (e.g. Cairo National Library, Ms. *Hadith Taimur* 450, p. 14: *wa sami'a ahl bilbais min nāḥiyat miṣr dajjatan hā'ilatan*, 'the people of Bilbais in Egypt heard a fearful roar', compared with the printed text, *wa sumi'a bi-tinnīs dajjatun hā'ilatun*, 'a fearful roar was heard in Tinnis'. Bilbais is also given by Lyons (1907), p. 283 and Sieberg (1932b), p. 188. The earthquake in Syria is also mentioned by al-Ya'qubi, II, 601, who makes no reference to Egypt.
3 See e.g. Sieberg (1932a), p. 802, followed by Kallner-Amiran (1951), p. 227, and Ben-Menahem (1979), p. 278, who say the shock was damaging in Jerusalem. See also Alsinawi and Ghalib (1975), p. 542; Poirier and Taher (1980), p. 2189.

873 September?   *late 259 H*   **Hejaz**

An earthquake occurred in the Arabian desert, which caused the bedouin to flee to the protection of the Prophet's tomb in Medina and of the Ka'ba in Mecca. They brought in goods belonging to the pilgrims whom they had intercepted on the route (north of Mecca). The tribes particularly affected were the Banu Hilal and Banu Sulaim and other sub-tribes of Qais, from the region round Medina; they reported that a great number of bedouin had been killed in the desert.[1]

The association of this earthquake with the pilgrimage and the reference to the plundering of the pilgrims on their way south suggests the shock occurred in the month before Dhu'l-Hijja 259 (began 28 September 873) and that news of the event was provided by the returning pilgrims. It can be inferred that Mecca and Medina were not themselves damaged by the shock, which probably chiefly affected the region between Taima and Medina, to the east of the main pilgrim route.[2]

1 Al-Ya'qubi, II, 624; Taher (1979), pp. 32–3/149–50. Poirier and Taher (1980), p. 2189, have 10 Sha'ban 259 (11 June 873), which is the date of an earthquake the previous year in Iran. The account of the earthquake by Ambraseys and Melville (1989) needs refining, although we retain our views on its general location.
2 For the tribe of Sulaim b. Mansur (Qais b. 'Ailan), with stations around Khaibar, and at Harrat Sulaim, Harrat al-Narain ('the Two Fires'), Wadi al-Qura and Taima, see Kahhala (1968), II, 543–4; and III, 1221, for the Hilal b. 'Amir (also of Qais b. 'Ailan), further to the south, round Mecca. See also al-Biladi, II, 213. It is difficult to be precise about the territories of these tribes. For the pilgrim routes, see Brice (1981), map 22, Cornu (1985), map VII, and the account of these places in al-Wohaibi (1973), esp. pp. 293–4 re. Wadi al-Qura.

879 February 9   *12 Jumada II, 265*   **Gulf of Oman**

A strong earthquake occurred in Suhar. The shock happened during the morning, but is reported without further details.[1]

1 Al-Salimi, I, 166, says it was a Sunday; 9 February was a Monday.

881   *267 H*   **Hellenic Arc**

A major earthquake probably originating in the Hellenic Arc was felt throughout the Middle East, including Egypt.[1] No details are given and no early sources of information have yet been identified.

On 22 Shawwal (26 May) this year, a destructive earthquake in the western Mediterranean was reported in Cordoba and in North Morocco.[2] Ibn al-Athir alludes to this by including Spain and North Africa within his general account of the earthquake, and Moroccan authors also mention that the eastern Mediterranean was affected. However, it is improbable that these two earthquakes are in fact connected.[3]

1 Ibn al-Athir, **VII**, 120. Al-Ya'qubi's account (**II**, 621) of an earthquake at dawn on 22 Rajab 257/15 June 871, apparently affecting Egypt, may have been misdated by later authors. The 257/871 event is not reported elsewhere, but is dubious (see below, under unidentified earthquakes, p. 106).
2 Ibn 'Idhari, **II**, 104–5; Ibn Abi Zar', p. 60, trans. Beaumier, pp. 132–3; al-Nasiri, **I**, 164; Roux (1934); Poirier and Taher (1980), p. 2189.
3 The connection is also made by modern cataloguers, e.g. Taher (1979), pp. 33–4/33. Sieberg (1932b), p. 188, puts a shock in Egypt on 6 May 887, probably referring to the same event. Kallner-Amiran (1951), p. 227, and Ben-Menahem (1979), p. 287, refer to a tidal wave off the coast of Acre in 881, ultimately on the unreliable authority of Tholozan (1879).

885 November–   *Jumada II, 272*   **Lower Egypt**
December

A strong earthquake in Egypt destroyed houses and the Friday mosque (i.e. the mosque of 'Amr) in Misr (Fustat). At least 1000 people were killed.[1]

The epicentre of this earthquake must have been in or very near Lower Egypt: no effects are reported from elsewhere in the region.[2]

1 Al-Tabari, **III**/4, 2110. Ibn al-Jauzi, *Muntazam*, **V**/2, 85, gives the month as Jumada I, but otherwise follows al-Tabari. Ibn al-Jauzi, *al-Shudhur*, fol. 34vo; Ibn al-Athir, **VII**, 140; Elias of Nisibin, p. 89. Sa'id b. Bitriq (Eutychius), **II**, 71, puts the event in 273 H. It is possible that the new mosque of Ibn Tulun (completed 879) in the Qata'i' district to the N.E. of Fustat is intended. See also Creswell (1940), p. 390; Volkoff (1971), p. 36; Taher (1979), p. 34/229.
2 One late source, Muhammad b. 'Ali, Paris Ms 1507, fol. 177ro, duplicates the shock under 272 and 274 H and says that Syria was also affected, but there is no contemporary evidence for this.

[899 November 14   *7 Dhu'l-Qa'da 286*   **Egypt]**
Al-Makin (d.1273) mentions that a strong earthquake was felt in Egypt during the night; shocks lasted till morning and there was a widespread fall of meteorites.[1] Not only is this an intrinsically dubious report for a genuine earthquake, but it is strange that no earlier authority mentions the event.[2] Eutychius, who is almost certainly al-Makin's source, mentions the fall of meteorites under 288 H but does not mention an earthquake.[3]

Furthermore, it is worth noting that al-Makin's text is almost identical to his account of similar events in 323/935 (see below), and he might have confused the two.

1 Al-Makin, ed. Erpenius, p. 181/trans. Erpenius, **II**, 17.
2 The earthquake is not mentioned by al-Tabari, al-Mas'udi or Ibn al-Athir.
3 Sa'id b. Bitriq (Eutychius), **II**, 72–3. Both authors refer to a great storm in 284 H.

912   *299 H*   **Egypt**

An earthquake shock was felt in Egypt this year.[1] It is not certain whether this originated from the central Mediterranean region, where an earthquake felt in Qairawan caused some damage on the Tunisian coast.[2]

1 Al-Mas'udi, *Muruj*, **VIII**, 282; he also refers to the appearance of Halley's comet this year, mentioned in other Muslim sources that do not report the earthquake, cf. Grumel (1958), p. 472. al-Mas'udi also mentions an earthquake in al-Kufa this year, Ambraseys and Melville (1982), p. 38.
2 Anon., *Kitab al-'uyun*, **IV**/1, 161; Ibn al-Athir, **VIII**, 22; Ibn 'Idhari, **I**, 166. See also Taher (1979), p. 35/262; Poirier and Taher (1980), p. 2190.

935 October 4   *3 Dhu' l-Qa'da 323*   **Lower Egypt**

A destructive earthquake with an epicentre somewhere in or near Egypt. There is some ambiguity as to whether the shock was damaging in Old Cairo (*misr*) specifically or Egypt generally, but some houses are said to have been destroyed.[1] Taher incorrectly implies that it was due to the earthquake that several leading members of society fled to Barqa (Barce in Cyrenaica).[2]

A fall of meteorites is also reported during the event, inviting comparisons with the alleged earthquake of 7 Dhu'l-Qa'da 286/14 November 899 (see above); Eutychius's text is very similar to al-Makin's account of the earlier shock. Al-Makin gives the date of this 'second' earthquake as 11 Dhu'l-Qa'da (12 October), during the reign of al-Radi (reg. 322–9/934–40).[3]

1 Sa'id b. Bitriq, **II**, 87; Ibn al-Dawadari, *Kanz*, Ms. cited by Taher (1979), p. 36/229, under 321 H. See also Lyons (1907), p. 284, and Sieberg (1932b), p. 188, under 934 AD. Poirier and Taher (1980), p. 2190, mistakenly cite al-Ya'qubi as their source and treat the earthquake as two separate events, which in turn misleads Guidoboni (1989), p. 715.
2 Taher, *l.c.*, has extracted the account of the earthquake in such a way as to misrepresent the purpose of Muhammad b. Tughj (the Ikhshid)'s movements first to Barqa and then to Alexandria. See also Lane-Poole (1901, repr. 1968), p. 81.
3 Al-Makin b. al-'Amid, ed. Erpenius, p. 208. A marginal note incorporated into Erpenius's text of al-Makin says that a similar event occurred in the 1240s, but this has not been verified (see below, Section 2.3).

950 July 25   *6 Safar 339*   **Lower Egypt**
A series of earthquakes destroyed most of the houses in *misr* (i.e. Old Cairo) and a portion of the old mosque

(of 'Amr) fell down.[1] The inhabitants fled into the open.[2]

There is some disagreement over the correct date in the sources, and room for confusion with the earthquakes that followed (see next entry).[3]

The damaging nature of the earthquake suggests a shock of local origin.

1 Anon., *'Uyun*, IV/2, 464; fuller details in al-Nuwairi, *Nihayat*, XXVI, fol. 18, Ms. cited by Taher (1979), p. 37/229–30. See also Poirier and Taher (1980), p. 2190. This was the first of a series of earthquakes in Egypt, mentioned by Lyons (1907), p. 284, and Sieberg (1932b), p. 188, under the year 954, which was actually the date of a great fire in Old Cairo: Lyons refers inaccurately to Petrie (i.e. Lane-Poole, 1901), p. 88.
2 Ibn al-Dawadari, V/3, 323 (under 338 H), Ms. quoted by Taher, *l.c.* See also al-'Umari, fol. 41ro (under 336 H).
3 Ibn al-Dawadari cites 'Sahib *al-Barq al-Shami*' for the year 338 H, i.e. 'Imad al-Din (d.597/1201) and cites al-Quda'i (an Egyptian author, d.464/1072) and Ibn 'Asakir (d. 571/1176) for an earthquake in Egypt in 340 H (see next entry), implying that they all refer to the same earthquake: which may be the case.

951 September 15   *Sunday 10*   Lower Egypt?
             Rabi' II, 340

Yahya b. Sa'id, who does not record the previous earthquake, records a damaging shock in Egypt (*misr*), with casualties. The lighthouse at Alexandria was damaged and in a number of places new springs of water appeared as a result of the earthquake.[1]

This may be a reference to the previous event, given the chronological uncertainties in the sources, but the damage to villages and other details suggest this was again a destructive shock local to Egypt, possibly just offshore from the Nile Delta.[2]

There might, nevertheless, be some confusion with another earthquake in 340 H that was locally destructive in southeast Anatolia, in the region of Aleppo and al-'Awasim.[3] This is unlikely to have been damaging in Egypt, but might have been felt there, particularly in Alexandria, and the two earthquakes might have been conflated into one account.

1 Yahya b. Sa'id, ed. Cheikho, p. 113 (ed. Vasiliev, p. 770). His date is very precise: 'the night before Monday 15 September'. But the detail about Alexandria, which suggests a distant epicentre, may refer to the earthquake of 344/956 (see below). Yahya (of Antioch) was the continuator of Eutychius's chronicle; he died in 1066.
2 Anon., *'Uyun*, IV/2, 467, a contemporary author who specifically refers to earthquakes in Egypt in 339 *and* 340 H. It is not clear whether Old Cairo (also *misr*) was affected, or only village districts. Al-Nuwairi, *Nihayat*, I/2, fol. 464, Ms. quoted by Taher (1979), pp. 38–9/230, clearly follows the *'Uyun*, but does not mention Egypt. Both say the shocks lasted three days. See also Poirier and Taher (1980), p. 2190.
Ibn al-Dawadari, V/3, fol. 324, Ms. quoted by Taher, p. 39 n.1, says the earthquake of Safar (see previous entry) was followed by another in Rabi' I (? 340), a month early. He repeats that the inhabitants left their houses to live in the open, and reports that the ground fissured open in places, revealing putrid water, echoing the account of Yahya b. Sa'id (and also al-Mas'udi's account of the earthquake

of 344/956 in Khurasan, cf. below). He also remarks that shocks continued for six months, which perhaps reflects the later seismic activity elsewhere in 340 H (see next note).
3 Yaqut, II, 791; a-Dhahabi, Ms. 1581, fol. 164ro; Ibn Taghribirdi, III, 305; Sibt b. al-'Ajami, p. 7. These are all relatively late sources. Nowhere is the *month* of the Anatolian shock mentioned, increasing the difficulties of identification. Some of the Syrian sources mentioned by Ibn al-Dawadari (see note 3 above) may refer to this event, rather than a separate shock in Egypt. We may also note an earthquake in the central Mediterranean, which is said to have caused damage in Sicily, reported on uncertain evidence in Amari (1854), I, 10, II, 127, and Baratta (1901), p. 17. The exact date is unclear.

956 January 5   *Saturday 18*   Eastern
             Ramadan 344   Mediterranean

A large offshore earthquake was felt in Syria and Egypt, where the shock was of long duration and caused the collapse of the upper 22 metres of the lighthouse in Alexandria. In Old Cairo the earthquake caused great concern but insignificant damage.[1] Later sources mention only Egypt and say that the earthquake destroyed some houses there. The shock lasted three hours, and the people turned to God in fear.[2]

This earthquake should not be confused with the earthquake of 7 Jumada I, 344/29 August 955, which affected Cordoba in southern Spain and perhaps parts of the Maghrib.[3]

1 Thirty cubits (*dhira'*) fell according to al-Mas'udi, *Tanbih*, pp. 48–50, followed by al-Maqrizi, *Khitat*, ed. Wiet, III/2 (1922), p. 120. Al-Mas'udi (d. 956) was in Fustat, and an eye-witness of this event, which he dates very precisely in a variety of calendars. He says the shock occurred at midday and lasted half an hour; he also describes a deep convulsion in the ground. It must be noted that al-Mas'udi does not mention the previous earthquake(s) of 339 and 340 H, and that his inclusion of Syria (which is not mentioned by any other source) might seem suspect (see also note 3 below).
2 Al-Dhahabi, Ms. 1581, fol. 193ro, repeated by al-Suyuti, p. 29/15; Taher (1979), p. 39. Even if exaggerated, this detail confirms the long duration of the shock and the likelihood of a distant epicentre; it also suggests the possibility of aftershocks, as in the case of the 340/951 event (three *days*).
3 Ibn 'Idhari, II, 220. Al-Mas'udi implausibly includes the Maghrib within the compass of the Egyptian event. For the earthquake in Egypt, see also Poirier and Taher (1980), p. 2190, who incorrectly have 1 January 956. Mallet (1853), p. 16, under 965 or 967, is probably referring to this event. Al-Mas'udi also mentions earthquakes in Khurasan this year, destroying villages and turning waters foetid. The similarity of Ibn al-Dawadari's wording concerning 340/951 adds to the sense of confusion surrounding these events, which cannot easily be resolved at this remove in time.

963 May 12   *14 Rabi' II, 352*   Hellenic Arc?

An earthquake was felt in Egypt during the night.[1] The shock was possibly associated with an earthquake that was alleged to have affected Sicily and Syria on 22 July 963,[2] and which might have originated in the Hellenic Arc.

Al-Suyuti states that earthquakes lasting six months occurred in Egypt during the reign of Kafur the Ikhshid

(355–7/966–8), which may refer also to this earth-quake: Kafur had been effective ruler of Egypt since 344/956.[3]

There is room for doubt over the true nature of this event: Yahya mentions that there was also a great storm that night, and the sun appeared red all the following day.

1 Yahya b. Sa'id, ed. Cheikho, p. 121; ed. Vasiliev, p. 791. Yahya's text reads Rabi' II in both printed editions, but Vasiliev's translation gives Rabi' I. See also Taher (1979), p. 46/231.
2 Baratta (1901), p. 17. The reference to a shock in Syria has not been confirmed in local sources; the report must be viewed with great suspicion. An earthquake in north Syria is reported in 362/972, al-Maqrizi, *Itti'az*, I, 132.
3 Al-Suyuti, p. 30/16. Taher, p. 38/230, appears to associate this with the events of 340/951 (*q.v.*, note 2). See also Section 3.2.

997   *387 H*   Upper Egypt

An earthquake is reported in Qus, along with other phenomena this year, particularly violent storms. Five hundred palmtrees were uprooted and a number of heavily laden boats sank.[1] This is probably a dubious seismic event; Ibn al-Dawadari, for example, mentions only the storm with no location, under the following year, 388 H. Nevertheless a fourteenth-century author, al-'Umari, is quoted as reporting that earthquakes occurred in the reign of the Fatimid Caliph al-Hakim (386–411/996–1021) and no other (or earlier) refer-ences have yet been found.[2]

The mention of Qus and not of Cairo suggests that the epicentral area of this shock, if genuine, must be sought in Upper Egypt, possibly in the vicinity of the Red Sea.

1 Al-Maqrizi, *Itti'az*, II, 16.
2 Ibn al-Dawadari, VI, 262; al-'Umari, *Masalik al-absar*, quoted by al-Suyuti, pp. 31–2/17–18; Ambraseys (1961), p. 24. Taher (1979), pp. 50–1/232 applies these verses to the year 407/1016. See also under 352/963, and Section 2.3.

1033 December 5   *10 Muharram 425*   Dead Sea

A damaging earthquake in the Jordan Valley was felt as far south as Gaza and Ascalon and probably in Egypt and the Negev (see Figure 2.7).[1] It was associated with a seismic sea-wave on the Mediterranean coast of Pales-tine and it was followed by many and strong aftershocks until 17 February 1034.[2] The correct date, 12 Teveth (5 December), is given by an eye-witness Hebrew source.[3] The variety of other dates available has led to some confusion in secondary sources.[4]

There is no specific evidence of the effects of the shock in Egypt and the northern Red Sea area.

1 Egypt is specifically mentioned by Ibn al-Athir, IX, 151; Ibn Shakir al-Kutubi, *'Uyun*, Ms. Or.3005, fol.112vo; al-Suyuti, p. 32/17. See also Taher (1979), pp. 51–3/35–8.

2 Cedrenus, pp. 730, 732, 737/II, 503, 707; Glycas, p. 315/587; Sawirus b. al-Muqaffa', II/2, p. 157/trans. 237–8; Ibn al-Jauzi, VIII, 77; al-Dhahabi, Ms. Or.49, fol.18ro.
3 Solomon b. Yehuda, p. 176 ff; trans. Mann (1920), I, 156–8. See also Gil (1992), pp. 399–400. Both Solomon and Yahya b. Sa'id (see next note) mention that the shock occurred on Thursday evening; December 5 was a Wednesday.
4 Nasir-i Khusrau, p. 19, who visited al-Ramla 13 years later, dates the earthquake 15 Muharram (10 December); Yahya b. Sa'id, ed. Cheikho, p. 272, gives 10 Safar/4 January 1034, which is followed by Poirier and Taher (1980), p. 2190. Al-Fariqi, p. 161, puts the earthquake in 439/1047, at the time of Alp Arslan's victories in E. Anatolia, i.e. probably intending 460/1068 (see below). Hoff (1840), p. 207, confuses this event with one reported in Constantinople on 6 March 1033 (Cedrenus, II, 500), which has misled later cata-loguers, e.g. Sieberg (1932b), p. 193. Ben-Menahem (1979), pp. 260, 287, lists the earthquake under 5 December 1033, 6 March 1032 and 4 January 1034.

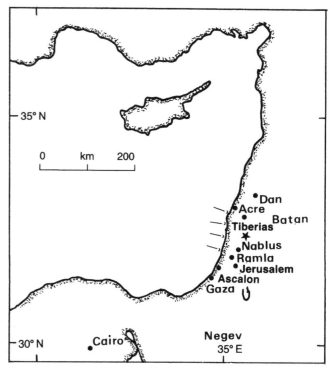

Figure 2.7. 1033 December 5, Dead Sea.

1068 March 18   *Tuesday 11*   Northern Hejaz
     Jumada I, 460

A major earthquake in the Hejaz and northwest Arabia occurred during the morning and was reported to have killed in all about 20 000 people. Aila at the head of the Gulf of Aqaba was completely destroyed with all but twelve of its inhabitants. In Tabuk, three new springs of water appeared at a place called al-Qur, and in Taima the ground was split open to reveal buried treasures of gold and golden jewellery. Near here a permanent and productive spring of water gushed out. The earthquake was felt at Wadi al-Qura, Khaibar, al-Marwa, Medina (where the shock brought down two decorative crestings of the mosque of the Prophet), and Wadi al-Safra, Badr and Yanbu', to the southwest of Medina (see Figure 2.8).[1]

In Sinai, the earthquake was strong enough to cause alarm at the monastery of St Catherine.[2] The shock was also experienced in Egypt, where Tinnis was apparently damaged, but not in Alexandria.[3] In Cairo, the only damage reported was to one corner of the mosque of 'Amr in Fustat.[4] Water rose in wells in Egypt and Palestine, and the retreat and return of the sea on the Mediterranean coast of Palestine drowned a large number of people (see below). The earthquake was felt as far as al-Rahba and al-Kufa on the Euphrates, the water of which was reported to have overtopped its banks;[5] it was also felt in Baghdad.[6] Two more shocks followed within an hour.[7]

The epicentral area of this event must be sought in the sparsely inhabited region between Aila and Taima, where the available evidence points to faulting but where information about the few inhabited villages, even if destroyed, would not attract much attention.[8]

It is generally reported that the earthquake was also destructive at al-Ramla (Ramallah) in Palestine, causing the ruin of many houses with loss of life. One account speaks of 15 000 casualties, including 200 lads at one school. Damage extended to Baniyas, where about 100 people were killed, and the same in Jerusalem.[9] There is no other evidence of damage in Jerusalem, though it is reported that the roof of the Dome of the Rock was displaced and then returned to its former position.[10] The assertion that the inhabitants of al-Ramla migrated to Jerusalem after the earthquake also suggests damage there was slight.[11] There is some archaeological evidence to suggest earthquake damage at the citadel of Amman in Jordan, possibly from this event.[12] The absence of information from Damascus and Aleppo to the north, and the silence of contemporary Byzantine historians, all argues against the 18 March 1068 earthquake having an epicentre near al-Ramla, which is 500 km from the region between Tabuk and Taima. In fact, a separate, locally destructive shock affected Palestine on 24 Rajab/29 May, two months after the earthquake in northwest Arabia,[13] and the two events have been amalgamated by later authors. The tsunami and effect on wells in Palestine possibly occurred during this earthquake, rather than in March.

The date of this earthquake is often misreported in earthquake catalogues, under 1067 April 20, 1067 November 11, and 1070 February 2.[14]

1 Ibn al-Banna, in Makdisi (1956), pp. 250–1, 256. He says it reached from Surair in Hejaz (a wadi near Medina, see Yaqut, III, 88–9) to al-Ramla and most of Syria. Ibn al-Banna's Wadi al-Safa must be equated with al-Safra near Medina; and al-Marwa refers to (Dhu) al-Marwa in Wadi al-Qura. See also Sibt b. al-Jauzi, Paris Ms. fol. 111vo–112ro, who says that all the places involved were shaken 'the same night', though he also records that the shock occurred at 'two and a half hours' of the morning (c. 8.30); al-Dhahabi, Ms. Or.50, fol. 4vo; al-Suyuti, pp. 34–5/20–1.
2 Anon. (1817), p. 125, under 1091 (see below); Eckenstein (1921), pp. 144–5, allows the possibility that the events in question occurred in 1068/9, but even if this is not the case, it is unlikely that the earthquake was not experienced in St Catherine's.
3 Sawirus b. al-Muqaffa', II/2, 182/trans. p. 277.
4 Ibn al-Jauzi, VIII, 256. He gives additional information about the effects of the earthquake in a second account, in which he dates the shock Tuesday 11 Jumada I, 462/25 February 1070. This misleads later authors and many modern cataloguers (see below). The date should clearly be 460 H, for Ibn al-Jauzi gives an alternative date of 18 Adhar (18 March in the Syriac calendar).
5 Sibt b. al-Jauzi, l.c.
6 Ibn al-Banna, p. 251.
7 Ibn al-Jauzi, p. 256; al-Suyuti, l.c.
8 Melville (1984a); Ambraseys and Melville (1989).
9 Ibn al-Qalanisi, p. 94, gives Tuesday 10 (sic) Jumada I; Ibn al-Banna, followed by Ibn al-Jauzi, VIII, 248, has 15 000 casualties; later authors have 25 000!
10 Ibn al-Banna, l.c.. Sibt b. al-Jauzi notes that other accounts say the Dome was cracked, cf. Ibn al-Athir, X, 20, and al-'Ulaimi, p. 270/trans. p. 69. The latter previously refers to the fall of a great lantern in the Dome, in 452/1060, which is mentioned by earlier authors who do not refer to the earthquake, see Le Strange (1887), pp. 286–7, 304: there may be some connection.
11 Ibn Shaddad, A'laq, p. 182.
12 A. Northedge, pers. comm. (1990). The chronological evidence is not very conclusive.
13 The contemporary Baghdad diarist, Ibn al-Banna, records the arrival of news early in Shawwal/August of an earthquake in Palestine in Rajab/May, Makdisi, p. 248. See also Gil (1992), p. 408. Unfortunately, Ibn al-Banna himself then repeats much of this information in his report of the Jumada I/March earthquake, misleading later writers, such as Ibn al-Jauzi. Contemporary correspondence between leaders of the Jewish communities in Jerusalem and al-Ramla appears to include no reference to the total destruction of the latter, under either date. It is possible that the seemingly exaggerated description of the earthquake in al-Ramla in Muslim sources echoes the earlier devastation of 1033, and refers to the later effects of famine and epidemic, which also affected Egypt, see Ibn al-Muyassir, p. 35, and Ibn Zafir, pp. 74–5, who make no reference to the earthquake. Sawirus, l.c., describes the depopulation of Tinnis and al-Ramla by an epidemic. Note also that Sawirus's account of the 1033 event (see above) could equally well apply to the 1068 earthquake and confusion between the two is clearly possible.
14 See Sieberg (1932a), p. 802, (1932b), p. 193; Kallner-Amiran (1951), p. 227; Ben-Menahem (1979), pp. 259, 287 (where the event is located offshore Yavne, a misidentification of Yanbu'). The date 1067 April 20 arises because Sieberg's late fifteenth-century source incorrectly gives the year 459 H; 11 November 1067 is the first day of the year 460 H, 25 February 1070 refers to the year 462 H, see above, note 4.

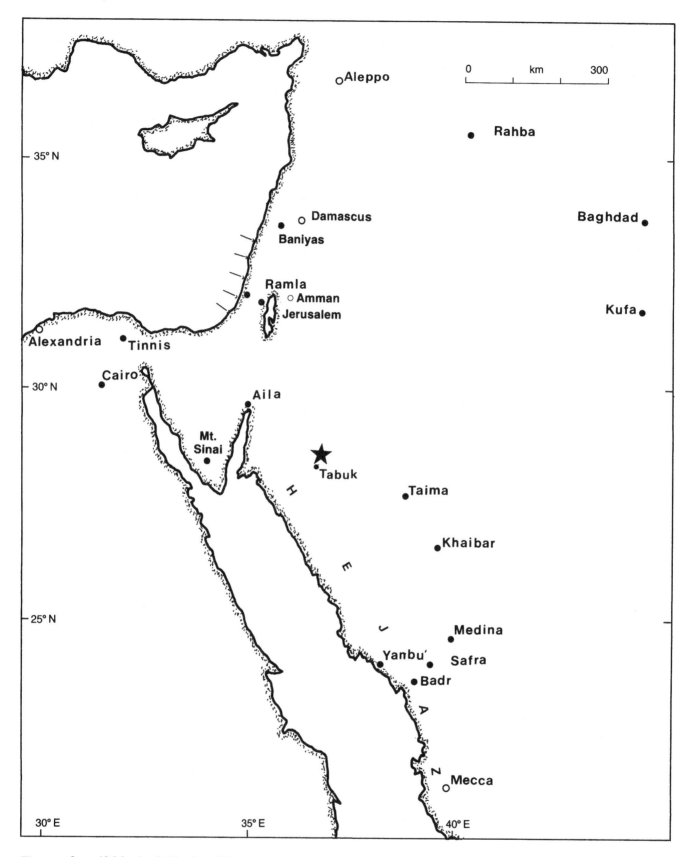

Figure 2.8. 1068 March 18, Northern Hejaz.

1072   *464 H*   **Yemen**

There was a great earthquake in San'a, Zabid and al-Mukha in the Yemen. Some houses were destroyed and about 50 people were killed in the wreckage.[1]

It is not clear where damage was concentrated, but the distribution of places mentioned suggests an epicentre to the northeast of Zabid, and a relatively large event (see Figure 2.9).

1 Al-'Umari, fol. 59ro; the shock is not mentioned in the BL manuscript. Al-'Umari (d. 1811) seems to have had a useful source of information on Yemen events, which has not yet been identified.

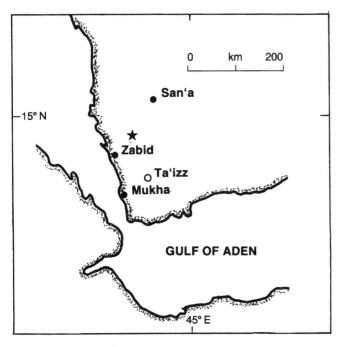

Figure 2.9. 1072, Yemen.

[1086   *479 H*   **Eastern Anatolia]**

A large earthquake in southeastern Anatolia was felt in Iraq, Mesopotamia and Syria.[1] One late author alleges that the shock was felt as far as Egypt,[2] though we have found no evidence of this.

1 Ibn al-Athir, X, 54; Abu 'l-Fida, II, 199; al-Suyuti, p. 36/22. See also Alsinawi and Ghalib (1975), p. 543; Taher (1979), p. 61/41.

2 Al-Qusi, fol. 94.

1091 February 12?   **Sinai**

A series of shocks was felt during the night at the monastery of St Catherine in Sinai. The earthquake is dated by reference to the death of the Archbishop John the Athenian, a date which is the subject of some dispute.[1]

1 Anon. (1817), p. 125, the sole author to mention the earthquake,

has 1091, but elsewhere 1061 or 1071. Eckenstein (1921), pp. 144–5, puts John's death in 1069, which raises the possibility of the 18 March 1068 earthquake, see above. Cheikho (1907), p. 416, quotes an Arabic manuscript that gives 1091 for his death, and this is followed by Rabino (1937), p. 82. No reference to the shock has been found in Arabic sources for this year (484 H), or the surrounding period, nor to the bedouin raids on the monastery that are said to have led to the Archbishop's death.

1111 August 31   **Lower Egypt**

A damaging earthquake in Lower Egypt affected Cairo and Fustat around 9.0 a.m. It was felt throughout the country and ruined a number of places.[1] It is alleged that the earthquake provided the opportunity for the destruction of the church of St Michael in the island of Rauda, on the orders of the wazir, al-Afdal.[2]

1 *Synaxarium Alex.*, p. 4/trans. p. 5; *Ethiopic Synaxarium*, trans. Budge, I, 11. This occurred on 3 Tut/Maskaram in the ninth year of the Patriarchate of Macarius II (1102–27).

2 Al-Makin (Ibn al-'Amid), ed. and trans. Erpenius, p. 298/369 reads 3 Tuba/29 December; fuller details in Sawirus b. al-Muqaffa', III/1, 5–7/trans. pp. 9–11. Sawirus's text has the year 818 of the Holy Martyrs (1101) instead of 828: See also Lyons (1907), p. 284, under the date Ramadan 504/March 1111 and Sieberg (1932b), p. 188 (under 26 May 1111). No reference to the earthquake has yet been found in the Muslim chronicles for this year (505 H). See also Taher (1979), pp. 64–6/233; al-Afdal was wazir until 1121.

1121   *515 H*   **Red Sea**

An earthquake in the Hejaz damaged the Yemeni (southern) corner of the Ka'ba in Mecca and also destroyed part of the mosque of the Prophet in Medina. According to the earliest source, Ibn al-Athir, other places were affected, including Mosul (?).[1] This may be a reference to a separate event.

No reference to the earthquake has so far been found in twelfth-century histories.[2] The facts that the shock was widely felt and that only slight damage occurred to relatively long-period structures located 330 km apart, suggest this was a large earthquake originating in the main trough of the Red Sea between Mecca and Medina.

1 Ibn al-Athir, X, 211; his account, without the statement about Mosul, is followed by all later authors, e.g. Abu 'l-Fida, II, 235; al-Suyuti, p. 37/23; cf. Taher (1979), p. 71/152. Poirier and Taher (1980), p. 2191, have 516/1122, misrepresenting their source, Ibn Kathir, XII, 188.

2 The early twentieth-century author, Rif'at Pasha (1925), I, 274, puts the earthquake in 559/1164, clearly referring to the same event. This date also finds no confirmation in Meccan sources.

1138 October 15   *8 Safar 533*   **North Syria**

It is said by a late source[1] that earthquake shocks were felt in Egypt, originating from the series of earthquakes that devastated northern Syria.[2] No contemporary evidence of the effects in Egypt has been retrieved, and the

the notice may be spurious, though the author is reliable and Egyptian sources for the period are lost.

1 Al-'Umari, fol. 72ro; see also Ambraseys (1961), p. 25, possibly intending the shock of 587/1191, see below.

2 For which, see Ibn al-Qalanisi, pp. 268–70; Ibn al-Athir, **XI**, 27–8; al-Suyuti, pp. 37–8/23–4; Taher (1979), pp. 74–8/43–5.

1145 September 15 *Saturday 25 Rabi' I, 540* **Yemen**

A strong shock in Yemen destroyed many houses, and many castles fell; the ground shook the people violently, but no-one was killed.[1]

The details given do not permit an accurate location of the event; in view of the absence of casualties, the shock might have occurred during the day and have been preceded by premonitory tremors.

1 Al-Khazraji, *al-'Asjad*, facs. edn, p. 134, Akwa' Ms. fol. 64ro. His source is probably the same as for the 1154 earthquake and it might have affected the same area, see below.

1154 September 11 *Sunday 9 Rajab 549* **Yemen**

A strong earthquake in Yemen affected the area between San'a and Aden. A great number of people died and many villages, forts and dwellings were destroyed.[1]

An unusually detailed break-down of the losses caused by this earthquake provides figures of casualties in many villages and fortified houses in the region round Ibb (see table below). In Ibb itself, 370 people were killed.[2] To the east, Hisn Habb, an important settlement in mediaeval times, was partly destroyed and eight people

were killed. In the same vicinity, Hisn 'Izzan was also partially destroyed and the two villages of Haqla 'Ulya and Sufla were completely destroyed along with their mosques. Many people were killed.[3] In Dhu'l-Maliki, further to the southeast, 14 people were killed and another four in the collapse of a dwelling.[4] In Raiman, a valley north of Jabal Habb, various dwellings were destroyed, with casualties, but further east, up to the head of Wadi Marar, a number of houses were destroyed but without fatalities.[5]

Between Ibb and Dhu Jibla to the southwest, the earthquake destroyed the villages of Dhi Hawal and al-'Aqayir, killing 70 and eight people respectively. Some months earlier, a meteorite impact had caused a shock in this area.[6] Damage was apparently most severe to the northwest of Ibb, towards Hubaish; heavy casualties were reported in al-Sahul and particularly al-Mahalla, with heavy losses in al-Shuwahit (near Khadid), Hisn al-Majma'a, Ghulas, Hisn Shu'aib, Akamat al-Hadda, Hisn al-Khadra and al-Shawafi.[7] Further west, towards al-'Udain, in the region of 'Anna, the shock caused damage but no casualties.[8] To the north, there were nine deaths in al-Mahsin, Rihab was destroyed with the loss of 26 lives, and 21 people were killed in Akamat al-Sumara, some 30 km north of Ibb.[9]

Many of the places mentioned cannot easily be identified, and the correct spelling is not always clear. Most are presumably located in the densely populated region

Places affected by the earthquake of 11 September 1154

Place	Details	Deaths
San'a	felt?	
Aden	felt?	
Hisn Habb	partially destroyed	8
Hisn 'Izzan	partially destroyed	
Haqla	two villages, completely destroyed	Many
Dala'a (?)	destroyed	14
Raiman	houses destroyed	
Manzil of Muslim b. (al-)Husain		5
Manzil of 'Isa b. Ahmad		3
Manzil of 'Ali b. Ahmad, his brother		3
Manzil of Ya'fur		5
Qasr of Muhammad b. Muslim		1
Dar of Ibn 'Abd al-Sami'		2
Ba'dan to wadi Marara	a number of houses destroyed	0
al-Sahul (Dar of Ibn al-'Arab/Gharb)		7+
Qaryat al-'Aqayir	destroyed	8
Qaryat al-Mahsin	destroyed	9
Qaryat Dhu 'l-Maliki (?)		14
Manzil of Dhi Qaifan		4
Akamat al-Ratisa (?)		15
Manzil Mu'mir		7
Dar of (Ibn) 'Abbas, partly ruined		5

Places affected by the earthquake of 11 September 1154 cont.

Place	Details	Deaths
Dar of Bani Mu'mir (al-Mahalla)		55
Qasr of Bani Mu'mir		7
al-Mahalla	partly destroyed	13
Akamat al-Hadda	below Ma'ri in al-Shawafi	67
Hisn Shuwahit		37
Hisn Dhi'l-Harbiyya (? Harasa)		40+
Akamat Sumara		11
Hisn al-Zafar	in al-Shawafi	8
Hisn al-Majma'a		35
Amir's mi'qab destroyed, with mosque		22
Rihab	destroyed	26
al-Manqal	in Sumara	3
Manzil al-Mahir (Makhir)	in Ghulas	7
Akamat al-Sahafi		7
Dar Ibn Misbah (?)	in Ghulas	36
Qasr Buhairan (?)	in Khadid	
Akamat in Masyuq (?)	house destroyed	14
Qasr Ibn Sabir		15
Hisn al-Khadra	in Uhada; destroyed	75
Hisn Yafuz		6
Hisn Shu'aib		30
Rusaiqa (? Rasagha)	2 manzils destroyed	2
Hisn Yaris		0
Qaryat al-Tafadi	no deaths	0
Dar al-Armad	below Hisn al-Hadda	5
Hisn Khabaz (? Habban)	in 'Anna, partly destroyed, with al- Buq'a and al-Maqra'a	0
Hisn Mushar (? Sar)		22
Akamat al-Ma		18
Akamat Munqidha	one survivor	22
Sumu' (?)	houses and mosques destroyed	23
Qaryat Warali (Wazali?)	partly destroyed	3
Ibb	destroyed (some say 300 deaths)	370
Manzil al-Rudaini (?) below Ibb		8
Manzil al-Mutahhara (al-Zahr?)		32
Manzil al-Kariyya		19
(Manzil) al-Hafif		12
al-Rasama (? Radama)	partly destroyed	8
Manazil Mawathir		12
Qaryat Dhi Hawal		70
Unamir	mostly destroyed	3
Dhu Jibla	houses and palaces shaken to pieces	2
Manzil al-Tab'i (?)		8
	house fell on Ibn Murad	4
Akamat al-Hamra		12
Manzil al-Muflih (?) in al-Mandam (?)		6
Qaryat al-Samra (Shamr ?)		30
Na'ima	Manzil of Ibn 'Abd al-Salam	5
Husun (Ibn) al-Maswad		3
Qatab	house fell on owner	
Qaryat al-Maqtah	completely destroyed	10
Qaryat 'Itab		
Qasr al-Tab'i (?)	partly ruined	0
al-Nawani (Tawani ?)	Dar of Yasin fell on him	12
	Total	**1345+**

around Ibb, rather than much further afield. The area worst affected lies northeast of Ta'izz (which, interestingly, was apparently not affected) and approximately halfway between San'a and Aden (see Figure 2.10).[10] The total number of victims must have been around 1400, and a great number of cattle and livestock were killed. In view of the large numbers of casualties, it is probable that the earthquake struck at night.

1 Yahya b. al-Husain, I, 308. A more detailed account is given by his cousin, author of the *Anba al-zaman*, of which two versions exist. Cairo Ms. 17075, I, 213, cited by Taher (1979), p. 80/178–9, is the only account (late seventeenth century) to mention the exact date of the shock. Both versions say the source of their information is a work called *al-Mawa'iz wa'l-'ibar* by the contemporary author, al-'Arashani (d. 557/1162), who, as noted by Taher, also wrote a book on earthquakes. Neither work has been traced, but al-'Arashani's account is certainly the basis for that of al-Khazraji, cited below.

2 The *Anba* Ms. 17075 is misquoted by Taher, who says the 300 people died in San'a. This is clearly incorrect: San'a is around 150 km north of the apparent macroseismic epicentral area. Both this and the Cairo Ms. 1347, fol. 51ro, say the 300 casualties were in Ibb. The latter version omits any reference to Aden.

The most detailed account of the event is preserved in al-Khazraji's *al-'Asjad al-masbuk*, facs. edn, pp. 132–4, Akwa' Ms. fols. 63ro–64ro, where it is reported that 300 people were killed in Ibb, 'but some say 370'. The higher figure is given by Ibn al-Daiba', *Qurrat*, I, 363, citing al-Khazraji. Al-Hibshi (1982), citing 'al-Khuraji's *al-Sahn al-masbuk* [sic], p. 132', follows the facsimile edition, but omits 24 lines of the text, including the reference to Ibb.

3 For Habb, 7 km east of Ibb, see Smith (1978), p. 153. Al-Waisi (1962), p. 44, vocalises this as Khabb. For 'Izzan (or 'Azzan) and Haqla, see the editor's notes in Ibn al-Daiba', *Qurrat*, I, 283, 336, II, 101. Haqla is perhaps to the southwest of Ibb, though localities of this name, and Haqlain, ('the two Haqls') can be found on modern maps to the S.E. See particularly the YAR (50) series maps, sheets 1344 A1 and A2.

4 Al-Hibshi (1982) reads Maliki, but both MSS of the *'Asjad* read Dhu 'l-Maliki (? Mulki). A village called Maliki exists S.E. of Ibb.

5 This is all part of the mountainous district of Ba'dan, wrongly located west of Ibb on the map in Kay (1892); cf. Redhouse (1908), p. 178 (note 1247). For this area, see YAR (50) sheet 1444 C4.

6 This occurred in al-Salahifa, a few kilometres west of Dhu Jibla (LA 9738). Conflicting dates are given in the sources: al-Khazraji, *al-'Asjad*, facs. edn, p. 132 has Fri. 6 Rabi' II (Sat. 20 June), reproduced by al-Hibshi (1982) as Fri. 26 Rabi' II (Sat. 10 July). Ms. Akwa', fol. 63ro has Fri. 6 Rabi' I (Fri. 21 May), followed by Ibn al-Daiba', *Qurrat*, I, 362. Both Mss. of the *Anba* have Fri. 16 Rabi' I (Mon. 31 May). It will be seen that the weekday only coincides in one instance, and the earliest date is to be preferred.

7 See editor's notes to Ibn al-Daiba', *Qurrat*, pp. 362–3 for identifications, pp. 336–9 for al-Sahul, Shuwahit (Shawahit?) and al-Shawafi, and other forts in the region mentioned by al-Khazraji, and p. 368 for Hisn al-Majma'a. Many of these places are also mentioned by 'Umara, *Tarikh al-Yaman*, see al-Akwa''s edn, pp. 88–91, 235–9, and al-Hamdani, *Sifat*, ed. al-Akwa', pp. 210–14. See also Redhouse (1908), pp. 61 (note 353), 79 (note 471), Smith (1978), pp. 197, 205 and 203, also Silvestre de Sacy (1798/9), p. 533, for al-Shawafi, between Ibb and al-Makhadir. Al-Sahul apparently denotes rather a large area, stretching partly to the south of Ibb. Khadid is a little to the N.W. of Ibb, see YAR (50) sheet 1444 C3. Al-Mahalla is tentatively identified as al-Mahall (YAR sheet 1443 D4, at LA 7864).

8 For 'Anna, see *Qurrat*, I, 335, and editor's note.

9 *Ibid.*, p. 363, for Sumara. Damage seems to have extended further in this direction than elsewhere.

10 It is difficult to discern a geographical pattern in the order in which places are mentioned in the text (as reproduced on the table), which appears, from such identifications as have been made, to be fairly random, though approximately anti-clockwise round Ibb, starting E./S.E. and finishing S. This should help the identification of the remaining localities, which must await further detailed investigation and preferably a field visit. However, many of the hill forts have probably been abandoned and many small settlements have probably moved or changed their name over the intervening centuries. The author, al-'Arashani, may himself have been compiling a detailed list of damages during a field trip for official purposes; he was anyway a regular visitor to the region (see Abu Makhrama, II, 136). Some of the co-ordinates found in the Official Gazetteer of the Yemen Arab Republic (Washington, 1976) appear to be at variance with the locations identified from the YAR 50 maps.

1170 June 29   *12 Shawwal 565*   North Syria

A catastrophic earthquake in northwest Syria was said to have been felt in Egypt.[1] There is no evidence of the effects there, nor has a reference so far been found in later Egyptian sources.[2]

1 Ibn al-Athir, *Bahir*, pp. 144–5; Abu Shama, I/ii, 467; Ibn Qadi Shuhba, p. 189.

2 For the effects of the earthquake in Syria, see al-Suyuti, pp. 44–5/30–1; Taher (1979), pp. 92–119/60–74. The event is mentioned in most histories of the Crusades, e.g. Runciman (1971), II, 389.

[1191 or 1196?   *587 or 592 ?*   Egypt]

A dubious and confused event in Egypt. Al-Suyuti correctly cites al-Maqrizi for a shock in Egypt in 587 H (1191); this has yet to be confirmed in an earlier source. Al-Maqrizi sandwiches his brief statement about the earthquake between an account of the high prices experienced in Egypt that year, and an account of a damaging simoom.[1]

Under 592/1196, al-Suyuti mentions a great wind throughout the world, which shook the Ka'ba a few times. The 'Yemeni' corner of the Ka'ba was damaged and there was an earthquake in Egypt. The 592 earthquake in Egypt (but not the destructive wind) is also mentioned by al-Maqrizi, who gives no details of the date or effects of the shock. Damage to the Ka'ba by the storm is also mentioned by Ibn al-'Imad, who does not refer to any earthquake.[2]

Possibly there were indeed two earthquakes, independent of the other phenomena reported. If there is some amalgamation of events, the later date may be preferred, since the storm winds of 592/1196 are also noted, for example, by Ibn al-Athir.[3] Another source puts both the earthquake in Egypt and the violent storm in 593/1197.[4]

The absence of information and the context in which they are recorded make all these earthquake reports dubious. The details of the Yemeni corner of the Ka'ba

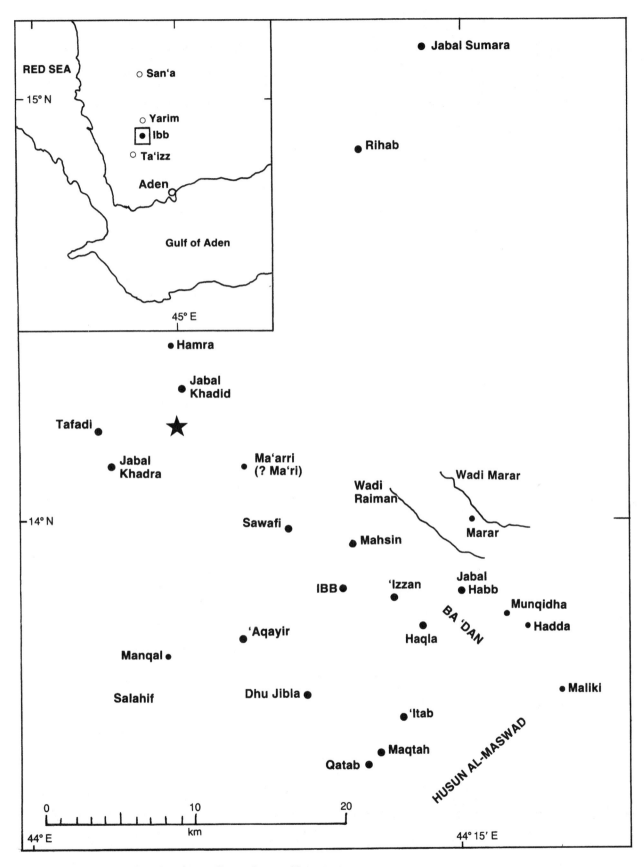

Figure 2.10. The earthquake of 1154 September 11, Yemen.
Inset shows the position of the epicentral area, round Ibb.

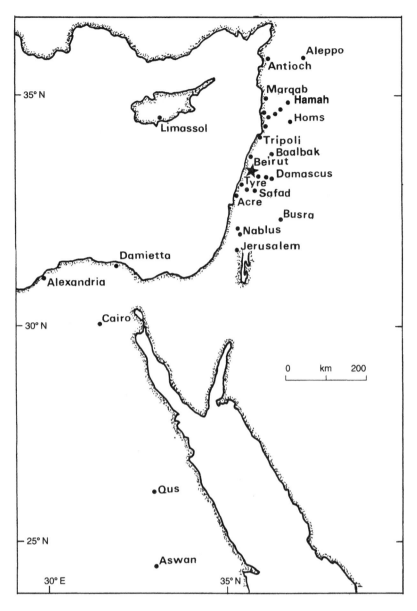

Figure 2.11. 1202 May 20, Dead Sea.

being affected are reminiscent of the shock of 515/1121 (see above).

1 Al-Maqrizi, *Suluk*, I/1, 108; al-Suyuti, p. 45/31.
2 Al-Maqrizi, p. 139; al-Suyuti, p. 46/32; Ibn al-'Imad, II, 308. See also Ambraseys (1961), p. 26.
3 Ibn al-Athir, **XII**, 48: the storms were in Iraq.
4 Anon., *Bustan al-Jami'*, p. 157. Ibn al-Dawadari, VII, 107, mentions a great shout 'which shook the earth' in 587 H in connection with events in Acre, but does not mention earthquakes in either that year or 592/3 H.

1195 April 4   *21 Rabi' I, 591*   **Hadramaut**
Three earthquake shocks are reported, without detail, in Hadramaut, south Arabia. The first, on 4 April, was followed by another on 11 Jumada I/23 April and a third on a Friday night in the middle of Jumada I/28 April.[1] The shocks possibly affected the hinterland in the Tarim–Shibam region.

1 Al-'Alawi, pp. 451–2, on the basis of local historical manuscripts. 'Friday night' was the equivalent of the evening of Thursday: so we have taken the middle of Jumada I to be the 16th (Thursday 28 April) rather than the 15th.

1202 May 20   *Monday 26 Sha'ban 598*   **Dead Sea**
A major earthquake in the upper Jordan and Litani Valleys was responsible for tens of thousands of casualties in the eastern Mediterranean region (see Figure 2.11). The shock occurred early in the morning and was felt throughout Egypt, causing great concern but little damage. In Cairo, the shock was of long duration and aroused sleepers. Three violent shocks were reported, but only tall or vulnerable buildings were particularly affected, along with those on high ground, which threatened to collapse. A lesser shock was felt about midday the same morning.[1]

The main shock was felt from Sicily to Azerbaijan in N.W. Iran, and from Constantinople to Aswan.[2]

1 'Abd al-Latif, pp. 264–73/trans. de Sacy, pp. 414–15.
2 Ambraseys and Melville (1988) give a detailed account of this earthquake, and of the duplications found in modern catalogues, e.g. Ben-Menahem (1979), pp. 260 (20 May 1202) and 273 (July–August 1201).

1203 October 23   *14 Safar 600*   **Ethiopia**

The effects of a volcanic eruption from Mts 'Awan and Dubbi in Ethiopia were widely experienced in the Tihama and inland Yemen, so that the year 600 H was known in Yemeni chronicles as the 'year of the ashes'.

The eruption started on the night of 14 Safar 600/ 23 October 1203, with a thick black cloud blotting out the horizon. This continued that night and the whole of the next day; on the second night something hard like particles of extremely white ice fell from the sky, accompanied by a profound darkness and convulsions. Confused winds blew over the whole coastal region and inland as far as Dhamar.[1] In Zabid, white ashes were followed by black ones and earthquakes and tremors occurred. The people went in fear of destruction.[2] The fall of ashes in Zabid was so heavy that it started to 'flow like a stream'.[3] In the valley of al-Ahjur (or Ahjir), W.N.W. of San'a, a large landslide occurred, possibly caused by earth tremors associated with the eruption.[4]

In Ethiopia, where the eruption originated, it is reported that Jabal 'Awan remained in eruption for several days, and that Jabal al-Dubb (Dubbi) was obliterated.[5]

On the night of 6 Rabi' I, 600/13 November 1203, three weeks later, there was a similar though lesser event, and stuff rained down like particles of coal, but only for an hour.[6] Further activity, possibly associated with the eruption, is reported in Yemen in early Sha'ban 601/late March 1205.[7]

1 Ibn Hatim, pp. 110–12; he cites two authorities for the account, and alludes to a third, for which see Smith (1978), p. 98.
2 Al-Khazraji, *al-'Asjad*, facs. edn, p. 176, Akwa' Ms. fol. 86ro; Ibn al-Daiba', *Bughyat*, p. 77 (cf. ed. Chelhod, p. 85). Al-Khazraji cites the authority of al-Janadi (d. 732/1332).
3 Yahya b. al-Husain, *Ghayat*, I, 381–2, and his namesake, in *Anba al-zaman*, Ms. 1347, fol. 65, record an eye-witness account that speaks of the darkness in the Tihama lasting three days, from Wednesday to Friday. October 23 was a Thursday. In Zabid and also San'a, prisoners were released and the people turned (briefly) to repentance. Both cite al-Khazraji's story illustrating the depth of the darkness in Zabid; Ibn Hatim tells another. The *Anba* also refers to similar earth tremors in Egypt and Syria, but these are unsubstantiated.
4 Yahya b. al-Husain, *l.c.* The village in question was called Marihin, which has not been identified (see Wilson, 1989, p. 291). There is not necessarily, however, a direct connection between these events.
5 Ibn Hatim, *l.c.* The eruption is not mentioned by Gouin (1979).
6 Ibn Hatim, *l.c.*

7 Yahya b. al-Husain, *Ghayat*, I, 387, refers to shooting lights and earth tremors in Yemen. See also Taher (1979), p. 136; Poirier and Taher (1980), p. 2192. Al-Hibshi (1982) follows this account, but has early Ramadan/starts 22 April 1205.

1212 May 1   *27 Dhu'l-Qa'da 608*   **Dead Sea**

A damaging earthquake in south Palestine also affected Egypt, where it was strongly felt in Cairo and Fustat and destroyed a number of houses. At al-Shaubak and al-Karak, towers and houses were destroyed, killing a number of women and children.[1] The earliest account says it was strongest in the part of Aila (Eilat) that is by the sea.[2]

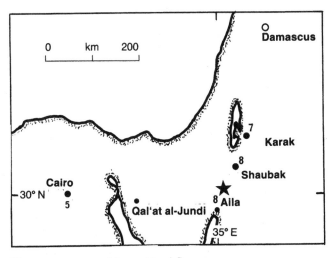

Figure 2.12. 1212 May 1, Dead Sea.

In the Sinai Peninsula, the shock caused severe damage to the monastery of St Catherine, destroying a major part of its fortifications and damaging the church. The northern wall of the monastery, with its northwest and northeast corners, collapsed. Of the cells, some were destroyed completely and others lost their roofs, apparently without loss of life. This was possibly due to the fact that the earthquake, which struck at dawn, was preceded by a foreshock at sunset the previous evening (Monday 30 April), which might have served as a warning. The shock triggered rockfalls from the mountains behind the monastery, and was followed by aftershocks that continued for a year. The date of the earthquake was commemorated as a day for prayers and penitence.[3] The walls and cells were almost immediately rebuilt with the assistance of the metropolitan, Gabriel of Supaki (Petra), whose caravan of builders originally sent to rebuild the church at Agia Koryphi (Jabal Musa), damaged by a previous shock, arrived at St Catherine's six days after the earthquake.[4] The earthquake is also said to have caused serious damage at Qal'at al-Jundi in Sinai,[5] but the evidence is inconclusive.

The location of the earthquake should be sought in the Gulf of Aqaba or south of the Dead Sea. No details of the shock are recorded for Syria or Damascus (Figure 2.12).

1 Ibn Kathir, **XIII**, 62; al-Maqrizi, **I**/1, 175; al-Suyuti, p. 49/35.
2 Abu Shama, *Dhail*, p. 78; also Taher (1979), p. 137/82, 238. Strong winds and shooting stars were noticed the previous day.
3 Nectarios (1758), p. 197, (1677), p. 198 (see also next note) and Eustratiades (1932), p. 125, quote an Arabic synaxarium written at the monastery in February 1214. The day for prayers, 1 May, is the definitive date of the shock; Muslim sources all give 27 Dhu'l-Qa'da, which is 2 May.
4 This information is recorded by Nectarios, a Sinaite archibishop writing in 1658. He gives the date as 1312 instead of 1212, while Papamichalopoulos (1912), p. 29 places the earthquake on 1 May 1608 on the authority of Ioasaph, which is not consistent with the other chronological elements in the account. These two sources have misled Ben-Menahem (1979), p. 258, who has 1312 (or 1608): the date should read 1212 (608 H). See also Melville (1984a), p. 99.
5 Ben-Menahem refers to Tamari (1977), who has a summary of the epigraphy relating to Qal'at al-Jundi's construction history. One might infer from the last inscription, dated 1201 (?) and a footnote referring to an edict of al-Malik al-Kamil dated 1221 (?), mentioned by Ibn al-Dawadari, that the fort was damaged by an earthquake during the intervening period. However, there is no such inference made by Tamari himself, and Ibn al-Dawadari's chronicle sheds no further light on the issue.

1222 May 11   Mediterranean

A destructive earthquake in Cyprus, associated with a seismic sea-wave, was felt in Egypt.[1] Paphos and Limassol were completely destroyed, with great loss of life.[2]

1 Oliverus, **II**/7, 279, has May; the exact date is given in Schreiner (1975), p. 199, (1977), p. 191. See also Ambraseys (1965); Ben-Menahem (1979), p. 289. No confirmation has been found in Arabic sources for an event this year (619 H). Oliverus was an exact contemporary and the similarity of this date to that of the previous entry is coincidental.
2 See sources in Alexandre (1990), pp. 173–4.

1256 June 30   *5 Jumada II, 654*   Hejaz

A series of shocks, precursory to a volcanic eruption at Medina, began on 1 Jumada II/26 June, accompanied by noise but causing no damage. The shocks became more violent on the 29th and continued during the day with increasing force, causing the collapse of a number of houses and damage to walls in Medina. Smaller shocks continued at intervals until next morning, when the extrusion of the lava dome began.

The eruption began on 30 June, east of Medina. The exact location of the crater is not known. Huge clouds of smoke and fire were emitted for a number of days, visible at Mecca, Yanbu' and as far as Taima. It is alleged that at night the glow of the eruption could be seen from as far as Busra in Syria, more than 900 km to the north, and the reduction in the strength of the sun's light caused concern in Damascus until the reason

was known.[1] One account states that 'pebbles and rocks were thrown out on all sides', and no-one could approach the scene of the eruption because of the intensity of the blaze.[2] The main eruption occurred at midday and did not cause any significant damage at Medina. The lava flow took a northerly direction and stopped at Jabal Waira in the valley of Wadi al-Shathat, which is a little to the east of Jabal Uhud (4 km from Medina). The flow covered an area 19 km in length and 6 km in breadth, up to a depth of 2.5 m. The whole sequence lasted about three months. The valley of Shathat became choked up; the lava carried whole rocks before it and heaped them up at al-Sadd (i.e. 'the dam or barrier'). The plain of Harra, which is on the pilgrim route from Iraq, was completely blocked.[3] Damage to man-made structures in the immediate vicinity of the lava field seems to have been insignificant, and there was no damage to the *haram* (Sacred Enclosure) in Medina.

The earthquakes associated with the eruption were not felt very far, perhaps not outside the Tihama.[4]

Not surprisingly, this eruption is widely reported in the Islamic sources,[5] in view of the importance of the Holy City; the mosque of the Prophet in Medina was destroyed later this year in a fire started by accident and unconnected with the eruption.

1 Abu Shama, *Dhail*, p. 192; this contemporary author has a detailed account on pp. 190–4.
2 Ibn Hatim, pp. 321–2 (also a contemporary). Other accounts do not mention projectiles; there is no specific reference to a crater.
3 *Ibid.*; writing at the end of the century, Ibn Hatim says that men and camels could still not cross the lava flow.
4 The fifteenth-century Persian author, Fasih Khwafi, **II**, 323, mentions an earthquake in Tihama in 653/1255, following an eruption in 652/1254 (*sic*), but his account cannot be regarded as reliable.
5 The fullest account is that of the sixteenth-century author, al-Samhudi, pp. 139–55, see also Taher (1979), pp. 139–65/153–68. The eruption is mentioned by al-Suyuti, pp. 49–50/35–6. Accounts may also be found in Burton (1893), pp. 60–1, and Neumann van Padang (1963), pp. 9–10. The eruption has recently been fully analysed by Camp *et al.* (1987).

[1259 June 6   *12 Jumada II, 657*   Lower Egypt]

Great earthquakes occurred in Cairo and the other towns of Egypt in 657 H.[1] This preceded the news of the Mongols' advance and is possibly a reference to the political situation in Egypt, which was undergoing a prolonged dynastic upheaval.[2] Some sources refer only to the strength of disquieting rumours (*arajif*) flying around at this time: the root meaning of r-j-f being to tremble or shake.[3]

Numerous problems surround the reporting of this event in the secondary sources.[4] Al-Maqrizi's Arabic text, which is usually cited for the date given above, merely reads 'in this year there were numerous earthquakes in Egypt'; and it goes on to record another event, which occurred on 12 Jumada II.[5] In other words, this date does not belong to the earthquake at all. The shock, if genuine, must in fact have occurred sometime shortly before 6 June this year.

1 'Imad al-Din, *al-Raud al-nadir*, fol. 156vo; Ibn Duqmaq, *Nuzhat*, fol. 117ro; al-'Aini, *'Iqd*, I, 224. None of these are contemporary authors.
2 Irwin (1986), esp. pp. 26–36.
3 See e.g. Ibn al-Dawadari, VIII, 44.
4 Lyons (1907), p. 284, duplicates the event, first under 28 May 1260, and again on 12 Jumada II, 657, which is erroneously converted as 21 February 1263. Lyons cites Quatremère's translation of al-Maqrizi (I/1, 89) for the first event, and simply al-Maqrizi for the second. Quatremère (I/1, 89) in fact only has the 12 Jumada II, 657 event: but the translation is misleading (see text). Neither al-Maqrizi nor Quatremère mention an earthquake on 28 May 1260, which would be the equivalent of 15 Jumada II, 658 H. Lyons's erroneous dates are repeated by Sieberg (1932b), p. 188. There may be some confusion with the earthquake of 20 February 1264 (see below).
5 Al-Maqrizi, I/2, 420. The text reproduced in Taher (1979), p. 166/238, is not al-Maqrizi's but al-'Aini's.

The printed text of al-Suyuti (*Kashf*, p. 50/36, *Husn*, II, 209) names his source as Ibn al-Athir (d. 1234); this should probably read Ibn Kathir. Ibn Kathir's text (XIII, 218), however, does not support the interpretation that this is a genuine seismic event.

1259 November 22 and December 10   *4 and 22 Dhu'l-Hijja 657*   Yemen

A non-damaging shock in San'a was followed a fortnight later by an earthquake to the west, which threw down the mountains and destroyed many places.[1]

There is no indication of the precise location of the shock, nor of how widely it was felt.[2] The earthquake was not destructive in San'a itself.

1 Al-Khazraji, I, 128/trans. I, 154; *idem.*, *al-'Asjad*, facs. edn, p. 234, Ms. Akwa', fol. 108ro-vo. Taher (1979), p. 166/180 and Poirier and Taher (1980), p. 2192, incorrectly date this 659/1261.
2 Freya Stark (1939), p. 481, refers to a manuscript history according to which Husn al-'Urr was destroyed in 657 H (which she wrongly converts to AD 1298), though not necessarily as the result of an earthquake. Al-'Urr is situated over 600 km east of San'a, so any connection with the San'a earthquake must be coincidental.

1264 February 20   *Tuesday 20 Rabi' II, 662*   Egypt?

A very strong earthquake destroyed houses[1] in a number of places.[2] Few reports of the event are available. Only al-Suyuti specifically states that the earthquake occurred in Egypt: the two earliest authors are both Syrian, and it might otherwise have been assumed that the earthquake affected primarily Syria.[3]

1 Al-Yunini, I, 553; the shock is also mentioned by al-Kutubi, *'Uyun*, Ms. Cairo 1497, fol. 234ro-vo (we owe this reference to Bill Tucker).
2 Al-Maqrizi, I/2, 508.
3 Al-Suyuti, p. 50/36. No report of the earthquake has been found in earlier Egyptian chronicles (such as Ibn al-Dawadari and al-Nuwairi). Taher (1979), p. 167/238, refers to al-Maqrizi's text in such a way

as to suggest that Egypt is specifically mentioned, but this is not the case. In his translation Taher mentions also al-Maqrizi's account of an earthquake in 693/1293 (see below).

Al-Qalqashandi (*Ma'athir*, II, 114–15) states that there was a strong earthquake in Egypt and south Palestine in 660/1262. Until further references to this event are found, it is regarded as dubious, since the details given suggest very strong parallels with the 702/1303 earthquake (see below).

The earthquake in Egypt on 21 February 1263, listed by Lyons on the authority of al-Maqrizi, may be a garbled reference to this event (see above, 1259 June 6).

1265   *663 H*   Yemen

An earthquake shock occurred in San'a, Yemen, this year.[1]

These details survive only in a late source. On the evidence available, the shock has to be located near San'a; it is not mentioned in the works of contemporary historians writing in the south of Yemen.

1 *Anba al-zaman*, II, fol. 366, Ms. 17075 quoted by Taher (1979), p. 167/180. The shock is not mentioned by the author's cousin, Yahya b. al-Husain, in the *Ghayat al-amani*, nor in the Cairo Ms. 1347.

1269 October 29   *1 Rabi' I, 668*   Hejaz

An earthquake shook al-Ta'if, about 60 km southeast of Mecca, during the fifth year of a great scarcity in the Hejaz.[1] No details of the shock are given, but it seems likely that had it been felt in Mecca, this would have been reported, so it was a relatively local shock.

1 Al-Fasi, p. 313. His source dates the onset of the dearth at the end of 663/1265. This was followed by an epidemic. See also Taher (1979), p. 168 (who cites al-Fasi, p. 363, and does not translate the passage into French); Poirier and Taher (1980), p. 2192, merely have 668/1270. Considering that al-Fasi devotes whole sections to the occurrence of floods, droughts and epidemics in Mecca, one would expect information on earthquakes to have been available, had they occurred.

1293 January   *Safar 692*   Dead Sea

A strong earthquake occurred in the region of Ghazza (Gaza), affecting al-Ramla, Ludd, Qaqun and al-Karak – particularly the latter, where three towers of the citadel were destroyed and many houses (see Figure 2.13).[1] In al-Ramla, the earthquake followed a destructive flood. The earthquake ruined many places in coastal Palestine and the minaret of the main mosque at al-Ramla fissured and fell; the minaret at Gaza also collapsed. Orders were given to assess the damage in both places, and a team was sent from Damascus to repair the damage to al-Karak.[2]

Al-Suyuti says that the following year, 693/1294, an earthquake that affected all Egypt caused some pillars in the mosque of 'Amr (in Fustat) to became partially detached, but this was less serious than what happened

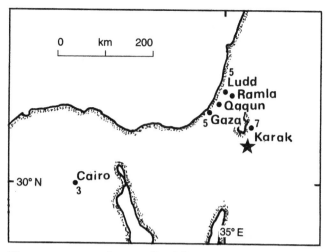

Figure 2.13. 1293 January, Dead Sea.

in the main mosque in Cairo.[3] It is very likely that the same earthquake is involved. No earlier confirmation of the earthquake in Egypt has yet been found.[4]

1 Al-Jazari, fol. 176–7/trans. Sauvaget, p. 29; al-Nuwairi, Ms. 1578, fol. 164ro; Ibn Kathir, XIII, 333; al-Suyuti, p. 50/36 (wrongly under 672 H).
2 Ibn al-Furat, VIII, 154–5; al-Maqrizi, I/3, 783. Amir 'Ala al-Din Aidughdi al-Shuja'i was put in charge of this expedition, accompanied by builders, engineers and stone-cutters, see al-Jazari, fol. 177.
3 Al-Suyuti, *l.c.* and *Husn*, II, 210. See also Taher (1979), pp. 171–2/238; Ambraseys (1961), p. 27.
4 Al-Suyuti's source is Ibn al-Mutawwaj (d. 730/1330), who was also extensively used by al-Maqrizi in his *Khitat* on the topography of Fustat, see Guest (1902), pp. 116, 125. The earthquake in Cairo is not mentioned under either 692 or 693 by contemporary Egyptian historians. It is possible that there is a confusion with the earthquake of 698/1299, see below.

1293   *692 H*   **Hejaz**

A minor volcanic eruption occurred near Medina this year, similar to that of 654/1256, but lasting only three days. It blazed, however, just as fiercely and consumed rocks but not palm trees. The eruption seems to have been more explosive than that of 1256, sending an eruption column high into the air.[1]

1 Ibn al-Fuwati, pp. 474–5; Ibn al-Kathir, XIII, 332; Taher (1979), p. 172/169.

1299 January 8   *3 Rabi' II, 698*   **Egypt**

A strong earthquake was felt in Egypt, such as had not been seen before. It followed a pair of weaker shocks earlier in the year, on 24 Safar 698 (1 December 1298), during the night, which were separated by a brief interval.[1]

There is no evidence of the effects of the second and stronger shock in Egypt, nor of events elsewhere in the

eastern Mediterranean with which these earthquakes might be associated, so this was probably a local event.[2]

1 Al-Jazari, fol. 280vo/trans. Sauvaget, p. 83. Al-Jazari (d. 1339) is a reliable author. He dates the first pair of shocks, which were separated by the time it takes to recite five verses of the Qur'an, as 5 Kanun I (December), i.e. using Kanun I as though it were the Coptic month Kiyahak (5 Kiyahak = 1 December). He also records hail and destructive rain in Egypt.
2 The shock is not mentioned by Egyptian chroniclers, which is surprising in view of the rich documentation of the period (though see above, 1293, note 4). There is some evidence of events in Italy around this date, see Baratta (1901), pp. 42–3, but nothing suggesting a large Hellenic Arc earthquake.

1303 August 8   *Thursday 23*   **Hellenic Arc**
                        Dhu'l-Hijja 702

A major earthquake in the Hellenic Arc caused serious damage in the south Peloponnese, Crete and Rhodes. Damage extended along the eastern Mediterranean coast from Cyprus to Palestine and Lower Egypt, and it was strongly felt in North Africa (see Figure 2.14).

Arabic sources report that the earthquake was strongest in Egypt, particularly in Alexandria, where many houses were ruined and much of the city walls was destroyed, killing a number of people. The lighthouse at Alexandria was shattered and its top collapsed.[1] The sea spilled over into the harbour, inundating the shore and then advanced, flooding Alexandria as far as the Sea Gate, carrying sailing ships and boats onto the land. Foodstuffs and merchandise stored along the shore were destroyed. Elsewhere in the Nile Delta there was widespread damage. Two villages in the Sharqiyya district were destroyed and Sakha, Abyar and Damanhur al-Wahsh were among the towns that were completely ruined.[2]

The shock was felt throughout Lower Egypt around sunrise and lasted several minutes.[3] Ground movements were slow, making people walk with difficulty, while those on horseback were thrown down. In Cairo, almost all the houses suffered some damage but relatively few collapsed. The earthquake caused panic, and women ran into the streets without their veils. Streets littered with fallen parapets and free-standing walls slowed down the evacuation of the city, whose inhabitants encamped that night outside Cairo between Bulaq and Rauda, leaving their houses to be rifled by looters. It appears that very few structures collapsed completely and that there were relatively few casualties, probably because when the earthquake struck, the streets would have been relatively empty. Most people returned to Cairo the following day, which was a Friday, and prayed in the mosques, to which it was apparently safe to return.[4]

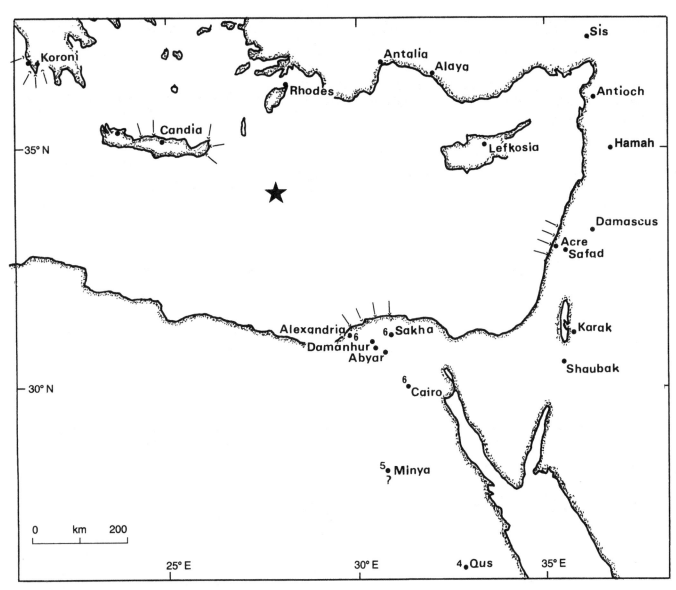

Figure 2.14. 1303 August 8, Hellenic Arc.

Many large public buildings in Cairo were, however, damaged and some collapsed. Minarets in the mosques of Cairo and Fustat were particularly affected. The mosques of al-Azhar,[5] al-Hakim,[6] and 'Amr b. al-'As at Fustat[7] partly collapsed and had to be pulled down and rebuilt. The minarets of the mosques of al-Hakim, of al-Fakkahin and of al-Salih b. Ruzaik outside the Bab Zuwaila, and of the madrasa al-Mansuriyya, were either destroyed or damaged to the extent that they had to be pulled down and rebuilt.[8] It is reported that after the earthquake, Cairo looked as though it had been wrecked by a conquering army; the Mamluk amirs spent large sums of money over the next two years on the repair and restoration of public buildings.[9]

Sporadic damage extended to Upper Egypt: at Munya

ibn Khusaib (Minya) rockfalls were reported from the mountains east of the Nile, and the bed of the Nile was briefly visible as the waters parted. The minarets of the mosque at Minya collapsed, along with houses and other buildings. Qus, further south, was also reported to have been 'destroyed'.[10]

In southern Palestine, the shock was damaging in al-Karak, al-Shaubak and Safad, where one side of the citadel collapsed. Further north, a wall of the Umayyad mosque in Damascus was fissured, and part of the walls of the citadel of Hamah was destroyed.[11]

According to Arab writers, aftershocks continued for at least 20 days.[12]

Sieberg wrongly places the epicentral region of this earthquake in the Faiyum, associates the event with faulting in this region, and says that in the Nile Valley ground movements were so severe that boats sailing on the river were cast on dry land.[13] In the first place, no source mentions Faiyum. Secondly, allusions to faulting and excessive ground motions responsible for setting up waves in the Nile large enough to cast boasts ashore, arise from a misinterpretation of al-Maqrizi's text. He describes the effects of very high winds on the day of the earthquake, which blew off the topsoil in Upper Egypt, in places exposing ruins of old buildings, and cast sailing boats a bow-shot onto the land.[14]

Damage was much more serious in Crete, where many fortified sites collapsed and 4000 people were killed. There was also widespread destruction and a damaging seismic sea-wave in Rhodes and along the coast of Palestine, particularly off Acre.[15] Clearly, this was a large earthquake originating in the Hellenic Arc and not local to Egypt. Not surprisingly, the event is duplicated under different dates in various catalogues.[16]

1 Al-Maqrizi, *Khitat*, ed. Wiet, III/2, 123, mentions the destruction of the lighthouse and its repair by the amir Rukn al-Din Baibars al-Jashnakir the following year. His statement that it was still standing in his day (mid-fifteenth century) is false. Al-'Aini, ed. Amin, IV, 265, gives precise details of the damage in Alexandria, and says repairs were carried out at the Sultan's expense.

2 The most detailed contemporary Egyptian accounts are those of Ibn al-Dawadari, IX, 101–8; Anon., ed. Zettersteen, pp. 126–8; al-Nuwairi, *Nihayat*, XXX, Ms. Leiden Or. 2–0, fol. 5ro–vo (cf. Taher (1979), pp. 176 ff); Mufaddal b. Abi'l-Fada'il, XX, pp. 86–90. The earthquake is reported in all the chronicles of the period. The fifteenth-century author, al-Maqrizi, *Suluk*, I/3, 942–5/trans. Quatremère, II/2, 214–22, has a very full account, including details not yet found in earlier sources. See also al-Maqrizi's contemporary, al-'Aini, IV, 260–5, citing various authorities. They both state that the earthquake was a punishment for the lax morals of the time; al-Maqrizi says the people gave up their unseemly behaviour. See also al-Suyuti, pp. 51–5/37–8: it was this event that led him to compile his earthquake catalogue.

3 Various estimates are given, e.g. a quarter of an hour (Ibn al-Dawadari, p. 101; Mufaddal, p. 86); less than one-third of an hour (Anon., ed. Zettersteen, p. 126); 5 *daraja* (al-Nuwairi, fol. 5ro; al-Maqrizi, p. 944). For the meaning of *daraja*, see p. xvii.

4 Al-Maqrizi, esp. pp. 942–3, al-'Aini, pp. 263–4.

5 Al-Maqrizi, p. 944. Repairs were carried out by Amir Salar, with Amir Sunqur al-A'sar; cf. al-Nuwairi, fol. 5vo (Taher, p. 180), al-Safadi, *Nuzhat*, fol. 75vo, al-'Aini, p. 265.

6 This mosque seems to have been particularly severely affected, Ibn al-Dawadari, p. 101. Al-Nuwairi, fol. 5ro, mentions the collapse of its minaret and part of its walls. It was restored by Amir Baibars al-Jashnakir in 704/1305, al-Muqri, pt. IV, fol. 71ro, and lavishly endowed, al-'Aini, pp. 264–5, al-Maqrizi, p. 945. See also Berchem (1891), I, 431–3.

7 The walls of this ancient mosque were fissured and it was repaired by Salar, see Ibn al-Dawadari and al-Nuwairi, also Corbett (1890), p. 784.

8 Ibn al-Dawadari, p. 101; Mufaddal, pp. 86–7. Al-Nuwairi notes that the Salihi and Zafiri (Fakkahin) mosques were repaired out of the Sultan's own purse. According to al-Maqrizi and al-'Aini, work on the Salihi mosque was under the supervision of Amir 'Alam al-Din Sanjar, while al-Safadi, fol. 75vo, has Saif al-Din Bektemür. The

Mansuriyya (i.e. the madrasa of Qalawun) was restored under the supervision of Amir Saif al-Din Kahardas: al-Nuwairi, al-'Aini, p. 265, al-Maqrizi, p. 944. See also Berchem (1891), II, 68, for the Salihi mosque and Combe, Sauvaget and Wiet (1944), pp. 242–5 for restoration inscriptions on the mosque of al-Hakim and madrasa of Qalawun.

9 Pious foundations (*waqf*) were drawn on and created to finance these operations. Al-Nuwairi, fol. 5vo mentions that 90 000 dirhams were spent on the repairs to the madrasa of Qalawun.

10 Anon. (ed. Zettersteen), was in Minya when the shock occurred. For Munya ibn (or Abi) Khusaib, see Yaqut, IV, 675. For Qus, see Mufaddal, p. 88, al-Maqrizi, p. 943; the latter tells a story about a man milking a cow, which seems to refer to Qus: both man and cow were lifted up by the earthquake wave.

11 Al-Nuwairi, fol. 5ro and Mufaddal, p. 86, mention Safad; Anon., *Jawahir al-suluk*, fol. 218ro, adds al-Karak and al-Shaubak. For Hamah, see Abu'l-Fida, IV, 50, and for Damascus, al-Maqrizi, p. 944. See also al-'Aini, pp. 261–2.

12 Mufaddal, p. 88; al-Maqrizi, p. 944. Al-Suyuti, *l.c.*, has 40 days.

13 Sieberg (1932a), p. 873, (1932b), p. 188.

14 Al-Maqrizi, p. 943, cf. al-'Aini, p. 263. Anon. (ed. Zettersteen)'s account of the bed of the Nile being briefly visible at Minya does, however, require explanation.

15 Pachymeris, II, 392–3; *Chronique d'Amadi*, p. 239; *Gestes des Chiprois*, p. 315; Florio Bustron, p. 134; Muratori (1723–51), IX, 254; XIV, 1123; XXII, 177; XXII, 772; Mallet (1853), p. 37. For the effects in Cyprus, see Ambraseys (1965) and Schreiner, I (1975), p. 203, and II (1977), p. 216. For the tsunami off Acre, see al-Nuwairi, fol. 5ro, al-Muqri, fol. 65vo. Several Arab authors were aware that the earthquake also affected Christian territory.

16 Taher (1979), p. 192/245 duplicates this earthquake under Dhu'l-Hijja 741/May 1341, on the false evidence of Ibn al-'Imad, VI, 127. Clédat (1923), p. 66, wrongly converts 702 H to AD 1324, which has misled later writers, including Sieberg (1932b), p. 188 (doubly inaccurate under 1326). For other erroneous dates, see Section 2.4.

1307 August 10 *Thursday 9 Safar 707* **Lower Egypt**

A shock of brief duration was experienced in Egypt during the night.[1] The shock is not widely reported and was evidently a small, local event.

1 Mufaddal, ed. Blochet, XX, p. 134. Cairo is not specifically mentioned, but it is likely that the shock affected Lower Egypt. Bonito, p. 538, refers to many earthquakes felt in Greece in 1307, but further evidence would be necessary before a firm association of these events with effects in Egypt could be made.

1313 February 27 *Monday 29* **Lower Egypt**
Shawwal 712

A shock of brief duration occurred in Egypt during the day.[1]

1 Mufaddal, ed. Blochet, XX, p. 227. In fact, 27 February was a Tuesday. This and the previous shock, in 707/1307, are reported in identical terms by Mufaddal.

1335 May 29 *5 Shawwal 735* **Lower Egypt**

Earth tremors were experienced in Cairo by many people during the afternoon and evening. This was possibly followed by another shock a year later.[1]

The evidence suggests that this was a local event,

though there may be some connection with an unidentified earthquake in Buhaira in 733/1333.[2]

1 Al-Muqri, pt. V, fol. 13vo. On fol. 10vo, the same author reports an earthquake in Egypt in the first 10 days of Shawwal 736 (began May 1336). This may be the same event; the order of folios in the manuscript appears to be disturbed. It is remarkable that none of the many other chroniclers of the period mention this event under either year.
2 See Section 2.3.

1336   *736 H*   Yemen

A powerful earthquake was experienced in Yemen this year. No details are given to allow an assessment of the event, but it might have occurred in the Zabid region.[1]

1 Al-Khazraji, *al-'Asjad*, facs. edn, p. 376. Someone was reported to have seen a tree bend over and touch the ground before springing back. 736 H began on 21 Aug. 1335. The earthquake is not mentioned in al-Khazraji's *al-'Uqud*, nor in Ibn al-Daiba''s *Qurrat*.

[1344 January 3   *Saturday 16*   Eastern Anatolia]   *Sha'ban 744*

A large earthquake in southeast Anatolia was said to have been felt as far as Egypt.[1] However, there are no details of its effects south of Damascus, where the shock was only slight.[2]

1 Ibn Habib, *Tadhkirat*, III, 58–60, has the rather neat expression 'it made those sitting down stand up, and those standing up sit down'. See also Ibn al-Shihna, IX, 175–6; Anon., *Manah al-Rabaniyya*, Ms. 1536, fol. 63vo; al-'Aini, *Tarikh al-Badr*, fol. 48ro–vo, cf. Taher (1979), p. 196/88; Sibt b. al-'Ajami, trans. Sauvaget, p. 9; al-Suyuti, p. 55/38. Poirier and Taher (1980), p. 2193, have 1343 January 1.
2 Ibn Kathir, XIV, 211.

1347 December 8   *4 Ramadan 748*   Lower Egypt

Cairo was shaken twice in one hour.[1] The shocks were probably small events and they are not widely recorded.

1 Al-Maqrizi, II/3, 741, cited correctly by al-Suyuti, p. 55/38. Some confusion surrounds this event in secondary sources. Taher (1979), p. 197 states, incorrectly, that al-Maqrizi does not refer to this earthquake. Poirier and Taher (1980), p. 2193, cite Ibn Kathir as their source, but the latter (a Syrian author) makes no mention of an earthquake this year. Ambraseys (1961), p. 27, following a single Ms. of al-Suyuti's *Kashf* in the British Library, gives Ramadan 778 (January–February 1377) for this earthquake, but the printed edition restores the correct year.

[1349 November–December   *Ramadan 750* Yemen]

A great earthquake occurred in Zabid. A fearful shrieking and a noise like heavy thunder were heard, which frightened the people.[1]

This description is reminiscent of an account of a destructive flood in Zabid in Safar 744 (June–July 1343), which was accompanied by a 'terrible roar and an earthquake', heard during the evening by the inhabitants of Zabid.[2] The flood is put by other sources in 743 H.[3] It is therefore possible that an amalgamation of

these separate events has occurred, although apart from the year, different months are given for the earthquake and the flood. The likely association of the earthquake with a flood throws doubt on whether it was a genuine seismic event.

1 Al-'Umari, *Athar*, fol. 130vo–131ro. Although he is a late author, al-'Umari generally provides reliable macroseismic data.
2 Al-Hubaishi, p. 119.
3 Al-Khazraji, II, 75/trans. II, 63; Yahya b. al-Husain, II, 513: neither refers to an earthquake.

1353 October 16   Hellenic Arc

A shock is reported to have occurred in Cairo during the last evening prayer, during Ramadan 753/October 1352.[1] This probably refers to the same earthquake in the Hellenic Arc that caused panic in Crete, also felt in Africa, the following year.[2]

1 Al-Maqrizi, II/3, 876; Taher (1979), p. 197/246. Poirier and Taher (1980), p. 2193, thus give the correct date (1353) by accident!
2 Coronelli (1693), p. 313. Kriaris (1930), I, 266, says the shock occurred the second day after the establishment of the Venetian Republic in Crete under Mark Genadius.

1359   *760 H*   Yemen

A great earthquake in Yemen affected Zabid, San'a and Aden. Shocks lasted from noon till late afternoon. Many houses were destroyed. In Zabid 51 people died and about 10 children in San'a and Aden.[1]

The large area over which the casualties have been reported implies an earthquake of considerable magnitude, with an epicentre somewhere east of Zabid.

1 Al-'Umari, *Athar*, fol. 132ro. The unusually precise details reported have not yet been traced to an earlier source, but lend his account great authenticity. This shock occurred ten years after the previous event reported by the same author (see 750/1349), which might suggest a duplication of one earthquake. The picture is confused by the fact that other, earlier, sources mention destructive rainfall with 80 casualties in Zabid in Ramadan 760 but again do not refer to the earthquake (al-Khazraji, II, 110/II, 92; Anon., *Tarikh al-daulat al-Rasuliyya*, p. 62; Ibn al-Daiba', *Bughyat*, p. 93; al-'Alawi, p. 676). This raises the further possibility that al-'Umari is turning floods into earthquakes. But al-'Umari himself puts the rains in 759 H, the previous year, and is clearly relying on a source not as yet identified (fol. 131vo–132ro; he says 30 people and 20 animals were killed). His account evidently applies to an earthquake rather than a flood, which could not affect three such widely separated localities.

[1361   *762 H*   Lower Egypt]

It is alleged that a great earthquake one Wednesday night shook Egypt, causing the collapse of a minaret of the madrasa of Sultan Hasan, with numerous casualties.[1] No contemporary Egyptian source mentions an earthquake this year, but the collapse of the minaret is recorded, on 6 Rabi' II/13 February (a Saturday). Around 300 children and others attending school were killed.[2] The mosque was still under construction, and the new minaret was of an unusual design; there is no need to invent an earthquake to account for its failure, and the Ottoman report is certainly spurious.

1 Süheyl Efendi, fol. 50 ro-vo; not a contemporary source.
2 Al-Maqrizi, III/1, 60. This was regarded as an omen of the instant demise of the Sultan, which duly occurred the following month. See further details in Ibn Kathir, XIV, 277.

1373 October 19 *1 Jumada I, 775* **Lower Egypt**
A strong earthquake is reported in Cairo,[1] although some accounts say that it was a slight shock.[2]

The absence of further information suggests that the less dramatic account is nearer the truth, and the shock was probably a small local event.

1 Al-'Aini, *Badr*, fol. 88vo; cf. Ibn Hajar, I, 61 note 3. He does not mention where the shock occurred.
2 Ibn Hajar, I, 60, merely has the month, though an alternative reading supports the date 1 Jumada I. Taher (1979), p. 197, misquotes Ibn Hajar's text: only al-Suyuti, p. 56/39 specifies that the shock was felt in Cairo.

1381 *783 H* **Hadramaut**
A great shock occurred in Wadi 'Amd in Hadramaut, which affected everybody strongly. In Dau'an, a parallel valley to the west, the earthquakes destroyed about twelve houses and caused cracks in the mountain.[1] There is no evidence that the earthquake was experienced on the coast.

1 Al-'Alawi, pp. 680–1, citing a manuscript history, which also reports an epidemic in the area this year. The shock in Wadi 'Amd was, literally, so strong that everyone thought it was (centred) in his immediate vicinity. For the location, see Serjeant (1962), map 1.

1385 September 19 *Tuesday 13* **Lower Egypt**
  Sha'ban 787
One or two slight shocks were felt during the night in Old and New Cairo.[1] One source says the shocks occurred in the day.[2]

1 Ibn Hajar, I, 303; al-Maqrizi, III/2, 534; al-Suyuti, p. 56/39; Taher (1979), p. 198/246.
2 Al-Jauhari, I, 120.

1386 July 17 *Monday 18 Jumada II, 788* **Lower Egypt**
A slight earthquake is reported in Old and New Cairo around the fourth hour of the day.[1]

1 Al-Maqrizi, III/2, 546; al-Jauhari, I, 134. Less detailed accounts, not specifying a location, are given by Ibn Hajar, I, 315; al-Suyuti, p. 56/39 and *ibid.*, *Husn*, II, 307; Taher (1979), p. 198/247.

1387 September 5 *20 Sha'ban 789* **Yemen**
Strong earthquakes shook Aden up to the end of Ramadan (14 October 1387).[1] Some houses in Aden fell down,[2] but there were no casualties there. The shocks extended to Hajar and al-Atlal (?), where the earthquake

obliterated the traces of monuments; a great number of people were killed in those districts.[3]

The correct identification of the area affected (? north of Aden) remains conjectural, and it is not possible to estimate the extent of the earthquake.

1 Anon., *Tarikh al-daulat al-Rasuliyya*, p. 96, ed. Yajima, p. 46.
2 Al-Khazraji, II, 193/trans. II, 170; al-'Asjad, facs. edn, p. 451. He says the shock lasted some days and people turned to reciting the Qur'an and reading the traditions of the Prophet in Bukhari (the *Sahih*, see *E.I.²*, I, 1296–7). For other examples of this practice in an emergency (such as plague or drought), see Dols (1977), p. 247.
3 Al-'Umari, fol. 139ro. He puts the event at the beginning of Sha'ban 779 (*sic*)/December 1377. Although the date in contemporary accounts is more likely to be accurate, al-'Umari is clearly once again using an unidentified independent source. *Al-atlal* means 'the ruins' and perhaps refers to some of the many masonry ruins in S. Arabia. It is also mentioned in connection with the earthquake of 830/1427 (see below). There are many places in Yemen called Hajar (which means a large village), including one N.E. of Zabid in the Wasab district; none exactly fits this context.

1394 March–April *Jumada I–II, 796* **Yemen**
An earthquake swarm lasting about two months in the Mauza' region, where about 40 shocks were felt in one day. This is reported on an eye-witness authority.[1]

1 Al-Khazraji, II, 267/trans. II, 238, and al-'Asjad, facs. edn, p. 482, citing the report of the jurist Abu Bakr b. Sulaiman. A briefer version of the same account is given by al-'Umari, fol. 145ro, under the following year, 797 H.

1400 February 22 *Saturday 25 Jumada II, 802* **Yemen**
A series of shocks is reported in the Yemen this year, connected either with meteorite impacts or with volcanic activity.[1]

Around midday on 25 Jumada II/22 February there was a great shock, which was associated with the fall of a great star that resembled a moon (probably a meteor). Many places in the mountains were destroyed.[2]

1 Sieberg (1932a), p. 795, regards this as genuine seismic activity; cf. Ambraseys and Melville (1983).
2 Al-Khazraji, II, 309/trans. II, 280 (wrongly, 20 February); al-'Asjad, facs. edn, p. 501. Redhouse (1908), p. 224, considers that this violent concussion must have been an earthquake, for a meteorite could hardly have given rise to ground motions strong enough to cause the damage reported, unless the places were ready to fall of themselves. Al-Khazraji reports at least two instances of large houses collapsing without the help of an earthquake, one in San'a in 1283 and another in Aden in 1346, both causing casualties (see also above, p. 17). For a previous meteorite impact in Yemen, in Shawwal 792/September 1390, see Ibn al-Daiba', *Qurrat*, II, 112.

1400 April 6 *10 Sha'ban 802* **Yemen**
A similar event occurred on 10 Sha'ban/6 April in the Zabid valley, following the impact of a great star which had come from west to east. It landed between two hills and fire broke out in the place where it landed. One of

the peasants was up a palm tree at the time, and nearly fell.[1]

1 Al-Khazraji, II, 310–11/trans. II, 281; al-'Asjad, pp. 501–2.

1400 July   *802 H*   **Ethiopia**

Volcanic activity in Ethiopia is reported in the latter half of 802/summer 1400. Seafarers who had come from Ethiopia reported that tremendous earthquakes lasted for just under ten days, as a result of which a number of places and many hills fell down. This was followed by the eruption, which caused the inhabitants to flee the region. The fire continued for some days and a group of hills was formed in that locality, where none had previously been known.[1]

This describes one of the eruptions of the volcanic complex of Dubbi on the Ethiopian side of the Red Sea.[2]

1 Al-Khazraji, II, 315/trans. II, 285.
2 Gouin (1979), pp. 93–4.

1408 ?   *811 H*   **Hejaz**

An earthquake in Mecca destroyed an arch at the sacred mound of al-Marwa, *c.* 200 yards north of the Ka'ba, before it died away. The arch was restored to a better state than it was in before the earthquake.[1]

1 Al-'Umari, fol. 153ro, with additional material in BL Ms. 6300, fol. 173. The collapse of the arch at al-Marwa and its subsequent repair are mentioned by the Meccan historians al-Fasi, pp. 95–6, under 801 H and by al-Nahrawali, p. 198, under 811 H. Neither mention that an earthquake was responsible for the damage, which is perhaps surprising.

1413 December   *Ramadan 816*   **Yemen**

A series of earthquakes in the mountain districts of Himyar, lasting for about half of Ramadan (began 25 November). A tower of Hisn al-Khadra collapsed and a cistern was fissured and the water poured out. The shock damaged Hisn al-Safra, among other forts in the region. The shock was not felt in Zabid.[1]

Himyar is a large and rather ill-defined area stretching south and southwest of San'a and Yarim, with its capital at Zafar (al-Ashraf). Hisn al-Khadra is probably to be located near Khadid in Jabal Hubaish, northwest of Ibb.[2] It is not possible to estimate the size of the earthquake.

1 Anon., *Tarikh al-daulat al-Rasuliyya*, p. 168. News of the earthquake arrived in Shawwal/beginning 25 December, which indicates that the shock was not felt there.
2 Ibn al-Daiba', *Qurrat*, I, 337n, 363n (cf. 549/1154 above); Yahya b. al-Husain, *Ghayat*, II, 609; 'Umara, trans. Kay (1892), p. 17 and index. A second possibility exists in Jabal al-Khadra E.N.E. of Ibb, cf. al-Waisi (1962), p. 44. Localities called al-Khadra and al-Safa (*sic.*) are to be found on YAR (50), sheet 1444 C4, at MA 3352 and MA 3652 respectively. We are grateful to Professor R.B. Serjeant and Abd al-Malik Eagle for advice over these locations.

1422 June 28   *8 Rajab 825*   **Lower Egypt**

A light earthquake is reported in Cairo on this date.[1] There is no indication of the epicentral area of the shock.

1 Ibn Hajar, III, 273; al-Suyuti, p. 57/40; Taher (1979), p. 202/247. The shock is not widely mentioned.

1425 June 23 *Saturday 6 Sha'ban 828* **Gulf of Suez?**

A strong earthquake was felt in Cairo and Egypt on 6 Sha'ban,[1] but did little damage. Houses, dwellings and minarets were set in motion by the ground which swelled in three waves. The first shock occurred at sunrise and was followed by two others. The shock lasted several minutes;[2] had it lasted longer, houses would have fallen down. An eye-witness speaks of buildings and other things being shaken with a frightening motion – he saw a wall moving out of position before returning to where it was. He was told that a man riding was jerked in his saddle so that he nearly fell. There were no casualties, but the people fled from their houses to the suqs.[3]

The earthquake evidently caused alarm in Cairo; the next day the Sultan called for a three-day fast, but the people did not heed this and did not turn to God.[4]

The long-period shaking caused by the earthquake, combined with the absence of information from Syria or the eastern Mediterranean, suggests a possible source in the Gulf of Suez region.

1 Al-Maqrizi, IV/2, 690–1; al-Jauhari, III, 95, al-'Aini, Ms. 1544, fol. 174ro. This date is preferred, as al-Maqrizi's account is based on an eye-witness source, though the day of the week does not match: 23 June was a Monday, not a Saturday. There is unusual disagreement among the Arabic chroniclers over the date of the shock; Ibn Hajar, III, 348, has 17 Sha'ban (= Friday 4 July). The Continuator of Ibn Duqmaq, fol. 144ro, appears to date it later in the year, between Ramadan and Shawwal. Taher (1979), p. 202/247, follows Ibn Hajar; the event is not listed by Poirier and Taher.
2 Two *daraja* is equivalent to 8 minutes, see above, p. xvii. Al-Maqrizi says that the first shock lasted the time it took to read *surat Ikhlas* (Sura 112, the third shortest in the Qur'an). Ibn Iyas, II, 99, a late source, says the shock occurred at sunset.
3 Ibn Iyas, *l.c.* It is strange that people fled to the suqs, which are no safer than the houses; possibly the text is corrupt. Al-'Aini, fol. 174ro, says the shock was of short duration and that most of the people were not aware of it.
4 The Sultan was Barsbay (1422–38); public fasting was also proclaimed during plagues. Al-Maqrizi's verdict on the morals of fifteenth-century Mamluk society is characteristically pessimistic (cf. 1303).

1426 November   *Muharram 830*   **Persian Gulf**

An earthquake is recorded in the island of Daraht, near Hurmuz, in the region of Bahrain (i.e. off the Arabian shore of the Persian Gulf). The Sultan's stables were partly destroyed, as was also the house of the qadi. A nearby mountain opened up and it was said that rats the size of dogs emerged from the fissure. This news reached Damascus from a reliable observer.[1]

The 'Daraht' of the text should probably be read Brukht (on Qishm), where there was a royal palace.[2] The earthquake was therefore local to the Persian Gulf.

1 Al-Maqrizi, **IV**/2, 736; his Syrian source has not been identified. Taher (1979), pp. 202–3/180, appears to locate the shock in Yemen.
2 Ibn Majid, trans. Tibbetts, pp. 222–3; Aubin (1973), pp. 102–3, 129–31, for the 'Sultan' at this time, Qutb al-Din Firuzshah b. Muhammad Shah, formerly ruler of Hurmuz and Bahrain, Lahsa and Qatif, before his deposition in 820/1417; cf. al-Maqrizi, p. 988.

1427 ? *830 H* **Yemen**
An earthquake in Yemen destroyed al-Atlal and al-Diman; Zabid was shaken 14 times in one day. Many houses were destroyed and about 60 people perished in the wreckage.[1]

Neither al-Atlal nor al-Diman (both meaning 'the Ruins') has been identified as a place name; the area affected was probably quite large, lying in the region east of Zabid and north of Aden.[2]

1 Al-'Umari, fol. 159vo, the same in BL Ms. 6300, fol. 178: another detailed report by this late author, whose original source of information has not yet been identified. The year 830 H began on 2 November 1426.
2 Damage to al-Atlal is also mentioned in 779 or 789/1387 (see above). *Al-diman* means 'abandoned and ruined places' (or deserted campsites), effectively synonymous with *al-atlal*. Perhaps the shock destroyed the ruins of former, historic settlements.

1432 May 20 *Monday 19 Ramadan 835* **Red Sea**
An earthquake occurred in the early morning, felt in Zabid and throughout Yemen. No casualties or damage are reported.[1]

This may be the same earthquake as was felt in the Tigre region of Ethiopia sometime before 29 August 1432. No damage is reported.[2] This suggests an epicentre in the Red Sea.

1 Anon., *Tarikh al-daulat al-Rasuliyya*, p. 249. Al-Ahdal, *Tuhfat*, fol. 314ro, also refers to a shock in Zabid before the eruptions the following year, 836 H (see below). May 20 was a Tuesday.
2 Basset (1881), p. 119; Palazzo (1915), p. 298; Gouin (1979), p. 26.

1432 December– *Rabi' II and* **Yemen**
January 1433 *Jumada I, 836*
An earthquake in Rabi' II and Jumada I, 836/between 26 November 1432 and 22 January 1433, was felt in Surdud and elsewhere in the Tihama and the mountains;[1] this suggests a relatively large event.

The exact place of this earthquake in the sequence of events in 835–6 H is not entirely certain (see below).

1 Al-Ahdal, fol. 314ro. For Surdud, see Wilson (1989), p. 188.

1433 March–April *Sha'ban 836* **Red Sea**
A volcanic eruption lasting half the month of Sha'ban 836 (which began 23 March 1433) is reported in the

Zubair group of seven volcanic islands in the Red Sea, between Kamaran and Dahlak (see Figure 2.15).[1]

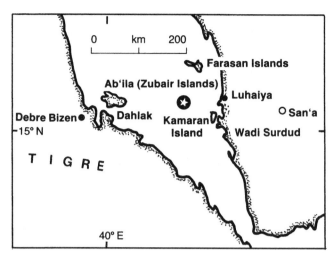

Figure 2.15. 1433 March–1434 September, Red Sea.

Al-Ab'ila was destroyed by the eruption, which then moved to an adjacent vent that was also destroyed. The eruption was audible in al-Luhaiya and the eruption column was visible in the Surdud region.[2] Similar effects were reported from Ethiopia.[3]

Some confusion of dating exists over this event, doubtless due to its having been part of a sequence. Al-Ahdal, an exactly contemporary source, records the eruption under Jumada II or Jumada I, 835 (February or January 1432), i.e. *before* the two earthquakes referred to above. The second earthquake might have just preceded the eruption and the date 836 H is preferred, particularly because the later date better matches the evidence from Ethiopia (see below, 1434).

1 Anon., *Tarikh al-daulat al-Rasuliyya*, pp. 261–2. The author was exactly contemporary.
2 Ibn al-Daiba', *Qurrat*, **II**, 131; Yahya b. al-Husain, **II**, 569. Both merely have the year, 835 H (*sic*), and paraphrase al-Ahdal's account, referred to below.
3 Al-Ahdal, fol. 314vo. He took the earthquakes as a fulfilment of Prophetic traditions on signs of the coming Resurrection. His account, which is chronologically quite confused, also confirms the correct spelling of Ab'ila, not Abghila.

1433 December 14 *1 Jumada I, 837* **Lower Egypt**
The Continuator of Ibn Duqmaq records an earthquake during the night, presumably in Cairo.[1]

1 Cont. of Ibn Duqmaq, Paris Ms. 5762, fol. 150vo. The shock is not mentioned by other sources, but the annals are exactly contemporary.

1434 September *Safar 838* **Red Sea**
The island of al-Ab'ila was again in an eruption heard as far away as al-Luhaiya and al-Hirda, i.e. along the

Yemen coast (see Figure 2.15). The eruption lasted several days.[1]

It is probable that some connection exists between this eruption and reports of numerous earth tremors in the Tigre region of Ethiopia during the year starting 29 August 1433, perhaps aftershocks of the relatively large magnitude earthquake of 1432 (see above). Unfortunately, the Ethiopian source does not refer specifically to volcanic activity and the Ethiopian dating does not resolve whether the year of the main eruption was 835 or 836 H, though it tends to favour the later date. The continuation of shocks (that would have accompanied the eruption) in Ethiopia until at least July 1434, appears to tally with the Yemeni account of this second phase of activity early in 838 H.[2]

A further renewal of volcanic activity on the remaining island of al-Ab'ila occurred in 839 (began 27 July 1435).[3]

1 Al-Ahdal, fol. 315vo; Ibn al-Daiba', *Qurrat*, II, 132; al-Hibshi (1982). Safar began on 6 September this year.
2 Palazzo (1915), p. 298; Gouin (1979), p. 26. The year is fixed by reference to a solar eclipse on 7 June 1434, also noted by Arab authors, e.g. al-Maqrizi, IV/2, 918–19 (Monday 29 Shawwal 837); cf. Chaine (1925), p. 272.
3 Ibn al-Daiba', *Qurrat*, II, 133.

1434 November 6 *Saturday 3 Rabi' II, 838* Lower Egypt

An earthquake is reported in Cairo shortly before noon on a Saturday morning.[1] Houses shook in various places, and it would have caused damage if it had lasted longer. But the shock was slight and passed quickly, causing no harm.[2]

Another shock occurred at the end of the month, 25 Rabi' II/28 November.[3]

1 The day is specified by Anon. *Damas. chron.*, fol. 150ro; this preferred date is also given by al-Maqrizi, IV/2, 935, and al-Jauhari, III, 308. Ibn Hajar, III, 546, gives 4 Rabi' II/7 November. The Continuator of Ibn Duqmaq, fol. 150vo, gives 4 Jumada II/5 January 1435, which was a Wednesday. All these authors are exactly contemporary. See also al-Suyuti, p. 57/40; Taher (1979), p. 204/247. Poirier and Taher (1980), p. 2193 (under 7 November 1437).
2 Ibn Hajar, *l.c.*; echoed by Ibn Iyas, II, 161.
3 The Continuator of Ibn Duqmaq is the only source to mention this. He puts the second shock on 25 Jumada II/16 January 1435, but as it apparently occurred in the same month as the first, we should read 25 Rabi' II.

1438 February 25 *17 Sha'ban 841* Hellenic Arc

Two very slight shocks were reported at the call for afternoon prayer, shaking the house of the historian al-Maqrizi in Cairo.[1] Most authors mention only Cairo; one has Cairo and 'Misr', which could mean Egypt in general or simply Fustat.[2]

This is probably the same earthquake as one felt in the eastern Mediterranean, but dated 2 February in the afternoon, according to a marginal note.[3]

1 Al-Maqrizi, IV/2, 1029; al-Jauhari, III, 402. Ibn Iyas, II, 181 has Rajab 841, which is a month early. See also, al-Suyuti, p. 58/41; Taher (1979), pp. 204–5/247.
2 'Abd al-Basit, fol. 4vo.
3 Lampros (1910), p. 157; Maravelakis (1939), p. 122.

1444 January 20 *Sunday 29 Ramadan 847* Red Sea

The effects of a volcanic eruption are reported in the Tihama of Yemen. White ashes fell during the night, during which earth tremors and a roaring sound were experienced. The ground was covered with ashes from Aden to the Hejaz, also parts of the inland mountain ranges. In Zabid, 847 H was called the year of the ashes.[1]

1 Ibn al-Daiba', *Bughyat*, p. 77.

1455 March 5 *Wednesday 16 Rabi' I, 859* Lower Egypt

A light shock was felt in Cairo and its surroundings, which shook the ground more than once. A few nights later, another weaker shock occurred.[1]

1 Ibn Taghribirdi, *Hawadith*, VIII/2, 225; 'Abd al-Basit, fol. 105ro; Ibn Iyas, II, 323.

1458 November 12 *Sunday 5 Muharram 863* Dead Sea

A damaging earthquake in southern Palestine was weakly felt in Cairo (see Figure 2.16). It destroyed parts of the citadel of al-Karak, including towers and parts of the walls, as well as the governor's palace and many houses.[1] It is reported that 100 people were killed in al-Karak.[2]

The shock destroyed minarets in al-Ramla, Ludd and Hebron (al-Khalil).[3] In Jerusalem, the top part of the

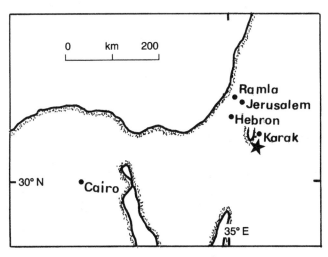

Figure 2.16. 1458 November 12, Dead Sea.

49

minaret over the Zawiya al-Darkah was thrown down and a dome near the Church of the Holy Sepulchre was destroyed.[4]

1 Ibn Taghribirdi, *Hawadith*, VIII/2, 319, dates the event Thursday 9 Muharram/16 November. In his *Nujum*, XVI, 127/trans. Popper, VI, 82, he has a briefer account dated Wednesday 1 Muharram 863/ 8 November. In view of this uncertainty, the date given elsewhere (see note 4) is preferred. Kallner-Amiran (1951), p. 229, relying on various early catalogues, duplicates this shock under 1456 and 1458, apparently identifying the latter with the Van earthquake of 861/23 April 1457.
2 Al-Suyuti, p. 58/41: his source has not been identified, and might not refer exclusively to al-Karak. Ibn Iyas, II, 350, merely says many houses were destroyed between Jerusalem and Hebron.
3 'Abd al-Basit, fol. 122vo, is the only author to mention Ludd.
4 Al-Sakhawi, *Dhail*, fol. 4ro; al-'Ulaimi, pp. 398, 400/trans. Sauvaire, pp. 165, 170. A story in al-'Ulaimi, pp. 599–600, makes it clear that the dome (*qubba*) was contiguous with the south side of the Holy Sepulchre (al-Qumama). The noise of the Christians 'reading their books' there sometimes could be heard as far as the Dome of the Rock, to the irritation of the Muslims, who saw the destruction of the chapel as retribution from God. The Christians were prevented from rebuilding the chapel. Both authors date the event 5 Muharram (12 November). According to al-Sakhawi, the damaged minaret in Jerusalem was situated over the Bab Asbab, i.e. the Bab Asbat, situated in the northeast corner of the sanctuary of the Dome of the Rock, see al-'Ulaimi, esp. p. 381/trans. Sauvaire, p. 129, and Burgoyne and Richards (1987), pp. 415–18; the top of the minaret was also repaired after the earthquake of 1927.

1463　　　*867 H*　　　Yemen

A series of damaging earthquakes occurred in Zabid, Yemen, where 50 houses were ruined. The tremors continued for three days, five times a day. Ten people were killed.[1]

1 Al-'Umari, fol. 168ro. This is again a very particular account; the original source has not been identified.

1466 December 15　*Monday 7 Jumada I, 871*　Yemen

A great earthquake at dawn caused alarm in Zabid. A lesser shock occurred before noon two days later.[1] There is no evidence of damage there or elsewhere.

1 Ibn al-Daiba', *Bughyat*, p. 140.

1467 December 15　　*Tuesday 17*　　Lower Egypt
　　　　　　　　　Jumada I, 872

A light earthquake at night in Cairo caused a few old places to collapse.[1]

1 Ibn Taghribirdi, *Hawadith*, VIII/3, 616; 'Abd al-Basit, fol. 189ro-vo; Ibn Iyas, II, 471.

1476 October–November　*Rajab 881*　Lower Egypt

A fearful shock occurred in Cairo during the night, the noise of which was heard in the buildings; several decayed buildings that had been vacated collapsed. Many people reported the earthquake. Had it lasted another minute, it would have caused great harm and

been extremely frightening.[1] Al-Suyuti calls it a slight shock.[2]

1 'Abd al-Basit, fol. 273ro; Ibn Iyas, III, 121/trans. Wiet (1945), p. 137.
2 Al-Suyuti, p. 58/41; Taher (1979), p. 205/248.

1481 March 18　　*Sunday 17*　　Mediterranean
　　　　　　　　Muharram 886

On 28 Muharram (29 March), news arrived in Damascus of a shock in three or five villages in Anatolia (al-Rum). Later, further news arrived of the death of the Qadi Sharaf al-Din b. 'Id in Egypt because of this earthquake, when a decorative cresting on the Salihiyya madrasa in Cairo fell on him and another man who died with him.[1] In the qadi's obituary, he is said to have been struck by a falling portion of the edging on the Hanbalis' porch in the madrasa, in which he was sitting when the shock struck. Poems were written about the earthquake and the qadi.[2]

The most detailed accounts of the effects in Egypt report that the earthquake was frightening in Cairo and Misr (Fustat) and the surrounding region. Reports came from Alexandria and even Anatolia. In Cairo, the shock struck about an hour after the afternoon prayer on 17 Muharram. The shaking was severe, and the minarets were seen to sway and undulate. Estimates of the duration of the shock vary from two to six *daraja*.[3] Many houses and weak buildings collapsed. The shock was accompanied by a grinding noise in the earth and caused panic: women ran into the streets with faces uncovered. Some people ran naked from the bath-houses. The earthquake was worse than anything that had been experienced in their time, or in that of their elders, and people thought the Day of Resurrection had arrived. Other casualties of the earthquake were 'Izz b. 'Id,[4] and al-Zaini Abu Bakr, who died ten days after the earthquake from a heart attack caused by the shock.[5]

It is possible that this earthquake also damaged certain parts of the nilometer structure and its foundations, which were repaired later the same year.[6]

The shock in Cairo is clearly a distant effect of a larger event in the eastern Mediterranean, most probably the earthquake of 15 March that ruined Rhodes and Cyprus, setting off a damaging sea-wave on the coast of Adalia. This was followed by further earthquakes on 3 May and 18 December 1481.[7]

1 Ibn Tulun, I, 33–4. He mentions that four others are said to have died with them.
2 Al-Sakhawi, *Dau*, X, 180–1; for the location and history of the important Salihiyya madrasa, see Parker and Sabin (1974), p. 46, Petry (1981), pp. 330–1. Other sources cited below give further details: the Sultan, Qayitbey, attended his funeral and allowed him

to be buried in the court of his own mausoleum. See also al-Suyuti, p. 58/41, and Ambraseys (1961), p. 28, who notes that Sprenger has 880 for 886 H.

3 Between 8 and 24 minutes: 'Abd al-Basit has 2 *daraja*; Ibn Iyas has 3, and Anon., *Jawahir al-suluk*, has 6.

4 'Abd al-Basit, fol. 310vo–110ro; al-Sakhawi, *Dhail*, fol. 79ro–80vo. These accounts complement without duplicating each other. 'Izz b. 'Id may be the 'other man' who died with the qadi.

5 Al-Jauhari, *Inba al-hasr*, pp. 509–10; Ibn Iyas, III, 178–9/trans. Wiet (1945), pp. 196–7; he says the earthquake was on a Wednesday, which is incorrect. See also in Anon., *Jawahir al-suluk*, fol. 391ro.

6 Popper (1951), p. 27.

7 Al-Sakhawi, *Dhail, l.c.*, reports that the earthquake was destructive in Rhodes. Ibn al-'Imad, VII, 344, a late author, reports an unprecedented shock in Mecca (*sic*.) on 17 Muharram 886, which has yet to be verified. It is probably an echo of the fire in Medina this year, started by a thunderbolt. The fire, which caused great damage on 13 Ramadan/5 November, is widely mentioned by contemporary Arab and Christian writers, e.g. Khatjikyan (1955), III, 40, 43; al-'Ulaimi, ed. Najaf, I, 191–2 (see also below, under 1483). Al-Samhudi, *Wafa*, III, 638, says the news caused the inhabitants of Rhodes to rejoice, only to be struck down by an earthquake: this must be a reference to the Rhodes earthquake of 18 December. See also Taher (1979), pp. 205–8/169–71, 248, for the Arabic texts: it should be noted that a report by Ibn Iyas quoted in Taher (p. 208/248) under the year 886 H should in fact refer to 896 H, and the entry in Poirier and Taher (1980), p. 2193, under 1481 July is thus redundant (see below).

For the earthquake in Rhodes and Cyprus, see Coronelli and Parisotti (1688), pp. 158–61; Ambraseys (1965). Confused entries on events in 1481 are found in Kallner-Amiran (1951), p. 229, and Ben-Menahem (1979), pp. 282, 292, based on earlier catalogues.

1483 June 7 *Saturday 1 Jumada I, 888* **Yemen**

A great earthquake during the night is reported in the Yemen, following the fall of a large star that travelled from west to east of Zabid.[1] Its association with what might be a meteorite impact makes this a dubious seismic event.

The report of an earthquake in Medina this year (888 H) can be taken as an inaccurate reference to the fire in Medina in 886 H.[2]

1 Ibn al-Daiba', *Bughyat*, p. 172.

2 Sani' al-Daula, *Muntazam*, II, 79: a nineteenth-century Persian work. He also mentions a flood in Mecca and a total eclipse. Grumel (1958), p. 469, lists two total lunar eclipses in 1483, and the flood in Mecca is put in 887 H (Ibn al-'Imad, VII, 346) or 888 H (Wüstenfeld, 1861, pp. 296–7). An alleged earthquake in Mecca is put in 886 H by Ibn al-'Imad, VII, 344, who is probably Sani' al-Daula's source, but we consider this to be a reference to the fire in Medina that year, which he does not mention; see above, 1481.

1483 June 15 *Sunday 9 Jumada I, 888* **Lower Egypt**

A perceptible though light shock of short duration was felt in Cairo during the night.[1] Had it lasted longer, it would have caused problems.[2]

This shock is very near in date to an earthquake in Yemen, which started a series of events lasting till 890/1485 (see above).

1 'Abd al-Basit, fol. 327ro, just has the month; the full date is given

by al-Suyuti, p. 58/41, who does not mention a place; cf. Taher (1979), p. 208.

2 Ibn Iyas, III, 201/trans. Wiet (1945), p. 223; he puts the shock a month earlier in Rabi' II, 888/May 1483 and implies it lasted less than a *daraja*. He uses stock phraseology here; cf. his account of the 881/1476 event.

1484 May 9 *Thursday 12 Rabi' II, 889* **Yemen**

There was a great earthquake at noon in Zabid; the roofs of the houses shook and people ran out in fear for their lives; people also fled from the suqs. Shocks lasted till sunset.[1]

There may be some connection with the report that during the previous night a massive star had fallen, moving from east to west.

1 Ibn al-Daiba', *Bughyat*, p. 172; 9 May was a Sunday this year.

1485 March *Rabi' I, 890* **Yemen**

A series of strong earthquakes affected Zabid and a wide area around the town sometime between Rabi' I and Ramadan/March and October this year. The tremors, which were also felt in other towns and caused great apprehension, were continuous until one Friday after the noon prayers there was a strong shock which caused people to flee the suq in fear, abandoning their treasures and merchandise. Those in their houses heard a strong vibration in their roofs. Everyone coming into Zabid from surrounding districts in the next few days said that they had experienced the earthquake.[1]

There is no indication from this account that the swarm of shocks or the main shock caused damage or loss of life, though they were evidently quite widely felt.

1 Ibn al-Daiba', *Bughyat*, p. 174/trans. Johannsen (1828), p. 223.

1486 October 11 *Wednesday 12* **Lower Egypt**
Shawwal 891

A frightening shock occurred in Cairo around midday,[1] moving the ground once or twice.[2]

There is no reference to a shock elsewhere in the region on this date and it was probably local to Egypt.

1 Al-Sakhawi, *Dhail*, fol. 103ro.

2 'Abd al-Basit, fol. 357ro; he calls it a light shock.

1491 April 24 *Monday 16* **Mediterranean**
Jumada II, 896

Two slight shocks a week apart were reported from Damascus, Cairo and Crete. Both earthquakes caused heavy damage in the island of Cyprus, where the forts at Limassol, Paphos and Famagusta and buildings in Nicosia were destroyed.[1]

In Damascus, the first shock, which was not widely felt, occurred after the sunset prayers on 16 Jumada II/

evening of 25 April; the second was before sunrise on 22 Jumada II/1 May.[2] In Egypt, the earthquake was alarming, shaking buildings and lasting a *daraja* or more.[3] The second shock was slight.[4]

1 A document in Archivo Ducale Sforzesco. Milan, Miscell. C. 646, dates the earthquake in Cyprus 24 April at midnight. Other sources have Sunday 24th or during the night of the 25th, and the damaging aftershock on May 1: Anonymous Pilgrim, fol. 331ro; Dietrich, p. 187/ 210–13; Darrouzes (1958), p. 245. Ben-Menahem (1979), pp. 257, 290 does not link the two events.
2 Ibn Tulun, I, 138–9; 16 Jumada II = 26 April, but the Muslim day started at sunset on the previous evening. There still appears to be a difference of one day between the Arabic and Occidental sources, perhaps due to the use of a different calendar (see pp. xvii–xviii). Ibn Tulun mentions the name of one of those who experienced the shock. He correctly dates the second shock 1 Ayyar (May).
3 Al-Sakhawi, *Dhail*, fol. 183vo. 'Abd al-Basit, fol. 402vo, describes it as a slight shock; it appears to come under Jumada I, not Jumada II (his chronicle ends this year).
4 Al-Suyuti, p. 59/41 correctly has Sunday 22 Jumada II. Ibn Iyas, III, 281–2/trans. Wiet (1945), p. 317, says the second shock occurred the morning after the first (note that in Taher, 1979, p. 208/248, Ibn Iyas's text is found under 886 H by mistake). Both he and 'Abd al-Basit say a solar eclipse occurred between the two shocks; Ibn Tulun mentions a lunar eclipse. There was, in fact, a solar eclipse on 8 May 1491 (Grumel, 1958, p. 469)/Sunday 29 Jumada II, as correctly noted by al-Sakhawi, who does not mention the second shock.

1498 September–October *Safar 904* Lower Egypt

A slight earthquake was experienced in Cairo.[1] There are no details of an earthquake elsewhere in the region at this time.[2]

1 Ibn Iyas, III, 399/trans. Wiet (1945), p. 441. Safar began on 18 September. The shock was considered a bad omen for the Sultan (Nasir al-Din Muhammad), who was indeed murdered the following month, cf. Ibn al-Himsi, fol. 30vo.
2 The murder of the Sultan on 15 Rabi' I/31 October provides a *terminus ante quem* for the shock, which is otherwise only vaguely dated. Oruç, ed. Kreutel, p. 106, reports a strong earthquake in Istanbul in the late evening of Sunday 20 Rabi' I/4 November, but this seemingly occurred too late to be relevant.

1500 July 24 *Friday 27 Dhu'l-Hijja 905* Hellenic Arc

A light shock in Cairo after the evening prayer, lasting half a *daraja*. Had it lasted longer, it would have been more of a problem.[1] The earthquake is also described as a serious shock, which destroyed many houses and caused a great commotion among the people.[2] These accounts are not necessarily incompatible, but reflect different attitudes to an event that probably caused only minor damage. The earthquake is perhaps to be associated with a shock that affected southern Greece and the Peloponnese sometime during 1500.[3]

1 Ibn Iyas, III, 443/trans. Wiet (1945), p. 482, paraphrased by al-Suyuti, p. 59/41; Taher (1979), p. 209/248. This is the last shock recorded by al-Suyuti. Ibn Iyas employs his stock phrase again here. He also refers to shooting stars being seen.
2 Ibn al-Himsi, fol. 57vo; he puts the event on 28 Dhu'l-Hijja.

3 Thevet (1554), p. 363b, places this earthquake in Southern Greece after the fall of Methoni to Bayezid, i.e. after 10 August 1500.

1501 January 26 *Tuesday 6 Rajab 906* Yemen

There was a great earthquake at dawn in Zabid which was heard in all parts of the town. The word used for earthquake is *hizza* ('shock, convulsion').[1]

1 Ibn al-Daiba', *Fadl*, p. 146; this passage is not found in the manuscripts used in Chelhod's edition.

1502 November 13 *Saturday 11 Jumada I, 908* Yemen

During the night and following day of 11 Jumada I, there was an earthquake in Zabid and its districts; tremors continued day and night and the people were alarmed.[1] There is no evidence that these shocks were damaging.

An earthquake in Aden is reported in similar terms this year in connection with a great fire there.[2] The fire in Aden in fact occurred in Safar 908/August 1502, and there must remain some doubt as to whether Aden was indeed affected by the November earthquake. It is, however, quite plausible that Aden was within the area shaken by the earthquake of 910/1504 (see below).

This earthquake was followed by a long series of shocks reported by the same authors up to 917/1511 (see below).

1 Ibn al-Daiba', *Fadl*, p. 166; al-'Aidarusi, p. 51. November 13 was a Sunday.
2 Ibn al-'Imad, VIII, 36. Taher (1979), p. 209/180, gives the Arabic text of al-'Aidarusi, but translates the account of Ibn al-'Imad. Poirier and Taher (1980), p. 2193, therefore put the shock in Aden.

1502 November 17 *15 Jumada I, 908* Lower Egypt

A strong earthquake is reported in Cairo, which destroyed several places.[1]

This unique account may be exaggerated. There is unlikely to be a connection with the event in south Yemen a few days before (see above).

1 Ibn al-Himsi, fol. 89vo. See also his account of the 1500 July 24 event. The shock is not mentioned by Ibn Iyas.

1504 August 30 *Friday 19 Rabi' I, 910* Red Sea

During the night there was a great earthquake at Zabid, which was also very strong at Zaila'. Some houses fell down and the people ran out down to the shore; they didn't return to their houses till the morning.[1]

The fact that this earthquake was felt in the Horn of Africa as well as the Yemen suggests a relatively large-magnitude event with an offshore epicentre near the Bab al-Mandab Strait (Figure 2.17).

On Tuesday 30 Rabi' I/10 September there was a

further shock in Zabid, following the explosion of a shooting star.[2]

1 Ibn al-Daiba', *Fadl*, p. 181; al-'Aidarusi, pp. 52–3; Ibn al-'Imad, **VIII**, 44 (the latter says the inhabitants of Zaila' took refuge in the fields, rather than on the shore). August 30 was a Friday. Ibn al-Daiba''s editor puts the shock on 19 Safar 910/1 August (see index, p. 373), and this also appears to be the date given by 'Isa b. Lutf-Allah, *Rauh*, p. 10, and Raşit (1875), p. 10. However, it is clear from the context that Rabi' I is correct: Ibn al-Daiba' previously mentions a great flood in Wadi Zabid on Sunday 29 Safar, and the Sultan's return from a raid on San'a on Friday 12 Rabi' I (from both of which it is consistent that 19 Rabi' I was a Friday); the issue is clinched by the fact that a total lunar eclipse was reported on Monday 15, four days before the earthquake. 15 Rabi' I = 26 August 1510, i.e. eclipse no. 4188 in Oppolzer (1962), p. 366.
2 Ibn al-Daiba', *l.c.*

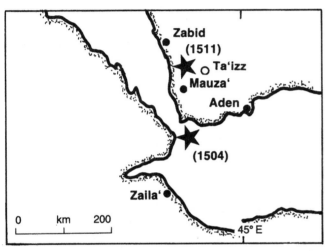

Figure 2.17. The earthquakes of 1504 August 30 (Bab al-Mandab Strait) and 1511 January 29 (betwen Zabid and Mauza'), Yemen.

1508 May 29 *Tuesday 30 Muharram 914* **Hellenic Arc**
A light shock in Egypt is reported during the night.[1] This was a result of the damaging earthquake in the Hellenic Arc on 29 May 1508, which ruined eastern Crete and many of the islands of the Aegean Sea. The shock was felt in Greece, Asia Minor and Syria.[2]

1 Al-Da'udi, p. 62/43; Taher (1979), p. 209/249; al-Hafiz (1982). The correct conversion yields Wednesday 31 May, unless Tuesday 29 Muharram is intended, which began on the evening of Monday 29 May.
2 Baumgarten (1704), **I**, 499, 502; Perrey (1850a), p. 21; Lampros (1910), p. 169; Maravelakis (1939), p. 131; Ben-Menahem (1979), p. 294. Despite the distant origin, however, the shock in Cairo was of relatively short duration: quarter of a *daraja* (1 minute) according to Ibn Iyas, **IV**, 132/trans. Wiet (1955), p. 128.

1509 April *Dhu'l-Hijja 914* **Hellenic Arc**
An earthquake in Egypt, only felt by a few people,[1] most probably from continuing activity in the Hellenic Arc.

1 Ibn Iyas, **IV**, 148/trans. Wiet (1955), p. 145.

1509 May 7 *Monday 17 Muharram 915* **Yemen**
There was an earthquake at dawn in Zabid; the floors cracked, loud noises were heard in the roofs and vases were overturned on shelves.[1]

1 Ibn al-Daiba', *Fadl*, p. 220; al-'Aidarusi, p. 91; Taher (1979), p. 209/181.

1509 September 10 *Jumada I, 915* **Turkey**
A destructive earthquake in the Marmara Sea was felt in the evening by a few people in Cairo.[1]

1 Ibn Iyas, **IV**, 160/trans. Wiet (1955), p. 157: Jumada I began on 17 August. Knolles (1603), p. 476; Perrey (1850a), p. 21; Hammer-Purgstall (1870), **II**, 349; **VI**, 90; Ambraseys and Finkel (1990).

1511 January 20 *Tuesday 20 Shawwal 916* **Yemen**
There was a strong earthquake in Zabid at dawn and again at midnight on Wednesday at the end of the month/29 January 1511. A third shock occurred the following morning. In the early evening of the same day a great star fell, moving from the east towards the north, which could be seen during the day and was followed by a powerful shock with a noise like heavy thunder. Then there was a great and unprecedented earthquake in Mauza' and its districts; it started at the end of Shawwal and shocks were still continuing at the end of Dhu'l-Hijja/March 1511, small and large shocks recurring day and night. These caused great harm to the people of the region, since large and solid houses were badly fissured and weakly built houses were totally destroyed. No house escaped damage; land prepared for cultivation was fissured; graveyards were disturbed and all the wells were affected.[1]

The continuing association of events in Yemen with meteorite falls or comets does not invalidate the genuine seismic activity reported here. Two distinct series of tremors are involved, one in Zabid and the other in Mauza', although many of the shocks (particularly that of 29 January) might have been felt in both places, as is implied in the case of the shocks in Zabid in February 1511 noted below. The absence of references to Ta'izz being affected is perhaps significant, since this was the summer capital of the Rasulids and might have provided local information (see Figure 2.17). The cumulative damage caused by the earthquakes in Mauza' is notable, as are the reported effects on underground water supplies.

1 Ibn al-Daiba', *Fadl*, p. 235; he quotes from a document sent by the qadi of Mauza', al-Faqih Muhammad b. Ahmad al-Mauza'i. This information is condensed by al-'Aidarusi, p. 93; 'Isa b. Lutf-Allah, *Rauh*, p. 13; Ibn al-'Imad, **VIII**, 72–3; Raşit (1875), p. 18 (who puts the shock on 13 Shawwal); Taher (1979), p. 210/181.

1511 February 27   *Thursday 29*   **Yemen**
Dhu'l-Qa'da 916

There was a great earthquake in Zabid during the afternoon and again the following night, 28 February.[1] These shocks were presumably part of the series of events experienced also at Mauza' (see above).

1 Ibn al-Daiba', *Fadl*, p. 236; 'Isa b. Lutf-Allah, *Rauh*, p. 13; Raşit (1875), p. 18.

1511 March 7   *Friday 7*   **Hellenic Arc**
Dhu'l-Hijja 916

A light shock in Egypt in the late afternoon shook the ground but caused the people little alarm.[1] This was probably associated with the widely felt earthquake in the Aegean Sea of that year.[2]

1 Al-Da'udi, p. 62/43, also in al-Hafiz (1982), p. 261, and Taher (1979), p. 210/249; Ibn Iyas, **IV**, 207/trans. Wiet (1955), p. 201.
2 Maravelakis (1939), p. 142.

1511 June 11   *Wednesday 14 Rabi' I, 917*   **Yemen**

At the end of the night there was a strong shock of earthquake in Zabid.[1] This is the last recorded shock in the sequence of tremors that began in 1501.

1 Ibn al-Daiba', *Fadl*, p. 242. The author died in 944/1537. This shock is dated by the death of a certain Shaikh al-Salih 'Ali b. Isma'il at dawn.

1513 March 28   *Monday 20*   **Lower Egypt**
Muharram 919

A slight shock in Cairo during the morning, lasting quarter of a *daraja*.[1] Three shocks were felt, which moved the ground perceptibly.[2] There is no indication of an epicentral area and the shock may have been local to Cairo.

1 Al-Da'udi, p. 62/43, reads one *daraja*, but not the text edited by al-Hafiz (1982), p. 261. Al-Da'udi has Monday 20 Muharram 918 (*sic.*)/7 April 1512, which was a Wednesday. This date is found in Taher (1979), p. 210/249 and Poirier and Taher (1980), p. 2193.
2 Ibn Iyas, **IV**, 297/trans. Wiet (1955), pp. 278–9 has 20 Muharram 919, which was a Monday, and his date is preferred. He sees the tremors as forerunners of the plague.

1518   **Ethiopia**

A series of shocks was felt during the year in Ethiopia, most probably in the Tigre region.[1]

1 Palazzo (1915), p. 298; Gouin (1979), p. 27.

1523 April 4 *Saturday 18 Jumada I, 929* **Lower Egypt**

A slight shock in Cairo about 10 *daraja* after the nighttime prayer, lasted about half a *daraja*, shaking walls and ceilings. The water in a dish was observed to shake according to one eyewitness account.[1]

There is no evidence of an epicentral area and the shock may have been local to Cairo.

1 Al-Shadhili, p. 63–4/44–5. Properly speaking, 'Saturday night'

began after sunset on Friday evening, 3 April. The shock thus occurred during the night of 3–4 April. The less detailed account by the other continuator of al-Suyuti, al-Da'udi, p. 62/43, perhaps reflects this problem by dating the shock Saturday 17 (*sic.*) Jumada I/3 April, also in al-Hafiz (1982), p. 261. Taher (1979), p. 210/249, gives 5 April, which was a Sunday.

1525 March 9   *Friday 14*   **Lower Egypt**
Jumada I, 931

A slight shock in Egypt (or Cairo) during the night.[1]

1 Al-Da'udi, p. 62/43, has Friday 14 Jumada II/8 April, which was a Saturday, cf. Taher (1979), p. 211/249. Since the night of Friday began at sunset on Thursday, al-Da'udi's text as edited by al-Hafiz (1982), p. 261, is preferred: 9 March was a Thursday. Poirier and Taher (1980), p. 2193, give 9 April 1525.

1527 July 14 *Sunday 15 Shawwal 933* **Lower Egypt**

A slight shock occurred in Cairo around dawn.[1]

1 Al-Da'udi, p. 62/43, also in Taher (1979), p. 211/250 and al-Hafiz (1982), p. 261.

1529 November 12   *Friday 10*   **Lower Egypt**
Rabi' I, 936

A light shock in Egypt (*misr*) which lasted about half a *daraja* towards the end of the night.[1] An eyewitness who was at the top of a minaret in Cairo at the time of the shock says the earthquake was extremely frightening and the minaret shook in such a way that he thought it was going to collapse. The oscillation lasted two or three *daraja*.[2]

1 Al-Da'udi, p. 62/43, identical in al-Hafiz (1982), p. 261, Taher (1979), p. 211/250. Poirier and Taher (1980), p. 2193 refer to source no. 45, not 46.
2 Al-Shadhili, p. 63/44. He was a muezzin and was presumably preparing for the dawn call to prayer: he says the earthquake occurred about 10 *daraja* (40 minutes ?) before dawn.

1532 July 10   *Wednesday 7*   **Lower Egypt**
Dhu'l-Hijja 938

A very slight shock reported in Cairo during the night.[1]

1 Al-Da'udi, p. 63/44; identical in al-Hafiz (1982), p. 261; Taher (1979), p. 211/250.

1534 March 23 *Monday 8 Ramadan 940* **Lower Egypt**

A slight shock was felt in Cairo after dawn.[1]

1 Al-Da'udi, p. 63/44; not in the Ms. edited by al-Hafiz (1982).

1537 January 8   *Tuesday 26*   **Eastern**
Rajab 943   **Mediterranean ?**

A very slight shock is reported (in Cairo ?) during the night.[1] This shock was also felt in Damascus.[2] There is no indication of an epicentral region for this earthquake.

1 Al-Da'udi, in al-Hafiz (1982), p. 261; this is the last shock mentioned by al-Suyuti's disciple, who died in Cairo in 945/1539. January 8 was a Monday; the shock occurred in the night of 8–9 January.

2 Al-Ghuzzi, in al-Hafiz, *l.c.* Another slight shock was reported in Damascus (but not in Cairo) on 27 Ramadan/9 March 1537.

1537   *944 H*   Eastern Mediterranean ?

Damietta was shaken by earthquakes that continued for four days, five times a day. These shocks may be associated with an earthquake reported to have shaken down many walls in Antioch the same year.[1] Alternatively, there may be some connection with the shocks reported in Cairo and Damascus, noted above under 943/1537.

Such associations, however, imply a large earthquake in or offshore from northern Syria, for which one would expect further details to be preserved, and it is more likely that the Antioch earthquake, at least, is a separate event.

1 Al-'Umari, fol. 188ro. He merely gives the year, 944 H (began 10 June 1537), for both events, which may be a year late.

**1546 January 14   *Thursday 10   *Dead Sea
                     *Dhu'l-Qa'da 952***

A series of shocks is reported from southern Palestine in 1546. The first occurred during the afternoon and strongly affected Jerusalem, al-Khalil (Hebron), Ghazza (Gaza), al-Ramla, al-Karak, al-Salt and Nablus and was felt as far as Damascus. It lasted a short while, but damaged or fissured most of the tall buildings in Jerusalem and al-Khalil. The madrasa of Qayitbey in Ghazza was destroyed and likewise southern, northern and eastern parts of his madrasa in Jerusalem. The top of the minaret at the Chain Gate was destroyed. Nablus was affected more than anywhere else and about 500 people were killed beneath the wreckage there.[1]

The Patriarch of Jerusalem reported the effects of the earthquake on the belfry of the Holy Sepulchre, which fell onto the Church of the Resurrection. The bell-tower of the church in Bethlehem also fell.[2] Effects on the rivers flowing into the sea near Jaffa are reported in contemporary accounts, as well as the damming of the river Jordan. In addition, the sea withdrew from the coast of south Palestine and returned as a tsunami, which drowned many people.[3]

Another shock, stronger than the first, occurred on the night of Saturday–Sunday 10 Muharram 953/13 March 1546 and another on Wednesday morning 12 Rabi' I, 953/29 September 1546, which some people felt more than others. These aftershocks followed a series of tremors over the preceding days and nights.[4]

There is no indication that the shock affected Egypt, although this was probably the case, by analogy with the earthquake of 12 November 1458, with which there are several points in common (see above). Modern writers

report effects in Tripoli in Syria and Famagusta in Cyprus, but this appears to be due to damaging storms that year, and an earthquake in Cyprus in 1547.[5]

1 Mujir al-Din, *Dhail* (ed. Mayer, 1931), pp. 86–90. Mayer fails to identify the correct date of the earthquake; see also Burgoyne and Richards (1987), p. 42, note 32. Duplications abound in Kallner-Amiran (1951), pp. 229–30. The earthquake is fully discussed by Ambraseys and Karcz (1992).
2 Dositheos (1715); Vincent and Abel (1922). The Muslim authorities apparently refused the Christians permission to rebuild these churches, according to a Hebrew source in Braslavskii (1938). For damage to Muslim monuments, see Burgoyne and Richards, pp. 119, 256, 272.
3 Anon. Wittenberg (1546), says 10 000 people perished in the tsunami, clearly an exaggeration. Further details of the drying up of the Jordan, and of losses sustained by the 'Ismaelites and the Gentiles', are provided by contemporary Hebrew sources, in Braslavskii (1938).
4 Mujir al-Din, *l.c.*
5 Sieberg (1932b), p. 193, under 14 January and 29 September. For Cyprus, see Enlart (1896), Ambraseys and Karcz (1992), p. 261.

1554   Ethiopia

Earth tremors were felt in the Tigre region of Ethiopia.[1]

1 Palazzo (1915), pp. 301–3; Gouin (1979), pp. 27–8. The shocks occurred in the Ethiopian year running from August 1553 to August 1554.

[1564   Hejaz]

Spanish diplomatic correspondence from Venice reports news that Mecca had been destroyed by storms and earthquakes, to the alarm of the Ottomans.[1] No contemporary confirmation has been found, and the reports might merely be rumours.

1 Archivo General de Simencas (Spain), Secc. de Estado, Legajo 1325, Corresp. de Venezia, 1564–6, fol. 1. The news came to Venice from Istanbul. Meccan historians of the period, such as al-Nahrawali (d. 990/1582), make no mention of such events, though he describes in detail building work done in Mecca at this date (e.g. pp. 350–3). AD 1564 corresponds to 971–2 H.

1573 February 4   *1 Shawwal 980*   Hellenic Arc ?

A slight shock is reported from Cairo. It was strongly felt in the al-Hakim mosque and the suqs of the city, but caused no damage.[1] This is possibly to be associated with the earthquake in the Hellenic Arc, felt in Cyprus and the Aegean Sea on 6 March 1573.[2]

1 Al-Shadhili, p. 64/45, gives a graphic eyewitness account. Also in Taher (1979), p. 212/250.
2 Klirides (1936), p. 122. Santorin was in eruption this year, see Richard (1657).

1576 April 1 *Monday 2 Muharram 984* Lower Egypt

A strong shock of earthquake was experienced in Cairo during the night, preceded by three weaker ones.[1] There may be some connection with the earthquake in Cyprus on Monday 28 January 1577, though the dates as reported do not coincide.[2]

1 Al-Jazzar, *Tahsin*, in Taher (1974), p. 17; also Taher (1979), p. 212/251–2. April 1 was a Sunday, and the shock occurred during the night of 1–2 April; it was not mentioned by al-Shadhili. It provoked a debate as to its causes and consequences for the people of Egypt.

2 Al-Dwaihi, p. 273, for example, mentions the Cyprus earthquake under the annal for 984/1576, though without referring to Egypt. He dates the shock 28 Kanun II (January), confirmed by other authors, Bonito (1691), pp. 716–17; Sathas (1873), **II**, 141; Bustronius (1989), pp. 160–1.

1583 ?   *991 H*   **Yemen**

A volcanic eruption blew apart a mountain in Yemen, following three days of earthquakes. Premonitory tremors occurred at the rate of twenty a day.[1] The eruption might have occurred in the region west of San'a.

1 Al-'Umari, fol. 197ro. No earlier or more local sources for this event have yet been identified. It is remarkable that al-'Umari does not mention any of the Yemen events recorded earlier in the century by Yemeni historians.

1588 January 4 NS   *Sunday 4*   **Northern Hejaz**
                         Safar 996

A little after midday, a strong shock of earthquake was felt in Cairo, where it was of long duration (about five *daraja*). The minarets shook, some of them losing their top; basins and water tanks tilted over, and one report speaks of an eye-witness running in fear from the bath-house, after the water sloshed violently in the pool.[1] Several quarters and houses of Cairo were damaged. The earthquake was destructive at the pass of Aila and caused rockfalls on the Egyptian pilgrim route to Mecca.[2] At Tabuk, on the Syrian pilgrim route, the shock was very strong and the castle collapsed on the pilgrims there.[3] Medina was also affected by the shock.[4] In Sinai, the mosque in the monastery of St Catherine's collapsed and, together with other structures, was later rebuilt.[5]

These details are consistent with an epicentre in the northern Red Sea area, to the east of the Gulf of Aqaba (see Figure 2.18).[6]

1 Al-Shadhili, p. 64. Note again his professional interest in minarets; their motion may have led him to overestimate the duration of the shock, see next note. January 4 was a Monday.

2 Al-Ishaqi, p. 154, and the more dramatic version in Digeon (1781), **I**, 136–7, which speaks of entire towns overwhelmed. Bedouin looted the goods of the pilgrims and *muhafizun* (escort?) stored in Aila. This was a regular hazard, cf. the pilgrimage of 872/1468, reported by Ibn Taghribirdi, *Nujum*, ed. Popper, **VII**, 748–9.

At the time of the shock, al-Ishaqi was in the house of the *naqib al-juyush* (Superintendent of the Army) in Fustat, and saw the walls of the courtyard swaying from side to side. Stones fell from a *qa'qa'a*

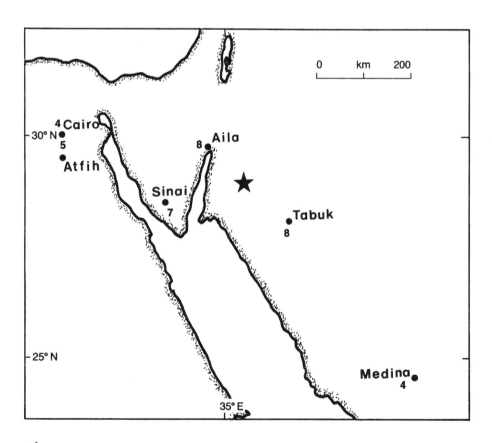

Figure 2.18. 1588 January 4–April 7, Northern Hejaz. Atfih was affected only in the event of 7 April.

(portico, *qaʿ?*) and the large lote-tree in the court shook as though in a violent squall. He says the shock lasted only a *daraja* and one-sixth, and also gives a chronogram written about the event. See Taher (1979), pp. 212–14/252–3, who wrongly gives the date as 14 Safar/14 January. The earthquake is duplicated under both dates by Poirier and Taher (1980), p. 2194.

3 Al-Ghuzzi, in al-Hafiz (1982), p. 260. The official pilgrim caravan had normally returned to Cairo by this date.
4 Al-ʿAidarusi, p. 443. He quotes a poem composed by the people of Mecca, which led Ambraseys and Melville (1989), to suggest that Mecca was also affected.
5 Ben-Menahem (1979), p. 258, without details, for which see Papamichalopoulos (1912), p. 242, quoting Zeki (1908), which we have not traced.
6 Melville (1984a), p. 99; Ambraseys and Melville (1989).

1588 April 7 NS   *Wednesday 10*   Lower Egypt
Jumada I, 996

A second earthquake was felt in Cairo soon after the previous event. It occurred at sunrise and lasted only a brief while. At Batnun to the east of of Atfih, in the Muqattam hills, three fissures opened and water poured out.[1] These details suggest that it was probably a local shock, though possibly connected with continuing activity in the northern Red Sea (Gulf of Suez) region.

1 Al-Ishaqi, p. 154; Taher (1979), p. 213/253; Poirier and Taher (1980), pp. 2194 (under April 9) and 2200. April 7 was a Thursday. Ambraseys and Melville (1989) associate these details with the earthquake of 4 January 1588. Batnun has not been located. The text translated by Digeon (1781), p. 138, puts these effects in the hills overlooking the market gardens near the citadel in Cairo, which is more plausible in view of the spectators who evidently witnessed the event.

1592 May   *1000 H*   Hellenic Arc

Three shocks of earthquakes were felt in Cairo this year.[1] These tremors should probably be associated with the large western Hellenic Arc earthquakes of May that caused damage in Zante.[2]

1 Al-ʿUmari, fol. 199vo. 1000 H began on 19 Oct. 1591.
2 Anon. (1592); Barbiani and Babiani (1863), p. 11.

1593 September   Ethiopia

An earthquake was felt in northern Ethiopia, presumably on the margin of the Plateau–Afar escarpment near Mai Chew. A second shock was reported from the same region during December, considered to be an aftershock.[1]

1 Gouin (1979), p. 28. The shock occurred during Meskerem 1586/began 29 August 1593. The year of the second shock (13 Tahsas/9 December OS) is not mentioned.

1608 December 23 NS   *Tuesday 15*   Ethiopia
Ramadan 1017

Secondary volcanic activity occurred in Aussa in Ethiopia. Smoke came out of the earth near the foot of

Mount Waraba on the western shore of Lake Abhe.[1]

1 Gouin (1979), p. 95, citing an Arabic Ms. Waraba is tentatively identified as Dama ʿAli volcano.

1609   Hellenic Arc

A destructive earthquake in the eastern Hellenic Arc in spring this year was strongly felt in Lower Egypt, Palestine and throughout most of Turkey. It caused great panic in Cairo and considerable concern in Alexandria, Rosetta and Damietta, but no damage.[1] The earthquake was particularly severe in Rhodes, where half the town, including the castle, was ruined.[2]

1 Anon. (1609); Winden (1613), p. 227.
2 Schöne (1940), sect. 20, citing *Zeitung aus Venedig*, 8 May 1609; BBA. *MD.* 78, fol. 698.

1613   *1022 H*   Gulf of Aden?

A great earthquake occurred in the town of Aden, during which the ground swelled in waves.[1]

1 Al-ʿUmari, fol. 205ro; Ambraseys and Melville (1983). Al-ʿUmari's source has not yet been identified; the information appears to be reliable. 1022 H began on 21 February 1613.

1613 June   *Rabiʿ II, 1022*   Hellenic Arc?

An earthquake in Cairo occurred shortly after the revolt of a newly arrived Ottoman regiment.[1] The exact date is not known but this may be associated with activity in the Hellenic Arc or the Ionian islands that occurred during the period.[2]

1 Al-Ishaqi, pp. 167–8; Digeon (1781), p. 203; Taher (1979), p. 215/253–4.
2 A large earthquake in the northwest of the Hellenic Arc occurred on 2 October 1613, preceded by a strong foreshock on 8 November 1612, Maravelakis (1939), p. 138.

1619 July   *Shaʿban 1028*   Yemen

An earthquake at midday was felt in Sanʿa and Kaukaban, and generally in most of the towns in the region (see Figure 2.19); it was followed by another, stronger earthquake during the night. Further shocks occurred on three successive nights, from Thursday to Saturday.[1] The earthquake was also strongly felt in Saʿda, where a great many houses were badly damaged. Among the buildings affected was the house of al-Mutahhar, son of the Imam, which was cracked by the shock, and one side of it was destroyed. Other ancient buildings in the city, thought to date from the period of Himyar, were severely shaken. Some of the villages were also affected, and the shock 'almost removed the foundations [of the mountains].'[2]

1 ʿIsa b. Lutf-Allah, *Rauh*, II, 99, gives the month; Yahya b. al-Husain, *Anba*, Ms. 1347, fol. 158, gives the days of the week. Precise dates are not given; the first Thursday in Shaʿban was 18 July (NS). Al-Hibshi (1982), who does not mention his sources but is clearly following the *Anba*, nevertheless has 1029 H (began 8 December

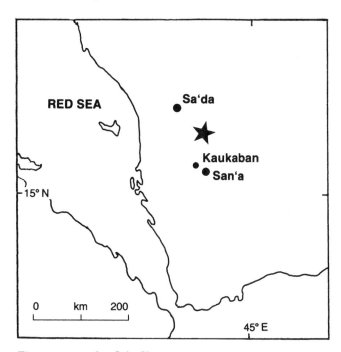

Figure 2.19. *1619 July, Yemen.*

1619). See also Raşit (1875), I, 221, who says there were three very strong earthquakes.

2 Al-Mutahhar was the grandfather of 'Isa b. Lutf-Allah, a native of Kaukaban. His chronicle ends in 1029 H.

1631 February 10 NS *Monday 8 Rajab 1040* **Ethiopia**
A series of earthquakes preceded a volcanic eruption, probably from Dama 'Ali in Ethiopia, on 12–13 Rajab/ 14–15 February. Shocks occurred almost continuously over a period of several months in Aussa, southern Afar. As a result, the town of Waraba was destroyed, with 50 casualties, and Aussa almost entirely burnt down.[1] A report from the Yemen mentions that 5000 people and many animals were killed.[2] The effects of the eruption were visible as far north as the Egyptian frontiers with Abyssinia.[3]

1 Gouin (1979), pp. 95–6. Further details are supplied by Yahya b. al-Husain, *Anba*, Ms. 1347. fol. 169ff. He cites reports from pilgrims coming to Jidda, who mention premonitory tremors around Y.f.r.s (?), before the eruption from Jabal Marmar, overlooking Aussa, on 13 Rajab.
2 See the briefer version by Yahya b. al-Husain, *Ghayat*, II, 834. The figures are probably greatly exaggerated.
3 Peresc (1633), fol. 264–5. He confounds this eruption with that of Vesuvius in December this year, cf. Gouin's discussion of a Portuguese account of the destruction of Aussa in December 1631, and Gassendi, trans. Rand (1657), p. 90 (who calls the volcano 'Semus'); Hoff (1840), p. 287.

1631 March *Sha'ban 1040* **Yemen**
A strong earthquake occurred in San'a and its district.[1]

1 Yahya b. al-Husain, *Anba*, Ms. 1347, fol. 169. We are grateful to Abd al-Malik Eagle for this reference. Al-Hibshi (1982) mentions

an earthquake in San'a in 1039 H, without identifying his source. Both the *Anba* and Yahya b. al-Husain, *Ghayat*, II, 832–3, report a meteorite impact in 1039 H. Another meteorite, two hours before noon on Tuesday 28 Rabi' I, 1042/Wed. 13 Oct. 1632 NS., was followed by an earth tremor, see Hasan b. Yahya, Ms. Or. 3330, fol. 222vo.

It may be noted that some European writers, such as Girardi (1664), the *Lettres Hist. et Polit.* XIV, 262, and Seyfart (1756), p. 30, refer to an earthquake in the Hejaz that caused damage in Mecca, particularly to the Ka'ba, in July 1630 (1039 H). Some later authors, such as Anon. (1693), Coronelli (1693), p. 321, Perrey (1850a), p. 22, and Sieberg (1932a), p. 795, state that both Mecca and Medina (where the Prophet was buried) were damaged by the earthquake. However, contemporary sources indicate that the cause of damage in Mecca was a flood, see Peresc (1633), fol. 264. Ibn Abi'l-Surur, p. 209, and Anon., *Zubdat al-ikhtisar*, fol. 5vo, give the correct date as 19 Sha'ban 1039/3 April 1630.

1632 June **Ethiopia**
An earthquake was felt in the province of Beghemder (Gondar) in Ethiopia.[1]

1 Palazzo (1915), p. 304; Gouin (1979), p. 28.

1633 November 5 OS **Hellenic Arc**
During the winter of 1632, at lunch time, a prolonged earthquake in Cairo caused houses to sway and waves to form in the Nile and in the fish-ponds in the city. The shock caused great concern but no damage.[1]

This report suggests a distant epicentre, and there may be some connection with earthquakes that caused great damage in the Ionian Islands the *following* winter, 1633.[2]

1 Peresc (1633), fol. 263. This is equivalent to 1042 H.
2 For the large Zante earthquake of 5 November 1633, see Coronelli (1693), p. 321 and Barbiani and Babiani (1863), p. 14. Berryat (1761), p. 563, also mentions earthquakes in Egypt in 1633.

1637 *1047 H* **Yemen**
A volcanic eruption is reported in an elevated location in Hajja, which is a district northwest of San'a and partly in Tihama. This continued for several days.[1]

1 Ibn al-Wazir, p. 59, mentions 'the appearance of fire'. For Hajja, see Wilson (1989), p. 129.

1644 September 22 NS *Thursday 20* **Yemen**
Rajab 1054
A great earthquake at al-'Ushsha in Wadi al-Ahjur (Ahjir) in the Yemen. The shock caused rockfalls of boulders from the mountains, which crushed flat the stones beneath.[1] It is possible that this was simply a landslide and not necessarily triggered by seismic activity. The collapse of the hill (Jabal al-Ahjur) brought down rocks and mud that choked up the surrounding cultivation and gardens.[2]

Wadi al-Ahjur is located in the Shibam and Kaukaban district, 25 km west of San'a.

1 Hasan b. Yahya, fol. 236vo, quotes an eye-witness account; al-Wasi'i, p. 54.

2 Ibn al-Wazir, p. 93; he mentions similar landslides in Hajja in 1071/1660-1 (p. 173) and 1074/1663-4 (p. 193); and one in al-Ahjur in 1076/1665-6, which cut the road to Kuhlan and 'Affar (p. 205).

1646 April 30 NS   *Tuesday 14 Rabi' I, 1056*   **Yemen**
A great earthquake was experienced by a number of reliable witnesses in the Yemen.[1]

An earlier source reports an earthquake in San'a and elsewhere in Dhu'l-Hijja 1056/January–February 1647.[2] It is possible that these are both references to the same event, or to a series of shocks.

1 Hasan b. Yahya, fol. 239ro; he says the shock was 'heard'; the region affected might have been round Kaukaban. April 30 was a Monday.

2 Ibn al-Wazir, p. 118. He comments on this, and on an earlier thunderbolt that crashed into the Friday mosque at San'a, killing two men who were at prayer, by referring to *Sura* 17, verse 59 of the Qur'an: 'And we do not send signs, except to frighten'.

1656   **Libya**
Towards the end of the year, an exceptionally strong earthquake in Tripoli in Libya destroyed almost half its houses and caused the loss of five pirate vessels in the harbour.[1] Later authors place this event in Tripoli in Syria.[2]

1 *Dressdnische* (1756), p. 122.

2 Hoff (1840), p. 305; Perrey (1850a), p. 23; Kallner-Amiran (1951), p. 230. The shock is sometimes put in February, which was the end of the Old Style year (March to March).

1664 November 20 OS   *1075 H*   **Hellenic Arc**
Three shocks of earthquake were strongly felt in Cairo, and caused fear. It is said that palmtrees, sycamores and other tall trees were uprooted by the earthquake, which was stronger than anything previously experienced.[1]

It is likely that this report refers to the distant effects of the earthquake of 20 November 1664, which caused great damage in Crete, particularly at Candia.[2]

1 Ibn 'Abd al-Ghani, p. 161. The year 1075 H began on 25 July 1664; the earthquake is mentioned between events in Rabi' II/10 November and Rajab/January 1665, and following the appearance of a comet. The details of the earthquake seem rather to point to storm damage (cf. al-Maqrizi, IV/2, 635, re. storms in 826/1423).

2 *Gazette de France*, 29 Jan. 1665, *Relationis Hist. Semestral.* (1665), p. 78, Bonito (1691), p. 790.

1666 November   *Jumada I, 1077*   **Yemen**
An earthquake and tremors occurred in San'a at dawn.[1]

1 Ibn al-Wazir, p. 216. He goes on to record a terrible plague of locusts, which spread from the Tihama as far as Anatolia.

1667 March 14 NS   *Sunday 18*   **Yemen**
Ramadan 1077
An earthquake shook San'a in the fourth hour of the night and some houses were fissured. Sleepers were awakened by the shock and the faint-hearted cried out in fear. The shock was felt throughout most of the Yemen.[1]

1 Ibn al-Wazir, p. 224; al-Hibshi (1982); Ambraseys and Melville (1983). Ibn al-Wazir then discusses the two shocks in 1077 H in the light of the dubious proposition, found in epics attributed to 'Ali b. Abi Talib, that earthquakes are a sign of cold weather, a rise in prices and a lack of rain – all of which occurred.

1667 October 22 NS   **Ethiopia**
An earthquake was felt in the region of Gondar in Ethiopia.[1]

1 Gouin (1979), p. 28. 15 Teqempt = 12 October OS.

1668 August 17 NS   *1079 H*   **Turkey**
A slight shock in Alexandria is reported to have destroyed a few walls.[1] The shock was most probably associated with the very large earthquake in central Anatolia on 17 August.[2]

1 Al-'Umari, fol. 216vo. He merely gives the year, which began on 11 June 1668.

2 Ambraseys and Finkel (1988). It may be noted that *Dressdnische* (1756) and later sources (e.g. Sieberg, 1932b, p. 188) refer to an earthquake in Alexandria that lasted intermittently for twelve days in March 1687. However, contemporary sources show that these shocks occurred in Alexandretta in North Syria (Anon., 1687).

1674 August   *Jumada I, 1085*   **Yemen**
Thirty shocks of earthquake were reported, particularly near Dauran, Yemen.[1]

1 Muhsin b. al-Hasan, *Tib al-kisa*, Ms. fol. 209, more detailed than Ibn al-Wazir, p. 311. No precise date is given, but the report comes after a notice of a lunar eclipse on 14 Rabi' II (18 July 1674; cf. no. 4450 in Oppolzer, 1962, p. 369) and news of events in Aden in late Jumada II (late September). Al-Hibshi (1982) also has 'about Jumada I'; Ambraseys and Melville (1983). Ibn al-Wazir again offers comments on the topical significance of these shocks, with an allusion to an earthquake in the Qur'an afflicting the prophet Shu'aib (*Sura* 7, verses 86–9).

1675 December 21 NS   *3 Shawwal 1086*   **Yemen**
A great earthquake occurred in San'a and elsewhere in the early dawn, waking those who were asleep; it was followed by another shock. During this period earthquakes and tremors also occurred south of San'a in Dauran; in one of them, most of the houses were fissured, among them the Dar al-Hasin.[1] Rocks from Jabal Dauran were strewn about, and the shock in Dauran lasted the time it takes to read the Qur'anic *Sura* 'Ya-Sin'.[2] The mind of some of those who experienced it became confused with fear; the Imam left for Ma'bar.[3]

The earthquake should be located a little to the north of Dauran.

1 Dar al-Hasin: i.e. probably the principal palace in Dauran, which was developed and strengthened at this period by al-Hasan b. al-Imam Qasim b. Muhammad, who named the town al-Hasin. It was then used as capital by his brother, the Imam al-Mutawakkil Isma'il (d. 1087/1676); see Zabara, *Nashr al-'arf*, I, 653–4.
2 *Sura* 36, containing 83 verses: a relatively long one, perhaps eight minutes to read.
3 Ibn al-Wazir, p. 320; al-Hibshi (1982), cites a similar account by Yahya b. al-Husain, *Bahjat al-zaman*, which was probably Ibn al-Wazir's source (see Sayyid, 1974, p. 265). This says that many with weak hearts became disturbed. Muhsin b. al-Hasan, *Tib*, fol. 211, says many lost their memories. The latter d. 1170/1757 and also wrote a continuation of Ibn al-Wazir's history (Sayyid, 1974, p. 270). Ma'bar is N.E. of Dauran, about 75 km south of San'a on the road to Dhamar; the Imam's departure there may be unconnected with the earthquake.

1679 September 21 NS *15 Sha'ban 1090* Red Sea
In the second half of Sha'ban (from 21 September) a volcanic eruption occurred in the Zuqar islands, N.W. of al-Mukha in the southern Red Sea.[1] It blazed like coal and threw sparks into the sea. The eruption was likened to a great minaret towering into the air, and it was seen by mountain dwellers in the Yemen as far away as Wasab (a district *c.* 30 miles northeast of Zabid). In the day, the eruption clouds were equally visible.[2] The eruption was followed by a series of earthquakes in al-Mukha, which was partially destroyed by a fire. The governor took to the sea in fear. The fire was extinguished by rainfall at the beginning of Shawwal (from 5 November 1679), i.e. after about six weeks.[3]

1 The text has Mount Suqar; for the location, between the Zubair islands (or al-Ab'ila) and the Bab al-Mandab Strait, see Tibbetts (1971), pp. 419–20.
2 Al-Hibshi (1982), apparently quoting from the *Tib al-kisa*, says the eruption was also visible from Raima (see Smith, 1978, p. 194), Hufash and Milhan, west of San'a (Wilson, 1989, pp. 140, 318). The author adds that in al-Mukha they could see the blaze from the volcano, which continued a long time, as though it were a sulphur mine.
3 Ibn al-Wazir, p. 362. He refers to the previous volcanic eruptions in 1256 and April 1433, also visible in Surdud, Kahfash (*sic*) and Milhan, so possibly al-Hibshi's quotation is inaccurate (see above).

1694 December 21 NS *1106 H* Egypt
An earthquake occurred in Egypt in the early morning.[1] People went into the open for three days until it ceased. Some houses were destroyed.[2]

No earthquake is reported elsewhere in the region in this year and the evidence suggests a local earthquake.

1 De Maillet (1735), p. 19, says the shock occurred around the same time of day as the 1698 event (see below). He implies that little damage was done.
2 A different impression is given by al-'Umari, fol. 221ro, who only has the year (1106 began 22 August 1694). He says it was a great shock. It is not clear whether the damage was caused in Cairo or elsewhere.

1698 October 2 NS Eastern Mediterranean?
An earthquake was felt in Cairo between 8.0 and 9.0 in the morning. The shock was also reported from Rosetta and Alexandria.[1]

1 De Maillet (1735), p. 18, adds that if there had been correspondence from Upper Egypt, no doubt similar reports would have been received. See also Lyons (1907), p. 285; Sieberg (1932b), p. 188. The date given falls in the Muslim year 1110 H.

1710 August 27 NS *Wednesday 2 Rajab 1122* Gulf of Suez?
A shock of earthquake occurred at eight o'clock in Cairo, where it was described as considerable. The people fled from the suqs and feared the houses would collapse. The shock is said to have lasted five *daraja*.[1]

There is possibly some connection between this event and a report from Venice that Mecca was destroyed by an earthquake.[2] This statement has not been verified and it seems likely that an earthquake of such size would have been more widely reported. Nevertheless, the long duration of the shock in Cairo suggests a distant epicentre, and the earthquake might have originated in the northern Red Sea or Gulf of Suez.

1 Ibn 'Abd al-Ghani, p. 228; A. N. (Paris), AE/BI/316 (Cairo); al-Jabarti, I, 106/trans. I, 90.
2 Luttrell, VI, 663, includes this in his Diary under the date Saturday 9 December 1710.

1733 November 30 NS Ethiopia
A severe earthquake on the Eritrean escarpment, northwest of Massawa, destroyed many houses, with casualties. The shock caused the collapse of trees and rockfalls from the mountains, which were visibly shaken. The epicentral area of this relatively large event must be sought between Asmara and Massawa. Aftershocks continued for a year.[1]

1 Gouin (1979), pp. 30–1; 23 Hedar = 19 November OS.

1741 January 31 NS *Sunday 14 Dhu'l-Qa'da, 1153* Hellenic Arc
A strong earthquake was felt during the night in Cairo; the shock was of long duration (estimates range from five to eight minutes) and caused terror; four or five mosques and a few houses were thrown down.[1]

This earthquake originated between Rhodes and Cyprus, where minarets fell and the church of Santa Sophia in Famagusta was damaged; the shock caused panic throughout both islands, and sporadic damage in Crete. Tsunamis detroyed several villages a kilometre inland from the coast northeast of Rhodes. Aftershocks continued until 9 February.[2]

1 Congreve, Ms. letter (1741); he gives the date as 19 January 1741

(OS) = 30 January (NS). The Muslim date given by *Takvimler* (1153 H.) corresponds to 31 January, which was a Tuesday (NS). In fact, the shock occurred shortly after midnight on the night of 30/31 January, see next note.

2 A.N. (Paris), AE/B1/952 (Rhodes 23 February 1741); BBA. *MMD*. 3609, fols. 516, 556, 568. See also Hasselquist (1762), p. 224; Cyprianos (1788), pp. 316, 740; Ambraseys (1965); Christophorides (1969), V, 90; Panzac (1985), p. 38.

1754 October   Gulf of Suez?

A locally destructive earthquake in Egypt affected Cairo, particularly the districts of Qarafa, Bulaq and part of new Cairo, where many houses were ruined with great loss of life. The shock was possibly felt in Alexandria, but there is no evidence that it caused damage elsewhere.[1] This earthquake is sometimes associated erroneously with the destructive shock of 2 September that affected northwest Anatolia, too far away to have caused the damage reported in Cairo.[2]

There is some evidence that at about this time it became necessary to repair the walls of the monastery of St Catherine in Sinai. The reason for their downfall is not given, except that earthquakes had occurred prior to 1771.[3]

1 The *Gazette de France* (1754), p. 568, mentions news coming on a ship from Alexandria, which Seyfart (1756), p. 135, may have misinterpreted. The report mentions thousands of casualties, which is probably an exaggeration, or else an amalgamation of the effects in Egypt with those of the Anatolian earthquake. Nevertheless, reports of damage in Cairo indicate a relatively local event. See also the Ms. *Chronicle of Augustins* (1754); Walther (1805), p. 113; Sieberg (1932b), p. 188. The shock is not mentioned by al-Jabarti. The Muslim year 1168 H began on 18 October 1754.

2 Berryat (1761), p. 626. Mallet (1854), p. 159, misquotes Seyfart, who merely has September instead of October. The vague dating of the Cairo shock makes confusion in the sources understandable.

3 Grigoriadis (1875), p. 142.

1756 February 13   Hellenic Arc

A large magnitude, intermediate depth earthquake in the eastern Mediterranean was felt in Alexandria and Cairo, where it lasted three minutes. It caused no damage in Egypt, but aftershocks continued to be felt for about 40 days. It was strong in Scarpantos, Rhodes and Satalia, where it caused damage. The shock was perceptible as far as Malta, Naples, Izmir, Corfu and Palestine, and caused alarm in Cyprus.[1]

1 *Gazette de France*, 3 April 1756; Donati (1759), fol. 67vo; Berryat (1761), p. 642; Cyprianos (1788), p. 316.

1775 June 3   *Friday 3 Rabi' II, 1189*   Yemen

A shock in the early evening was followed by another, and then a third; on the night of Sunday 5 Rabi' II/5 June, just before daybreak, another, strong shock occurred. This sequence was reported at first hand by someone who was in Hais at that time.[1]

1 Contemporary annotation at the end of BL Ms. Or. 3330, fol. 255ro. June 3 was actually a Saturday in 1775. We are grateful to Abd al-Malik Eagle for this reference. Hais (Hays) is about 30 km south of Zabid. Simkin *et al.* (1981), p. 39, list a volcanic eruption in southern Ethiopia this year.

1778 June 22   Upper Egypt

An earthquake was felt in Upper Egypt at Nag Hammadi and Tahta, followed by several aftershocks during the night. It caused no damage, but terrified the inhabitants of Nag Hammadi.[1]

1 Sonnini (1799), III, 173; Sieberg (1932b), p. 188.

1786 July 28   Iraq

An earthquake shock of 17 seconds' duration occurred at two o'clock in the morning, waking people in Basra. It caused no damage.[1]

1 Griffiths (1805), p. 391.

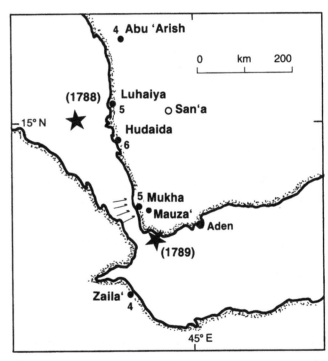

Figure 2.20. The earthquake sequence reported between 1788 November and July 1789, Red Sea. A general location near Zubair islands is given for the 1788 events, and a separate earthquake in 1789 is distinguished near the Bab al-Mandab Strait.

1788 November–   *Rabi' I –*   Red Sea
July 1789   *Shawwal 1203*

A series of shocks occurred along the west coast of Yemen and Arabia, from al-Mukha to Abu 'Arish (see Figure 2.20). Two or three shocks occurred a day and houses and buildings fell down, causing the people great fear and consternation. Shocks affected al-Mukha, al-Hudaida, Bandar Laja (?) and Bandar Laban (?) and places belonging to Bait al-Qasir in Bait Hindi (in the region of al-Hudaida) as well as Abu 'Arish. It was reported that at one place the ground was fissured and fire emerged, consuming many places and people left their homes.[1] This suggests that the earth tremors were associated with some volcanic activity.[2]

A small town in the interior, about 30 km inland from al-Mukha (possibly Mauza'), was also destroyed by an earthquake in 1789, which was strongly felt in al-Mukha itself.[3]

It is not possible to distinguish individual earthquakes within this sequence, though possibly one large shock with an epicentre in the Red Sea affected the coastal areas from Abu 'Arish to al-Mukha, and a separate event struck the region between al-Mukha, Aden and Zaila'.

1 'Abd al-Qadir, p. 157; it has not been possible to identify the two ports (*bandar*) mentioned in this account, from either late mediaeval or modern maps, but one of them might reasonably be associated with al-Luhaiya. The author was part of an embassy from Tipu Sultan of Mysore.
2 Simkin *et al.* (1981), p. 41, list a volcano called Jabal al-Nar at 13.3°N–43.7°E, but no dated eruption; Camp *et al.* (1987), Fig. 1, plot an eruption at Jabal al-Tair, dated 1750, with which there may be some connection, cf. below, 1796.
3 Degrandpré (1801), II, 267–9. He was in al-Mukha at the time, and left his house for fear it would fall. He also refers to the frequency of shocks in Aden, possibly connected with this series of events, and to the coincidence that shocks felt in al-Mukha were generally felt in Zaila' (in Ethiopia).

1790 May 26   *12 Ramadan 1204*   **Hellenic Arc?**
A light shock of earthquake was felt in Cairo in the sixth hour of the night (around midnight).[1] There may be some connection between this shock and an event reported in Crete during 1790.[2]

1 Al-Jabarti, IV, 118/trans. V, 76.
2 Panzac (1985), p. 38.

[c. 1796   **Red Sea]**
About this year, the volcanic islet of Jabal al-Tair (formerly known as Saiban) in the Red Sea, west of al-Luhaiya, was in eruption and was given the name 'Jabal Dukhan' (Mountain of Smoke) by the Arab pilots.[1] This eruption should probably be dated to the period 1788–9, mentioned above.

1 Perrey (1856), p. 26.

1799 August 18   **Ethiopia**
Before sunrise, an earthquake was felt north of Lake Tana, in Dembia and in Tigre, Ethiopia. It caused some panic, but no damage.[1]

1 Palazzo (1915), p. 304; Gouin (1979), p. 31. Sunday 13 Nahassie = 6 August (os).

[c. 1801   **Sinai]**
The fortification walls of the monastery of St Catherine in Sinai were repaired in 1801 after damage alleged to have been caused by an earthquake.[1] However, local sources of information, while confirming the completion of the construction of the walls in May 1801, state explicitly that the reconstruction work became necessary after the flash floods of 17 December 1798, which carried away the whole length of the north side of the walls of the monastery.[2]

1 Rabino (1937), p. 25; see also Ben-Menahem (1979), p. 258.
2 Grigoriadis (1875), pp. 126–7.

1801 October 10   *Sunday 2*   **Lower Egypt**
Jumada II, 1216
A shock of earthquake was felt in Cairo in the third hour of the night.[1] There is no evidence of an earthquake elsewhere in the Eastern Mediterranean on this date, and the shock was perhaps of local origin.

1 Al-Jabarti, V, 311/trans. VII, 65. 'Sunday night' according to the Muslim calendar is equivalent here to the late evening of Saturday.

1802 June 30   **Ethiopia**
Earth tremors were felt near Gondar in the province of Dembia, north of Lake Tana in Ethiopia.[1]

1 Gouin (1979), pp. 31–2. The date is 24 Senie, year 7294 of Creation (18 June os).

[1803   **Libya]**
A light earthquake is reported in Tripoli, Libya, without details.[1]

1 Sieberg (1932a), p. 872, but not mentioned elsewhere. This may be a duplication of the 22 November 1903 earthquake, see below.

1805 July 3   *Tuesday 5 Rabi' II, 1220*   **Hellenic Arc**
A strong earthquake was felt around sunrise in Egypt; in Cairo the shock is said to have lasted four *daraja*.[1] These effects are to be associated with a large event in the Hellenic Arc that caused damage in Crete. The earthquake was perceptible in south Italy and Greece, as well as Syria.[2]

1 Al-Jabarti, VI, 230/trans. VII, 382. The translation is misleading, saying four shocks occurred during the night of Monday 2 July. See also Lyons (1907), p. 285; Sieberg, (1932b), p. 188. Actually, 3 July was a Wednesday.
2 Perrey (1850a), p. 37; Mallet (1855), p. 61; Baratta (1901), p. 317.

1809 February 26   **Ethiopia**
An earthquake was felt in Gondar in Ethiopia.[1]

1 Gouin (1979), p. 32; on Sunday, 23 Yekatit, year 7301 of Creation (14 February os).

1810 February 17   *Saturday 13*   **Hellenic Arc**
                  Muharram 1225

At the beginning of the seventh hour of the night, three strong shocks were felt in Cairo with a total duration of about four minutes. People were awakened from their sleep and thrown into turmoil. Many left their homes and fled into the alleys, seeking to escape into the open country. In the morning, everyone discussed what had happened. Some walls and old houses collapsed in the shock; other walls were fissured and the minaret of Basus fell, as well as half the minaret of Umm Akhnan (Umm Khunan) in Minufiyya. During the late afternoon, there was another earthquake, which though weaker than the first, still caused fear and agitation. Many rumours later circulated that there was going to be a further, more destructive shock, but such fears proved groundless.[1]

These effects were produced by a large, probably intermediate-depth event with an epicentre off Crete, where many places were destroyed and many people killed. All the other islands in the southern part of the Archipelago suffered some damage. In Malta, the shock caused panic, and widespread minor damage in Valletta, where ships in the harbour were violently shaken. The effects extended to Syria; in Lattakia, it is related that ancient buildings that had previously been swallowed up were seen at the bottom of a chasm, before it closed up again. The shock was also felt in central Italy, Turkey, Cyprus and various parts of North Africa. Minor damage was reported from Rashid (Rosetta) and Alexandria, where the earthquake set up waves in the harbour and in canals.[2] The shock was probably also strongly felt in the Siwa oasis (see below, Section 2.4).

1 Al-Jabarti, **VII**, 90–1/trans. **VIII**, 239–40. Muharram 13 is 18 February (a Sunday), but as he gives the time as 'Saturday night', i.e. the night of Friday–Saturday, the correct date should be 16 February, as in the European sources (see below): 17 February local time. Al-Jabarti, **IV**, 132/trans. **V**, 89, records a previous earthquake scare in Cairo, where a disastrous earthquake was predicted by astrologers for 27 Jumada I, 1205/1 Feburary 1791. The population left Cairo during the night in question, but no shock materialised.
2 Al-Jabarti, **VII**, 95/trans. **VIII**, 245–6; this news arrived in Cairo in Safar/March. See also *Gentleman's Magazine* (1810), pp. 372–3; F.O. SP 105/131 no. 309 (Smyrna, 10 March 1810); French Diplomatic Archives, Correspondance Consulaire et Commerciale, Smyrne 33 (1810), and Archives Diplomatiques, Nantes: Correspondance Consulaire (Îles), Candia, 23 March 1810. Baratta (1901), pp. 330–1; Sami (1928), p. 553; Ambraseys (1965); Ben-Menahem (1979), p. 294; Ambraseys (1991).

c. 1810   **Yemen**

Jabal Yar, on the boundary between Yemen and the 'Asir, was probably in eruption at the beginning of the nineteenth century. Hot springs abound in the region.[1]

1 Lamare (1924); Neumann van Padang (1963); Simkin *et al.* (1981), p. 40. We have found no contemporary source for this event.

1814 June 27   *Monday 9 Rajab 1229*   **Gulf of Suez**

A strong earthquake was felt in Sinai. Near the convent of the Arba'in (Forty Martyrs) north of St Catherine's in the Wadi al-Leja, the shock caused rockfalls that blocked the valley. There is no indication that there was any damage to the monastery of St Catherine.[1]

It is very likely that this is the same as the earthquake reported in Cairo at the time of the night prayer on Monday 9 Rajab 1229. The shock, lasting two minutes, caused minarets to shake violently and brought down a decorative cresting from the mosque of al-Azhar. A lesser shock followed around the fifth hour of the night and another, weaker still, at sunrise the next morning.[2]

1 Turner (1820), **II**, 439, who was there in August 1815, says the shock had happened the previous year. For the location of the Arba'in, see Meinardus (1962), map V.
2 Al-Jabarti, **VII**, 283/trans. **IX**, 96; cf. the discussion in Creswell (1940), p. 338, concerning the alleged effects at the mosque of Ibn Tulun.

1818 June 30   **Ethiopia**

A violent earth tremor was felt in Adua, Ethiopia, but caused no damage.[1]

1 Palazzo (1915), p. 305; Gouin (1979), p. 32.

1824   **Red Sea**

A volcanic eruption is reported in Jabal Zubair (Saddle Island) in the Red Sea.[1]

1 Simkin *et al.* (1981), p. 38; no contemporary source has been identified.

1825 June 21   **Lower Egypt**

At about 9 p.m., four rather severe shocks were felt in Cairo.[1]

1 Schmidt (1879); Sieberg (1932b), p. 188, lists another shock on 21 August 1825, probably a duplication.

1832   **Red Sea**

During this year, the volcano Jabal al-Tair (or Saiban) was possibly in eruption, emitting smoke.[1]

1 Perrey (1872a), p. 11. Fuchs (1876), p. 225 refers to an eruption in 1834, for which we have found no contemporary information: this is possibly a reference to the same event. See also Simkin *et al.* (1981), p. 38, under 1833. Gouin (1979), p. 33, mentions earth tremors 10–15 km northeast of Halai, on 4 May 1832, felt by a caravan on the move.

c. 1832   **Arabian Peninsula**

Palgrave, who was at Hufuf in December 1862, was told by the inhabitants that slight shocks of earthquake were not at all uncommon there. An earthquake of unusual severity, and to which the rents and clefts in the high walls and upper storeys of several houses in Hufuf still bore witness, was said to have taken place about 30 years before his visit.[1]

1 Palgrave (1869), II, 355. He adds that earthquakes appeared to be wholly unknown to both the historical records and living traditions of upper Nejd.

1838 February 25   Ethiopia

From 25 February to the end of March, a series of earth tremors was felt at Massawa and Imakullu (Muncullo), 12 km inland from Massawa. Some of the shocks were accompanied by a loud noise and they were rather strong, but caused no damage.[1]

1 Gouin (1979), pp. 96–7.

c. 1839   Sinai

Rabino refers to the repairs of the fortification wall of St Catherine's monastery in Sinai during this year, following an earthquake that caused them some damage. Repairs to the 'museum' building in *c.* 1840 are also mentioned.[1]

1 Grigoriadis (1875), p. 48; Rabino (1937), p. 25; Ben-Menahem (1979), p. 258.

1842 December 8   Ethiopia

Strong shocks at Ankobar in Ethiopia triggered landslides and flowslides that buried many houses, destroying roads and part of the palace, killing a number of people. For many miles, landslides destroyed or heavily damaged some villages, ruined the crops and caused a number of casualties. As a result of this, the capital of Shoa was relocated at Debre Berhan.[1]

1 Gouin (1979), pp. 34–6.

1844 October 23   Ethiopia

An earthquake accompanied by a noise was reported from Imakullu, near Massawa in Ethiopia.[1]

1 Gouin (1979), p. 97. A further series of shocks starting on 1 August 1848 is listed by Perrey (1860), p. 121 cf. Gouin, pp. 97–8.

1845 February 12   Ethiopia

An earthquake occurred about noon and was felt throughout western Ethiopia. At Lasta a village was destroyed by landslides and at Gondar a few walls collapsed. A church at Gojjam was destroyed and three people were crushed in a crevasse at Wara. The shock was not felt in Bahrdar but it was perceptible from southern Eritrea to Shoa and Gudru in the south (see Figure 2.21). The meridional alignment of the sites for which we have information is due to the fact that much of this was collected along the route of a geodetic survey

being carried out at the time of the earthquake. The exact location of the epicentral region is not known; as a first approximation it may be placed in Wollo, east of Lasta and Wara.[1]

1 Gouin (1979), pp. 36–8. Earthquakes in the region round Lake Tana are also reported to have occurred eight years earlier, in 1836, *ibid.*, pp. 33–4.

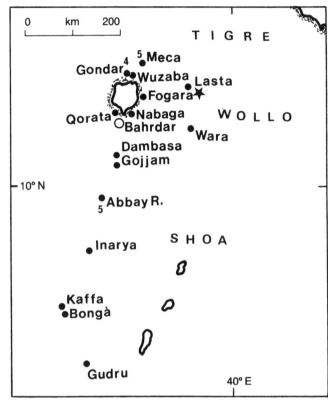

Figure 2.21. 1845 February 12, Ethiopia.

1845 October 31   Red Sea

The volcano Jabal al-Tair, located in the Red Sea between al-Luhaiya and Massawa, was in eruption this year.[1]

1 Itier (1848), III, 331; Perrey (1872b), p. 6; Sieberg (1932a), p. 795.

1846 March 28   Hellenic Arc

Parts of Lower Egypt, particularly Cairo and Alexandria, were shaken by the large-magnitude, intermediate-depth earthquake on this date in the Hellenic Arc. In Cairo the ground motions lasted for three minutes and caused considerable concern, but no damage.[1] The shock was felt in south Italy, Greece, western Anatolia and Syria, and caused widespread damage in Crete.[2]

1 Jomard (1848), p. 278.
2 *Echo de l'Orient* (Istanbul), 24 April 1846; *Journal de Constantinople*, 21 April 1846; *Ceride-yi Havadis* (Istanbul), 22 Rabi' I, 1262. See also Schmidt (1879); Agamennone and Issel (1894); Lyons (1907), p. 285; Ben-Menahem (1979), p. 294.

1846 June 15   **Lower Egypt**

Two shocks with a total duration of 40 seconds were felt in Cairo.[1]

1 Jomard (1848), p. 278; Sieberg (1932b), p. 188, incorrectly has 5 June.

1846 July 14   **Red Sea**

At about 10 a.m., an eruption of the volcanic islet of Saddle in the Zubair group (formerly known as al-Ab'ila) began with great noise and continued for half an hour.[1] This group is located about 90 km northwest of al-Hudaida in the Red Sea.

1 Perrey (1872a), p. 11, citing the *Nautical Magazine* (October 1846), p. 551. Sieberg (1932a), p. 795, refers to a seaquake on 14 August, probably the same event.

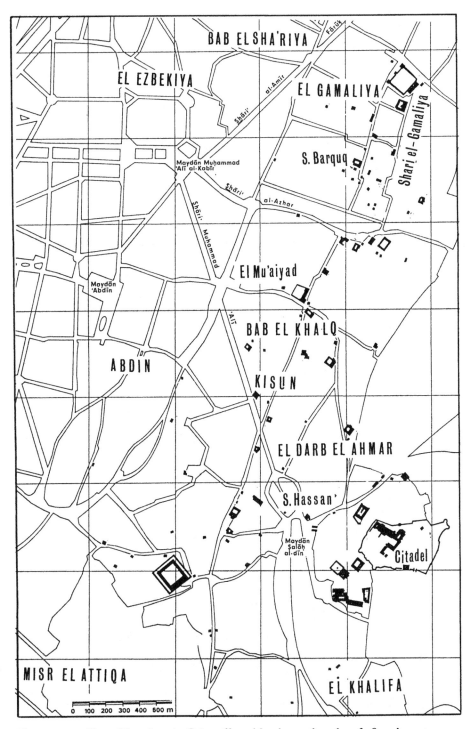

Figure 2.22. Plan of locations in Cairo affected by the earthquake of 1847 August 7.

1847 August 7   *25 Sha'ban 1263*   **Lower Egypt**
Shortly after 8 a.m. a strong earthquake shook Lower
Egypt, causing widespread damage to local houses and
to a number of public buildings. It was the strongest
shock with an epicentre on land in the last 150 years in
Lower Egypt and was felt as far as Asyut in the south
and Gaza and Jerusalem in the northeast.

In Cairo (see Figure 2.22), the shocks continued
intermittently for about a minute, causing panic and
considerable damage. In al-Azbakiyya 14 houses partly
collapsed and one was totally destroyed, together with
an underground cistern and a wall which collapsed and
killed a child. The tops of three minarets fell off, one
of them killing a woman. There was no damage to newly
built houses or public buildings in this district, and

altogether, damage to modern, European-style houses
was negligible. In the Bab al-Sha'riyya, seven houses
and one mosque were damaged. In al-Jamaliyya, three
walls collapsed and many shops were ruined along the
main commercial street. In the northern cemetery, to
the east, the upper part of the southern minaret of the
mausoleum of Sultan Barquq collapsed. In the Bab al-
Khalq, eight columns of the mosque of al-Mu'ayyad
were destroyed and its northern minaret was much damaged.[1] In the 'Abdin district four houses and a mosque
were partly destroyed and many houses damaged. In
Qisun, to the east, two houses and a dye-house partly
collapsed, killing a horse, and a small mosque and rooms
over a shop were destroyed. Near the citadel, several
large stones fell from the dome of the mosque of Sultan

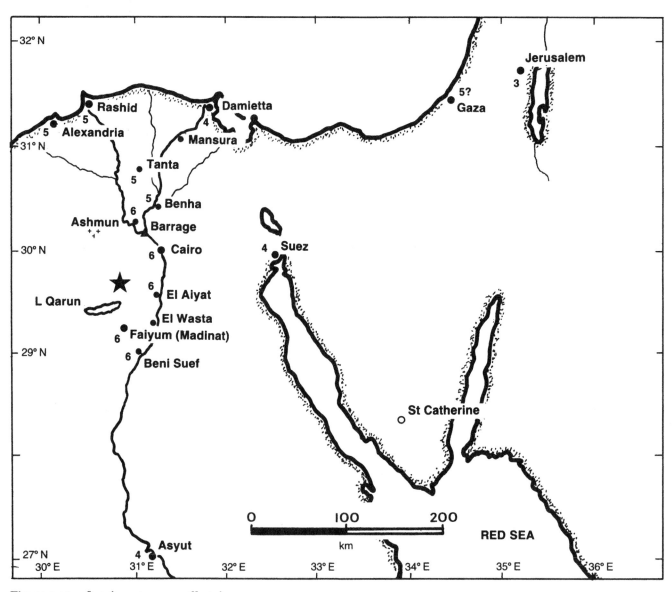

Figure 2.23. 1847 August 7: area affected.

Hasan. To the south of the city, in the Darb al-Jamamiz (in al-Sayyida Zainab district), 16 houses were severely damaged and one dwelling, two cisterns and two mosques were destroyed. In al-Khalifa (southern cemetery), 27 old houses partly collapsed, a mill and rooms over two shops were destroyed and some of the ruins of the old aqueduct crumbled. In Old Fustat, three khans and ten dilapidated houses were destroyed and one new house collapsed, with casualties. To the north, in the Bulaq quarter, built on the sands which the Nile left when it shifted its bed in the fourteenth century, ten houses were damaged and three smaller dwellings and a minaret were destroyed. There was no damage to the landing piers at Ramla or to warehouses. In all, 12 people were killed in Cairo and its suburbs, out of a population of about 250 000.[2]

Outside Cairo (see Figure 2.23) damage extended from Ashmun in the north to Faiyum and Beni Suef in the south. Official damage statistics give total losses for the administrative region from Cairo to Beni Suef as 2987 houses, 42 mosques, a cistern and 45 pigeon towers damaged or destroyed, and 37 men, 48 women and 56 animals killed, and 62 injured. This was mainly concentrated in the villages in the densely inhabited Faiyum oasis (pop. *c.* 200 000). Minor sporadic damage occurred in Madinat al-Faiyum and Beni Suef. In Asyut, a farmer turning a water-wheel fell off and drowned in an irrigation ditch. There is no information about damage to historical monuments in this region.[3]

The earthquake was strongly felt throughout the Nile Delta. In Alexandria the shock lasted about 35 seconds. It caused alarm by opening doors, shaking walls, setting bells ringing and stopping all pendulum clocks. Cracks appeared in the walls of a few old houses. In Mansura the shock caused great concern and some minor damage to a chimney stack and a minaret. The earthquake was reportedly strong in Damietta, Rashid and Suez and was distinctly felt in Jerusalem. In Gaza the earthquake caused some damage, ruining part of the house of the Mutazalim and the leper house.[4]

An aftershock was strongly felt in Cairo and Alexandria on the morning of 10 August 1847, but it caused no damage.

Sieberg provides a grossly exaggerated assessment of the damage and regards Faiyum as the epicentral region, associating the earthquake with a fault-break there (see Figure 2.24). In addition to the damage statistics above, he reports that in the region between Cairo and Asyut another 1000 houses and 27 mosques were destroyed, 25 people killed with the loss of five beasts of burden. He also states that destruction of houses and loss of life occurred also in Alexandria.[5] These details

arise from an amalgamation of the effects of the 1847 earthquake with those of the earthquake of 26 June 1926.[6] Sieberg's account, nevertheless, is generally followed by later authors.[7]

In fact, despite the relative severity of the 1847 event, it is important to recall that the shock had little effect on better-built houses and none whatsoever on more substantial engineering works. There is no evidence that public buildings had to be pulled down, nor that any relief measures were taken. Whatever damage was done was quickly repaired. It is remarkable that accounts left by travellers passing through Cairo, the Faiyum and the Lake Qarun region after August 1847 make no reference to the earthquake.[8] Similarly, there is no evidence that damage was caused to the Barrage on the Nile, started in 1835 about 20 km below Cairo, not considered structurally safe at the time of the earthquake. One of its designers, in 1856, could hardly remember the 1847 earthquake.[9]

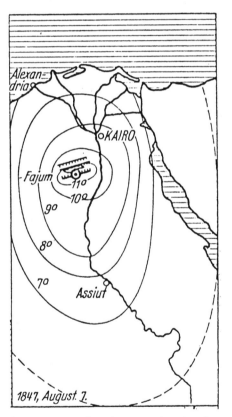

Figure 2.24. 1847 August 7: isoseismal map from Sieberg (1932b).

Despite the relatively full information available for this earthquake, it remains difficult to fix a precise epicentral location. The distribution of localities for which intensity data are available is artificially distorted by the concentration of population in the Nile Valley and it is impossible to construct a proper isoseismal map. A macroseismic epicentre in the region between Cairo and the Faiyum oasis, in the same area as the earthquake of 1992 October 12, remains the best estimate, although it must be emphasised that there is no evidence whatsoever for faulting in this area.

1 Lane (1896), pp. 96, 122. The Cairo press (see below) puts the mosque of al-Mu'ayyad in the Darb al-Ahmar district. It is situated by the Bab Zuwaila. 25 Sha'ban is the equivalent of 8 August, unless the astronomical hijri calendar (starting a day earlier, on 15 July 622) was being used.
2 The detailed survey of damage published in *al-Waqa'i' al-Misriyya* gives a total of 111 buildings damaged or destroyed in Cairo, see Sami (1928). A total of 12 deaths is given in National Archives, Cairo Citadel: European Archive, Italy no.22. See also F.O. 78/708, no.50 (Alexandria). For the European Press, see *Allgemeine Zeitung* (Hamburg), 24 August 1847, *Journal des Débats* (Paris), 29 August 1847. The earthquake is noted briefly by Perrey (1848), p. 450. The best account of Cairo at this period is by Abu-Lughod (1971), pp. 83 ff.
3 Details for the Faiyum are included in the French Correspondance Commerciale des Consuls, Égypte, 19, Cairo (no. 28, 17 August 1847) in the Bibliothèque Nationale, Paris. See also *Al-Waqa'i' al-Misriyya*, 3 Shawwal 1263/14 September, followed by *Ceride-yi Havadis* (Istanbul), no. 355, 7 Dhu'l-Qa'da 1263/17 October. Displacement of masonry walls and roofing stones of a ruined temple of the Thirteenth Dynasty near Lake Qarun (Brown, 1892, pp. 53–4), though suggestive of an earthquake, are not necessarily to be associated with the 1847 event.
4 *Ceride-yi Havadis*, no.349, 25 Ramadan/6 September; Neuville (1948); Neale (1852); Brehm (1862).
5 Sieberg (1929), (1932a), p. 873; (1932b), p. 188.
6 For full details, see Ambraseys (1991). The Cairo press took statistics of the 1847 event for comparison with those of 1926, but gave the impression that the figures all referred to the earlier earthquake; see *The Times*, 5 July 1926, p. 13.
7 For example, Kárník (1971); Maamoun (1979); Ben-Menahem (1979), p. 257.
8 Bayle St John (1849, 1850); Brehm (1862), p. 27.
9 Mayer (1856a, 1856b). It is remarkable, in fact, how little information is found in the consular correspondence; there is nothing, for instance, in F.O. 141/13–14 (Alexandria, Damietta, Aden); F.O. 142/15–16 (Cairo), in the PRO, Kew.

1849 July 23   Lower Egypt
Early in the morning, a light shock was felt in Cairo.[1]

1 Perrey (1850b), p. 220; Sieberg (1932b), p. 188.

1850 July 17   *7 Ramadan 1266*   Yemen
A large earthquake in the Yemen demolished 300 houses in San'a, with casualties.[1]

1 *Ceride-yi Havadis*, no. 499, Sunday, 9 Dhu'l-Qa'da 1266. The paper also reports a destructive fire in al-Mukha, according to news received from Egypt. It is not necessarily implied that there is any connection between these events.

1850 October 27   Upper Egypt
At about 9.30 in the morning, a strong earthquake was felt north of Asyut, lasting about 30 seconds. The shock was felt for many miles, and in many places near the Nile the earth was cracked a full inch.[1]

1 Melly (1851), I, 128.

1851 April 3   Hellenic Arc
A shock of earthquake on this date was felt slightly in Cairo but not in Thebes.[1] The earthquake was the largest aftershock in a series which began on 28 February in Makri (Fethiye), where the coast was flooded several metres above its normal level. The shock caused rockfalls and landslides inland and was strongly felt in Rhodes.[2]

1 Perrey (1852), pp. 361–4.
2 BBA. *IMV.*, 6790, 29 March OS (= 10 April NS) 1267 H (1851).

1853 August 5   Libya
Many earthquake shocks were felt in the region of Murzuq in the Fezzan, Libya, and again on 11 October.[1]

1 Vogel (1862), p. 140.

1854 January 3   Sudan
Late at night, a brief shock was felt in Khartoum, Sudan.[1] The claim that the shock was also felt in Upper Egypt is without foundation.[2]

1 Perrey (1855), p. 530; Schmidt (1879), p. 175; Ambraseys and Adams (1986a).
2 Lyons (1907), p. 285; Sieberg (1932b), p. 188.

1854 February 21   Ethiopia
A severe earthquake in the region of Tambien in Tigre, Ethiopia, was preceded and followed by many strong shocks lasting till July. Fissuring in the ground and landslides were reported over a large area, from Lake Ashangi to Betalihem along the Guf Guf graben (see Figure 2.25). The shock caused liquefaction at Lake Ashangi and changes in the underground water supply in the Doba Valley. The earthquake was felt at least 200 km away.[1]

1 Gouin (1979), pp. 39–42.

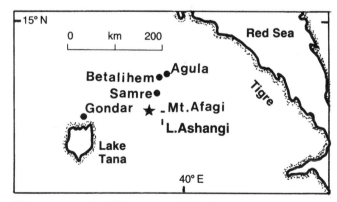

Figure 2.25. 1854 February 21, Ethiopia.

1856 October 12   **Hellenic Arc**

A major earthquake in the Hellenic Arc was felt in the Nile Delta. The earthquake occurred before dawn and lasted over two minutes. In Alexandria, a few old walls tumbled down; no substantial damage was done, but the people were thrown into a panic. Some old houses along Ra's al-Tin were damaged but only two lives were lost, from injuries. In Damietta, the shock was slight, but in other provincial towns of the Delta, such as Tanta and Damanhur, several minarets fell, killing some people. At Suez the shock was felt, but it was slight. Elsewhere in the Delta it caused panic, people finding it difficult to stand or walk, the ground movements causing furniture to move and water to slosh out of tanks.[1]

In Cairo the effects of the earthquake were more serious. Three successive shocks were felt, of one, one-half and two minutes' duration. The railway clock and all hotel clocks stopped and water in canals ran up the embankments; water from the Khalij canal was thrown over its sides and well-water rose to the surface. Only about 20 houses collapsed completely and about 200

were ruined, killing four people, but many houses of local construction were damaged. The mosques of Sultan Hasan and Da'ud Pasha fissured from top to bottom; the minaret of the Mahkama mosque collapsed and the structure of the building was badly cracked. The Catholic Church, off al-Muski, was also damaged and part of Shepherd's Hotel fell in. In the district of Bulaq, there was more debris than in Cairo. Here about 20 mosques were damaged, almost all through the collapse of their minarets, among them the mosque of Abu 'l-A'la which lost its top, with the loss of four lives. Panic was great throughout Cairo and a large proportion of the populace camped in open spaces for a day or two.[2]

The earthquake had no effect on major engineering works and it was hardly felt south of Helwan (see Figure 2.26). It was strongly felt by boats sailing off Egypt and as far as Central Italy. Almost the whole of the Adriatic Sea, Malta, Greece, western Anatolia, Cyprus and Palestine experienced the shock. Maximum damage was inflicted on Rhodes and Crete and neighbouring islands of Karpathos (Scarpantos) and Casos.[3]

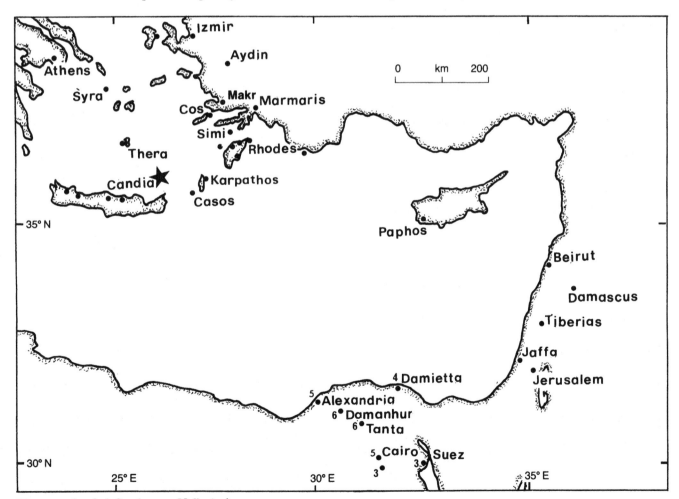

Figure 2.26. 1856 October 12, Hellenic Arc.

69

Aftershocks continued for a few days. Two shocks on 14 October were particularly strong, one of them at midnight causing the collapse of the minaret of the mosque of Da'ud Pasha in the district of al-Sayyida, and further damage to a few old houses in Bulaq.[4]

1 F.O. 78/1222 (Alexandria, 15 October 1856); Mayer (1856a, 1856b); Raulin (1869), pp. 428–30.
2 Neimann (1856); *Ceride-yi Havadis*, nos. 808, 810, 814, 29 Safar – 15 Rabi' II, 1273.
3 F.O. 78/1210, pp. 184, 190, 192, 200 (Canea, 14 October–8 December 1856); 78/1211 (Rhodes, 18 October 1856); Buist (1857); Fenech and Froud (1857); Sopwith (1857), p. 95; Lyons (1907), p. 285; Mazzarelli (1947); Goby (1955); Ben-Menahem (1979), p. 294.
4 Mayer (1856b), Schmidt (1879), pp. 47–54.

1857 April   Ethiopia

Frequent shocks were reported during the Spring, felt in Tigre and Eritrea. They were light in Massawa, but damage was reported in an unnamed town in Tigre.[1]

1 Gouin (1979), pp. 42, 98.

1858 June 13   *1 Dhu'l-Qa'da, 1274*   Persian Gulf

A strong shock was felt in Bushire (Bushahr) in the Persian Gulf, followed by aftershocks, probably associated with an earthquake in the Ganava plain.[1]

1 *Vaqa'i'-yi Ittifaqiyya*, no. 391, 1274 H; Pelly (1863).

1858 December   Lower Egypt

Near the end of the month, an earthquake was felt in Cairo.[1]

1 Perrey (1862a), p. 31.

1859 January   Yemen

Early in January a series of shocks affected al-Mukha and Aden and the intervening areas, probably from an inland earthquake. The shock was not felt at Oboc in Djibouti.[1]

1 Perrey (1862b), p. 41.

1860 August 6   Sudan

Shortly after midnight (local time), three shocks were felt in Berber in Nubia, as a result of which, two houses and a wall were damaged.[1]

1 Dinome (1862), p. 140; Ambraseys and Adams (1986a).

1860 December 22   Libya

A violent shock was felt at Matmata, on the limits of the Little Desert on the Tunisian border with Libya. People lying on the ground were made to roll. There were no man-made structures to be damaged.[1]

1 Perrey (1862c), p. 73.

1861 May 7   *Dhu'l-Qa'da 1277*   Ethiopia

An eruption of the volcano Dubbi in Ethiopia was accompanied by strong earthquakes felt over a distance of a least 200 km. In the immediate vicinity of the volcano, two villages were destroyed, animals killed, and 106 people lost their lives. The shock was felt in al-Mukha and al-Hudaida and there, as well as along the entire coastal plain of Yemen, dust fell for several days. Dust also fell as far as Walda in Wollo and in the Dahlak Archipelago, so thickly that ships were unable to continue their voyage. The eruption was heard as far away as Massawa, Perim and Aden (see Figure 2.27). Local tremors were felt for four months, ending with the final eruption in September.[1]

Reports from San'a confirm that the effects of the volcano were observed throughout Yemen, with the exception of San'a itself and part of the district of Yahsub, near Yarim. Explosions were heard for three days in Dhu' l-Qa'da (began 11 May 1861) and a kind of

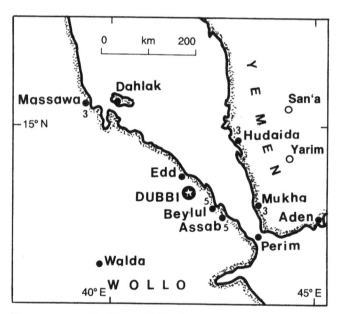

Figure 2.27. 1861 May 7, Ethiopia.

'lustrous ashes' fell. There was a recurrence of similar activity on 12 Shawwal 1279 (2 April 1863), i.e. two years later. The people heard terrible noises in the air like cannonfire, and saw smoke and lightning flashes that crossed the sky from the direction of Mecca towards Aden. These experiences caused great terror.[2]

1 Playfair (1861, 1863); Perrey (1864), pp. 95–9; Heuglin (1868), pp. 62, 328; Schmidt (1879); Gouin (1979), pp. 98–101. Lyons (1907), p. 285, lists earthquake shocks in the Red Sea on 24 September 1861, probably part of the same event; see also below.
2 Al-Harazi, pp. 90–1, 139. The 1863 activity is also reported by Simkin *et al.* (1981), p. 38, where it is associated with both Dubbi and Jabal al-Tair in the Red Sea.

[1861 July 19   **Gulf of Aden]**
Pumice seen floating near Steamer Point at the entrance
to the harbour of Aden was attribted to a submarine
eruption off Sira Island.[1] No shock was felt and it is
probable that this was the result of the Dubbi eruption
rather than a local phenomenon.

1 Perrey (1864), p. 86.

1863 April 22   **Hellenic Arc**
A destructive earthquake in the eastern part of the Hellenic Arc was strongly felt in the Nile Delta (see Figure
2.28). In Alexandria it caused some concern but no
panic, and it was also reported from Suez. In Cairo,
however, the earthquake, which happened late at night
and lasted for about 30 seconds, caused great panic,
most of the inhabitants spending the rest of the night
out in the open. No damage was reported in the city or
in other parts of Egypt.[1] The shock was very widely felt,
from the Ionian Islands to Palestine and from Istanbul
to Cairo. Maximum damage occurred in the islands of
Rhodes, Karpathos and Casos.[2]

1 Perrey (1865), p. 128; Chaplin (1883); Schmidt (1879), p. 207.
2 F.O. 195/758 (Rhodes). Perrey (1865), p. 126; Lyons (1907),
p. 285; Ben-Menahem (1979), p. 292.

1864 September 14   **Ethiopia**
A series of shocks was felt in Massawa between 5 March
and 21 October this year, including two tremors on 14
and 15 September. No damage is reported in any of
these events.[1]

1 Gouin (1979), p. 101.

1865 April 11   **Lower Egypt**
An earthquake shock was felt in Alexandria in the morning, lasting only two seconds.[1]

1 Perrey (1867), p. 63.

[1867 September 20   **Hellenic Arc]**
A large shock in the Hellenic Arc is alleged to have
been felt along the coast west of Alexandria.[1]

1 Only Lyons (1907), p. 286, mentions Egypt. Cf. Perrey (1870),
pp. 182–6; Schmidt (1879), pp. 103–8.

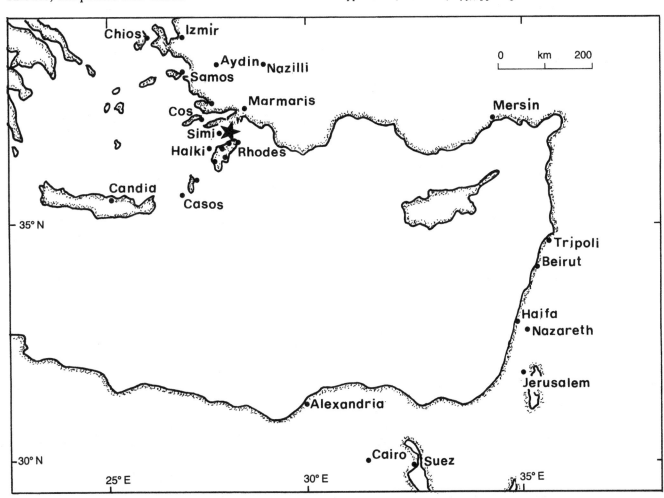

Figure 2.28. 1863 April 22, Hellenic Arc.

1868 February 20   **Eastern Mediterranean**

Two strong shocks were felt in Alexandria during the night. They were also felt in Cairo and Jerusalem, as well as along the Suez Canal, particularly from the line of works between Ismailia and Qantara. The shock was felt on board ships in the roadstead of Port Said, riding at anchor in bad weather, and at various works encampments of the Canal around Port Said.[1] Nowhere did the shock cause damage.[2]

1 *The Times*, 10 March 1868; Perrey (1872a), p. 66; Chaplin (1883).
2 It occurred at 03.15 on 20 February local time, see Lyons (1907), p. 286, and Sieberg (1932b), p. 188.

1870 June 24   **Eastern Mediterranean**

A large, probably intermediate-depth earthquake was felt throughout the eastern Mediterranean (see Figure 2.29). In Alexandria, three successive shocks, lasting about two

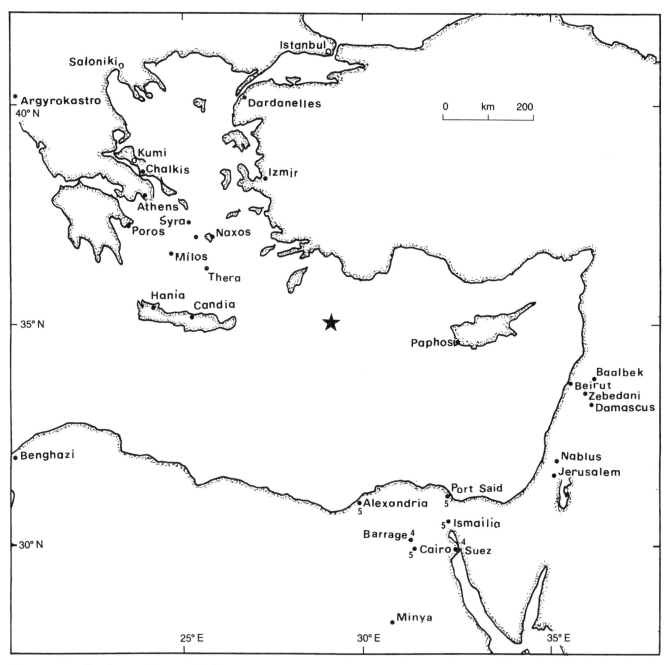

Figure 2.29. 1870 June 24, Eastern Mediterranean.

minutes, caused considerable concern but no damage, except in the New Port area where there was some sporadic cracking of plaster. Here some people ran out of doors and the sea flooded the quay. The shock was felt on board ships in both the Old and New Ports and offshore opposite the quarries of Mex. Everyone along the coast of the Nile Delta felt the earthquake and it was reported from Port Said and from construction sites along the Suez Canal.[1]

The earthquake was stronger, but of shorter duration, at Ismailia and Cairo, where it caused some slight panic and damage to a few houses. It is said that here, as well as in Alexandria the shock was presaged by domestic animals.[2] In Suez the earthquake was felt by some people but in general it was not strong. The same applies to the Barrage near Kalyub, which at the time of the earthquake was not holding any water, and to the region north of Cairo. There is no evidence that in Egypt the shock was felt south of Minya, but it was reported from the east coast of the Red Sea, in the Hejaz, according to the Turkish press.[3]

The earthquake was felt in a number of places in south Italy and Sicily, in parts of Albania, Greece, west Turkey, in the Greek islands, Cyprus, Lebanon, Syria, Palestine and at Benghazi in Libya. Nowhere did this earthquake cause any damage and, although it was reported from many places far apart, it also passed unnoticed in most other places in the eastern Mediterranean.[4]

This is one of the first earthquakes recorded instrumentally. The shock was registered by the primitive seismographs of the Observatory of Naples at 17h 16m 22s local time.[5]

1 Coumbary (1870a); Craveri (1870).
2 Craveri notes that both horses and dogs were disturbed in Alexandria.
3 *Selanik Vilayet Gazetesi*, 21 Rabi' II, 1287, p. 128. The Turkish press received reports from Aden that the shock had been felt on the east coast of the northern Red Sea. There is no suggestion that the shock was felt in Aden itself, see next note.
4 Perrey (1873), pp. 109–11; Schmidt (1879), pp. 274–8; Lyons (1907), p. 286 (two separate entries). Much of the information given about this event by modern writers is grossly inaccurate. Sieberg (1929) considers the earthquake to have been centred in Ismailia. Later (Sieberg 1932a,b) he places this large shock just off the Nile Delta, about 100 km north of Rosetta. His isoseismal map shows Cairo to have experienced intensities of IX and Alexandria and Ismailia VIII. Aden is included among the places where the shock was felt (see above). This in turn misleads Ben-Menahem (1979), p. 283, who assigns to the earthquake a magnitude $M_L = 7.2$, and a radius of perceptibility of 2500 km.
5 Baratta (1901), p. 450.

[**1872 April 2**   **Rhodes**]
An earthquake in Rhodes and elsewhere is alleged to have been felt in Egypt as well as Syria.[1] This is in fact the large earthquake of April 3 that was felt in Alexandretta.[2]

1 Lyons (1907), p. 286, quoting Schmidt, who also mentions an aftershock on 28 April (p. 307).
2 F.O. 78/2243, p. 131 (Beirut); Fuchs (1886), p. 484.

1873 January 12   **Mediterranean**
An earthquake was felt in Cairo in the early afternoon by many people, the oscillations of the ground persisting for about a minute and a half, but causing no damage.[1] This earthquake was apparently felt along the Palestinian and Lebanese coast and it could have originated from the southeast Mediterranean region.

1 *The Times* (London), 24 January 1873; Schmidt (1879). Sieberg (1932b), p. 188, lists an earthquake in Cairo on 14 February, thereby amalgamating the Cairo shock with an earthquake in the eastern Mediterranean on 14 February, for which see *Nature*, 6 March 1873, p. 351, Fuchs (1886), p. 485.

1873 October 9   *1290 H*   **Yemen**
A strong earthquake occurred in the district of Zabid in Yemen, at about 9.0 p.m., but caused no damage or loss of life.[1]

There is no connection with another earthquake reported in 1290/1873, in the region of al-Haima, west of San'a, which is probably not a genuine seismic event.[2]

1 *Basiret* (Istanbul), 23 November 1290. The date is given as 27 September in the Rumi calendar, the time as 3.10 Turkish time.
2 A large landslide occurred from a mountain overlooking Bait al-Nash; the course of a river was diverted and its waters turned red, and much damage was caused. The episode provoked disputes between the local communities affected. Al-Wasi'i, pp. 113–14; Ambraseys and Melville (1983).

1875 November 2   **Ethiopia**
A destructive earthquake occurred in north Eritrea, probably associated with faulting, with an epicentre possibly somewhere north of Keren (see Figure 2.30). The shock caused many casualties and damage extended to Adua, where landslides were triggered.[1] The earthquake was strongly felt at Suakin[2] and as far as Gebeit. It was also probably noticed at Jidda and on ships offshore.[3]

1 Gouin (1979), pp. 42–3.
2 Lyons (1907), p. 286. Sieberg (1932b), p. 189 duplicates this with a second earthquake between Massawa and Jidda on 11 November 1880; see also Ben-Menahem (1979), p. 255.
3 *Neologos Konstantinoupoleos*, no. 246, 3 December 1875.

1878 June   *Jumada II, 1295*   **Yemen**
A series of shocks and tremors occurred in the towns of Dhamar and Yarim and their respective districts.

Shocks lasted for three months up to Ramadan (began 29 August) and destroyed many houses.[1]

1 Al-Wasi'i, p. 118; Ambraseys and Melville (1983).

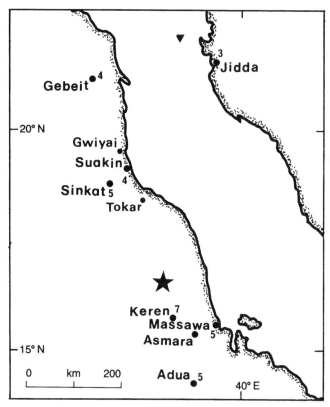

Figure 2.30. 1875 November 2, Ethiopia.

1878 November — Ethiopia

Earth tremors over a period of about six days were reported from Tigre. There is no indication of the location or intensity of the shocks. Another earthquake occurred on 30 June 1880 in Tigre, probably in the neighbourhood of Axum, causing panic but apparently no damage.[1]

1 Gouin (1979), pp. 43–4.

1879 July 11 — Gulf of Suez

An earthquake consisting of three moderately violent shocks was felt at night in Cairo. In the Bab al-Nasr quarter (near the mosque of al-Hakim) some isolated walls fell and an old, somewhat dilapidated, minaret suffered so badly that it had to be taken down. The shocks were felt also near Giza and it was rather strong in Alexandria.[1] At Tor, in Sinai, the earthquake was associated with a sea-wave that flooded the village. The shock was reported from some of the railway stations to Wasta and Madinat al-Faiyum.[2]

1 *Nature*, 7 August 1879, p. 353.

2 Kaiser (1922), p. 21; Sieberg (1932b), p. 189. Kaiser says he had experienced no earthquakes during his previous ten-year stay in Sinai.

1881 June 10 — Libya

A swarm of strong earthquakes caused extensive but minor damage in the region between Gabes and Meret in eastern Tunisia. The activity continued for about ten days, causing the partial evacuation of Gabes. None of the strongest shocks was felt over a radius of more than about 35 km. Smoke seen from Mt. Ain Buni was probably dust due to rockfalls triggered by the shocks.[1]

1 *Bollettino de Vulcanismo Italiano*, 8 (1881), pp. 47–8. *Comptes Rend. Acad. Sci.*, 93 (Paris, 1881), pp. 31, 105; Fuchs (1886), p. 508.

1881 June — Rajab 1298 — Yemen

A strong earthquake during Rajab cracked some houses, and was followed by another, weaker shock.[1] It is not clear from the context whether this occurred in San'a or Sa'da, but it was probably the latter.

1 Zabara, p. 13. Rajab began on 30 May 1881.

1883 March — Ethiopia

An earthquake in Ethiopia, probably originating in the Aussa district, was widely felt; there is no evidence of damage.[1]

1 Gouin (1979), pp. 101–2. Simkin *et al.* (1981), p. 38, list an eruption of Jabal al-Tair in the Red Sea this year; there may be some connection.

1883 August — Shawwal 1300 — Libya

The town of Ghadamis in Libya was shaken by an earthquake that caused great concern. The shocks were accompanied by loud explosions, and the shaking was felt within a radius of 350 km from the town,[1] implying a large-magnitude shock.

1 Duveyrier (1883). The date is given as late in Shawwal/began 4 August.

1883 October 13 — Arabian Peninsula

The Political Agent at Muscat, Oman, reported a double shock felt there at 3 p.m., causing doors and windows to rattle.[1]

This is perhaps the earthquake reported in the European press in March the following year, said to have been strongly felt in Muscat and Nazwa, in the vicinity of which nine villages were destroyed and the ground in places was deformed.[2]

1 I.O.R., R/15/1/193 (Bushire Diary, 27 October 1883); see also *Iran*, no. 528, 1300 H and no. 530, 1301 H.

2 Fuchs (1886), p. 489; Sieberg (1932a), p. 795.

1884 May 19   *24 Rajab 1301*   **Persian Gulf**
A destructive earthquake in the central part of Qishm island ruined 15 villages, 34 mosques and killed 238 people, injuring about 500. The shock caused cracks to open in the ground and many water reservoirs were destroyed, forcing the inhabitants to leave the island. The earthquake did not cause much damage in the town of Qishm. It was strongly felt in Bandar 'Abbas, Linga and Ra's al-Khaima. Aftershocks continued until July.[1]

1 *Iran*, no. 547, 1301 H; Lorimer (1915); Kababi (1963); Ambraseys and Melville (1982), p. 162.

1884 July 20   **Ethiopia**
A damaging earthquake occurred in Eritrea, with an epicentre probably offshore from Massawa. Many houses in Massawa and a few in Muncullo were ruined, without loss of life. High sea waves built up in the harbour at Massawa, especially in the bay between Taulud and Edaga Barai, and swept over the causeway. Ships in the harbour were seen to rock violently. Several times the sea flooded the land, leaving dead fish on the shore. The shocks were also strongly felt on the Dahlak islands. Aftershocks continued intermittently until October.[1]

Contrary to some catalogues, there is no indication that Cairo was affected by this earthquake.[2]

1 Gouin (1979), pp. 102–3. Several further tremors were reported in Massawa in the next ten years: 1886 May 8; 1891 February 12 and April 27; 1892 November 23; 1894 September 12; 1896 December 11 and 1897 September 30. For details, see Gouin, pp. 103–5.

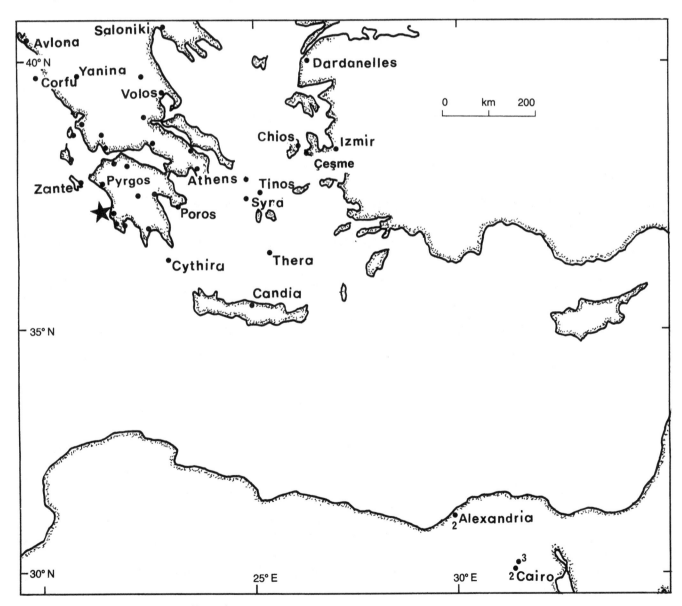

Figure 2.31. 1886 August 27, Hellenic Arc.

2 Sieberg (1932b), p. 189, lists an earthquake on 25 July 1884 that was felt in Saudi Arabia and in Cairo; cf. Sieberg (1932a), p. 886 under 23 July. His reference to Nejd and Muscat indicates some confusion of the Massawa earthquake with the Oman event of 13 October 1883 (see above). Cairo is located *c.* 1500 km from the epicentral area, but it is one of the centres of information from which press reports were derived.

1886 August 27   Hellenic Arc

A very large earthquake offshore from east Greece was strongly felt in Alexandria and Cairo, particularly Abbasiyya.[1] It caused no damage in Egypt, but inflicted heavy damage in southwest Peloponnese. It was felt as far as Izmir, Istanbul, Saloniki and Albania, and was perceptible in parts of Sicily, south Italy and the Adriatic coast (see Figure 2.31).[2]

1 *Nature*, 2 September 1886, p. 434; *Ankara*, 8 September 1886 (1302 H), p. 4. Sieberg (1932b), p. 189 mentions a slight shock in the Nile Delta in July 1886, an inaccurate reference to the same event.
2 Galanopoulos (1941, 1953).

1886 November 17   Lower Egypt

A slight shock was felt in Cairo at 4.30 p.m. The vibration lasted several seconds.[1]

1 *The Times* (London), 18 November 1886.

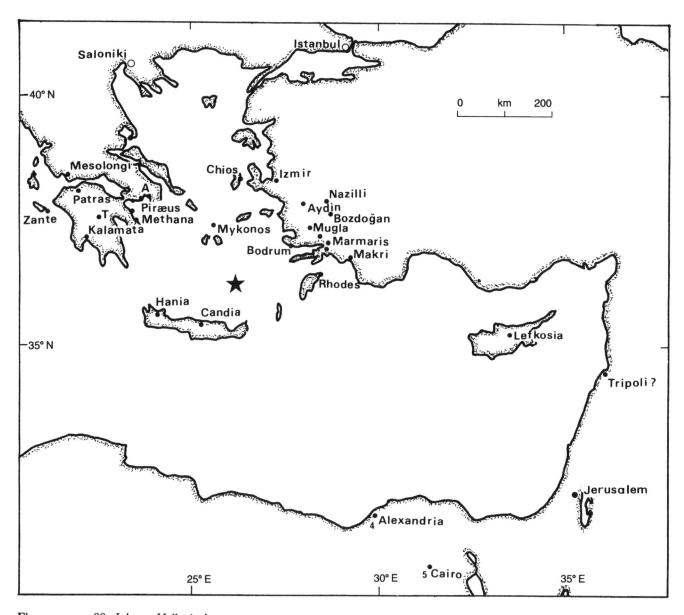

Figure 2.32. 1887 July 17, Hellenic Arc.

1887 April 1   Yemen

Earthquake shocks were felt repeatedly in Aden for four days, but no damage is reported.[1]

1 *The Times* (London), 6 April 1887.

1887 July 17   Hellenic Arc

A large, intermediate-depth earthquake in the Hellenic Arc was strongly felt in Egypt. In Alexandria the shock caused some concern but no damage. In Cairo, two successive shocks were felt, lasting 40 seconds, and causing some panic mainly in Bulaq, where people ran out of their houses. In Fam al-Khalij, near Fustat, three minarets of mosques were damaged and in Qasr al-Nil an old house partly collapsed, killing one and injuring three people. In New Cairo, the shock was felt by many people and clocks were stopped in a few hotels. No damage of any kind was reported from the city.[1]

The earthquake was reported from railway stations along the Nile Valley and it was perceptible at Luxor. We could find no evidence that the shock was felt along the Suez Canal, but it was felt in Jerusalem, Tripoli and Cyprus. Damage was reported from Crete, southwest Turkey and mainland Greece (see Figure 2.32), and the earthquake was felt in south Italy and part of Sicily.[2]

1 *The Times*, 18 July 1887; *al-I'lam* (Cairo), 26 Shawwal 1305; *al-Muktataf*, vol. 11 (1887), p. 703; Legrain (1900). Sieberg (1932a), p. 873, (1932b), p. 189, gives a grossly exaggerated description of the effects in Egypt, assigning intensity VII in Cairo. He claims the shock was felt as far as the Sudan, apparently following Lyons (1907), p. 286, who mentions that Suakin was also affected, but there is no evidence for this.
2 Baratta (1901), pp. 517–18; Kallner-Amiran (1951), p. 232; Galanopoulos (1953).

1889 October 4   Ethiopia

Earth tremors were experienced during the night in Keren. Further shocks in the same vicinity are reported in 1894 May 13 and July 4 and 1894 November 8.[1]

1 Gouin (1979), pp. 44–5.

1894 November 3   Ethiopia

Two strong tremors were felt during the night at Halai, 55 km southeast of Asmara. There are no reports of damage.[1]

1 Gouin (1979), p. 45.

1895 August 2   *Safar 1313*   Yemen

A series of shocks occurred at al-Mukha and Ta'izz in the Yemen. Local sources report two particularly strong earthquakes, followed by a weaker one, and then another strong one, all on the same night.[1] It is not known whether any damage was caused. Ottoman sources put the third shock in al-Mukha and the fourth in Ta'izz.[2]

1 Zabara, p. 179, citing the Imam of the mosque in San'a. Safar 1313 began on 24 July 1895. Al-Wasi'i, p. 163, has the same information under the year 1314 H/beg. 12 June 1896.
2 Agamennone (1896), reports the date and the time of these shocks; Ambraseys and Melville (1983).

1895 December 7   Lower Egypt

Two shocks, of two or three seconds each, were felt within five minutes in Cairo, as well as in Alexandria and Ismailia. They caused no damage but some concern in the Tanta–Zagazig region.[1]

1 *Egyptian Gazette* (1895); Agamennone (1896).

1896 June 29   Hellenic Arc

A slight earthquake lasting a few seconds was felt in Cairo. The shock originated off Cyprus and it was widely felt in Lebanon and Palestine, causing some damage on the south coast of Cyprus (see Figure 2.33). The shock was not reported from Alexandria, but it was felt by a few people on board ships in Port Said.[1]

The earthquake was recorded by almost all the seismographs in Europe.[2]

1 *Sabah* (Istanbul), 11 July 1896; Agamennone (1900), p. 165. Kallner-Amiran (1951), p. 232; Ambraseys (1965); Ben-Menahem (1979), p. 290; Ambraseys and Adams (1993).
2 Agamennone (1896, 1904).

1896 June–July   *Muharram 1314*   Yemen

An earthquake occurred in San'a after the night-time prayer.[1]

1 Zabara, p. 196. Muharram began on 12 June. The earthquake is reported on the authority of the Imam of the Friday mosque in San'a. The shock was followed in late Safar/early August by a great dust storm.

1897 February 20   Red Sea

An offshore shock caused the failure of the submarine cable between Perim, Assab and Massawa. Shocks were also reported from ports on the west coast of the Red Sea in December 1896 and September 1897.[1]

1 *British Assoc. Adv. Sci.* (1898), p. 252; Gouin (1979), p. 105.

1898 September   *Jumada 1316*   Yemen

A great earthquake in the Yemen is reported without details. It was followed by news of the fall of a meteorite which destroyed a village in the Tihama.[1] The two events are not necessarily connected, but this is the most likely interpretation.

1 Zabara, p. 256. He does not specify which month: Jumada I began on 17 September 1898, Jumada II on 17 October. The shock was presumably felt in San'a, over 100 km from the coastal Tihama. Usually earthquakes are reported after meteorites, as when the thundering noise of a meteorite was heard in the sky in late Rajab 1309/

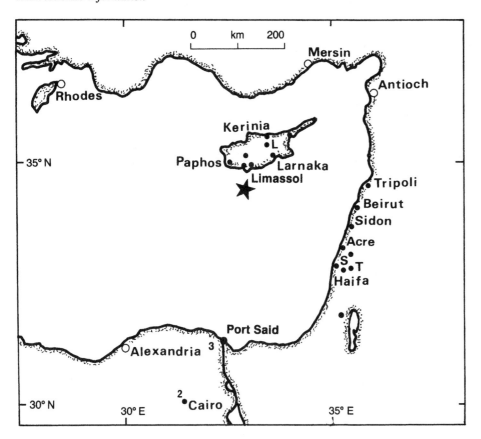

Figure 2.33. 1896 June 29, Hellenic Arc.

February 1892, accompanied by an earth tremor in the region of Khaulan, east of San'a. The meteorite landed on and destroyed some houses in Bani Siham; Zabara, p. 74. The sequence of events reported here also allows a logical connection to be drawn.

1899 February 11   Ethiopia

A strong earthquake fissured many houses in Djibouti. The shock was followed by another, weaker tremor on 23 November.[1]

1 Gouin (1979), p. 148.

1900 January 18   Gulf of Suez

The Cairo press reports a slight but quite perceptible earthquake felt by some persons within the city at 5.30 a.m.; the movement lasted five seconds and was undulatory. The only event recorded at Helwan was at 15h 29m at a time later than that reported from Italian and other European stations. It seems unlikely therefore that the recorded event was local. In view of other earthquakes this year in the Suez region, we have assumed that the early morning felt report was from a foreshock of the event on 1900 March 6.[1]

1 *British Ass. Adv. Sci.* [*BAAS*] (1900), no. 2, p. 40; *al-Mu'ayyad* (Cairo), 21 January 1900; *Egyptian Gazette* (Cairo), 20 January 1900.

1900 March 6   Gulf of Suez

This was an earthquake of relatively large magnitude (Ms 6.2), centred in the Gulf of Suez in the uninhabited region between the lighthouses of Zafarana and Gharib in Egypt, and Maghara in Sinai, where it caused rockfalls. At the monastery of St Catherine's, at Tor, at the Ashraf lighthouse and at Helwan, hanging lamps were set swinging, objects were displaced and glass rattled, but there was no damage. In Cairo the earthquake consisted of three consecutive shocks lasting in all about 30 seconds; they were felt by everyone and some people went out from their houses. The only damage they caused was the fall of a heavy chandelier on the first floor of the Grand Continental Hotel. The earthquake was felt slightly along the Nile Valley, where it caused nausea and was just perceptible as far as Alexandria and Luxor, but not at Aswan. It was widely felt in Gaza, where it lasted about 10 seconds, but not in Jerusalem (see Figure 2.34). The earthquake was recorded at Helwan where the boom of the Milne pendulum was thrown off scale. Instrumental readings from other seismograph stations up to 63° away confirm an epicentre in the Gulf of Suez.[1]

1 *Sabah* (Istanbul), 10 March 1900; *Egyptian Gazette*, 7–9 March 1900; *al-Mu'ayyad*, 7–8 March 1900; Lyons (1907), p. 286; Sieberg (1932b), pp. 184, 189.

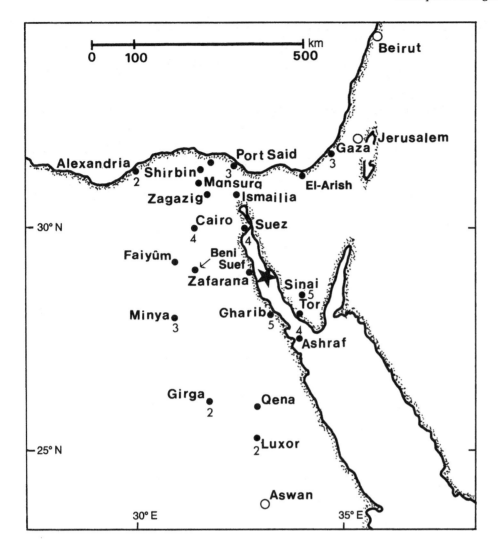

Figure 2.34. 1900 March 6, Gulf of Suez.

1900   **Iraq**

A contemporary field report states that in about 1900 an earthquake devastated Amara in Iraq, necessitating its rebuilding, but we have been unable so far to confirm this from other sources.[1]

1 Gubbins (1944), *Sub anno.*

1902 August 5   **Yemen**

At the end of Rabi' II, 1320, a middle-sized earthquake during the night was perceived in San'a by those who were awake.[1] There is no evidence of damage, and no instrumental recordings were reported.

1 Zabara, p. 368.

1903 July 19   **Mediterranean Sea**

An earthquake with an epicentre in the Mediterranean Sea off Egypt was felt in Cairo, where it lasted 45 seconds. The shock was felt in Tanta but not in Alexandria.[1] The shock was also perceptible at Athens and Mineo in Sicily.[2]

1 Rudolph (1905), p. 396; Sieberg (1932 b), p. 189.
2 Station Bulletins, Athens and Mineo.

1903 August 11   **Hellenic Arc**

A large-magnitude earthquake in the western Hellenic Arc was generally felt in Alexandria, and at a few other places in the Nile Delta (see Figure 2.35). In Tanta, Cairo and Ismailia the shock was felt by few people, but it was not reported from Port Said, Suez or in Sinai.[1]

1 *Petermanns Mitteilungen*, **49** (1903), p. 190; Rudolph (1905), p. 472.

1903 November 22   **Libya**

An earthquake was felt in Tripoli, causing some concern but no damage. The shock was more violent further inland in the region of Tarhuna.[1] The shock was not recorded by the early Italian network, but some of the European stations confirm that this was a genuine event that occurred at about 09h 45m. From a single reading of the Milne instruments at Kew the surface-wave magnitude of the event appears to be about 5.5.

1 *Osservatore Romano*, 27 November 1903; *Levant Herald*, 1 December 1903; Rudolph (1905), p. 596.

1904 October 3   Arabian Sea

This large-magnitude earthquake in the Arabian Sea was strongly felt on board a steamship, which was obliged to stop its engines to avoid damage. Less-violent shocks were felt one and a half and two and a half hours later, causing great concern. The earthquake, which was not reported felt on land, was well recorded by many stations and overloaded close ones. The instrumental

solution is 12° N, 58° E and the surface-wave magnitude 7.1 (± 0.3) from 19 stations. The BAAS location at 7° N, 61° E is not substantiated.[1]

1 Breitung (1905), p. 84.

1906 December 26   Egypt

An earthquake widely felt in Lower Egypt in the afternoon was preceded by a foreshock of almost equal strength in the morning. The main shock, which lasted five to ten seconds, was strong at the lighthouses of Ashraf and Shadwan, where it caused some minor

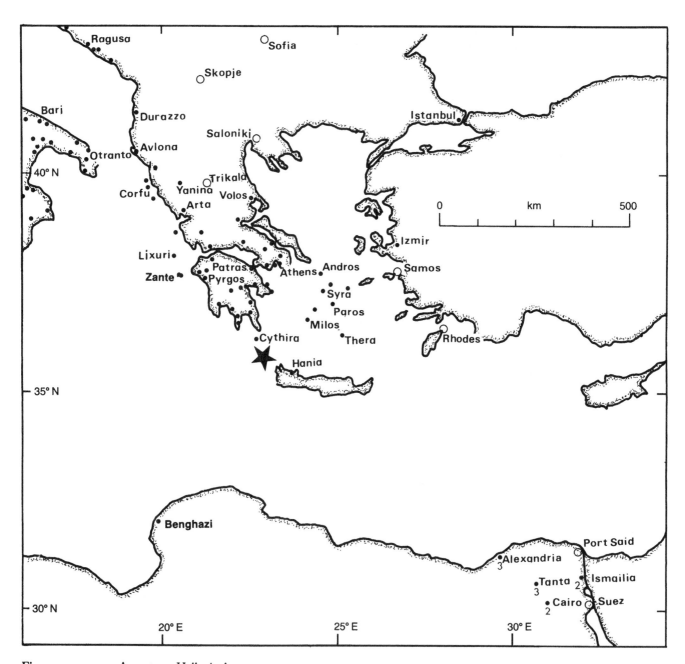

Figure 2.35. 1903 August 11, Hellenic Arc.

damage to the equipment, and at Semna, where it caused some rocks to move. It was widely felt at Asyut, Sohag and Nag Hammadi and it was perceptible as far as Aswan in the south and Cairo and Helwan in the north, but not in Alexandria or Ismailia. The main shock was not felt at Luxor, although the foreshock had allegedly caused the collapse of a ceiling in the Savoy Hotel, and a feeling of giddiness (see Figure 2.36).

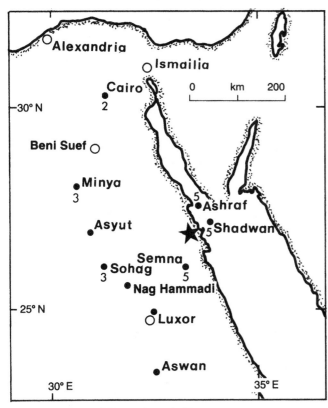

Figure 2.36. 1906 December 26, Egypt.

Both shocks were recorded at Helwan, but the seismogram of the foreshock was masked by a large South American earthquake at the same time.[1] The main shock was recorded by only three stations, insufficient for an instrumental solution, but adequate to determine a surface wave magnitude of 4.9 (± 0.2) and to confirm a rough location indicated by the macroseismic data.

1 Keeling (1906), pp. 182–3; *BAAS* (1906), p. 127; Sieberg (1932b), p. 189; Melville (1984a), p. 100.

1907 February 4   **Red Sea**

A rather severe shock was felt by the lighthouse keepers on the Brothers Island in the Red Sea. The shock lasted half a minute, and was accompanied by a low rumbling sound. The shock was not recorded by the seismograph at Helwan or any other station.[1]

1 *Cairo Scientific Journal*, 2 (1908), p. 60.

1907 July 20   **Gulf of Aden**

Two shocks of short duration were felt in Aden, causing insignificant damage.[1]

1 *Die Erdbebenwarte*, 7 (1907), p. 131.

1908 April?   **Yemen**

Several shocks were reported from various parts of the Yemen.[1] No precise dates are given, and it is likely that one or more of these events is connected with the earthquake of 25 January 1908, poorly recorded instrumentally.[2]

1 *Levant Herald*, 21 April 1908.
2 See below, Section 3.2.

1909 February 19   **Ethiopia**

An earthquake shock was felt in Djibouti at 6.0 a.m. local time.[1] Due to a misprint in a Yemeni chronicle, the earthquake is often associated with a damaging earthquake at Wainan in Yemen.[2] However, the earthquake in question is actually the Gulf of Corinth event of 30 May 1909, and the Yemeni shock should be deleted from catalogues.[3]

1 Gouin (1979), p. 148. This is one of a sequence of shocks reported in Djibouti between 1907 and 1912.
2 Ambraseys and Melville (1983); cf. *Naval Intelligence Division* (1946), p. 22, which associates Wainan with Wailan or Wa'lan. The ultimate source, al-Wasi'i, p. 224, lists the shock under Rabi' II, 1327 (beg. 20 April 1909), on the basis of the Egyptian newspaper *al-Mu'ayyad*, but 'Bilad al-Wainan' should read 'Bilad al-Yunan' (= Greece).
3 See Ambraseys and Jackson (1990), pp. 679–81.

1909 February 26   **Egypt**

A slight tremor was felt at Ras Gemsa on the Red Sea coast. The shock was sufficiently strong to dislodge fragments of plaster and to cause roof timbers to creak. The earthquake was not recorded at Helwan, nor at any other observatory.[1]

1 *Cairo Scientific Journal*, 3 (1909), p. 96.

1910 January 6   **Sinai**

A slight earthquake was felt at Nakhl in Sinai; it lasted about one second, and it was not recorded at Helwan.[1]

1 *Cairo Scientific Journal*, 4 (1910), p. 27.

1910 February 18   **Hellenic Arc**

An earthquake originating in the western part of the Hellenic Arc was felt in the Nile Delta, and was reported as slight in Alexandria.[1] Apparently the shock was not felt in Cairo or Suez.

1 *Cairo Scientific Journal*, 4 (1910), p. 51.

1910 March 6   **Red Sea**

An earthquake was felt by the lighthouse keepers on

Brothers Island in the Red Sea. The shock was recorded at Helwan.[1]

1 *Cairo Scientific Journal*, **4** (1910), p. 78.

1911 January 5   Egypt

A local shock of short duration was felt in the neighbourhood of Cairo. It was sufficiently strong to cause plaster to fall and crockery to rattle. The earthquake was felt at Helwan and also recorded there with very small amplitude.[1]

1 *Cairo Scientific Journal*, **5** (1911), p. 50; *Helwan Station Bulletin* (1911).

1911 January 26   Gulf of Suez

A series of shocks lasting 45 seconds was felt at the Ashraf lighthouse and at Tor in Sinai. The earthquake was recorded at Helwan but it was not felt at the station nor in Cairo. It is said that a shock was felt in Alexandria, but the local press does not confirm this.[1]

1 *Cairo Scientific Journal*, **5** (1911), p. 76; *Helwan Station Bulletin* (1911).

1911 May 13   Yemen

An earthquake in Yemen, followed by many small aftershocks in the district of Zabid, caused considerable damage and the collapse of a house in the town, without loss of life.[1]

1 *Sabah* (Istanbul), 15 May 1911.

1911 August 22   Egypt

This was a local earthquake in Lower Egypt centred on the west coast of the Gulf of Suez. At the monastery of St Paul it was very strong, but caused no damage, and it was severely felt at Karimat. In Cairo the shock was felt generally and it is said that in Bulaq some houses fell and walls and roofs were cracked. In a few isolated instances the shock displaced furniture, and in Cairo it is said to have disrupted the tramway service through failure of an overhead power line. In the Qasr al-Nil barracks some furniture was overturned, but most people in the city did not notice the earthquake. The shock was just perceptible at Suez, but it was not felt in Alexandria, Port Said or Minya (see Figure 2.37). The shock was followed six hours later by an aftershock, and it was recorded at Helwan, but not at any other station.[1] Shocks continued to be felt at Tor for the rest of the year.

1 *Cairo Scientific Journal*, **5** (1911), pp. 236–7; *al-Muktataf*, **39** (1911), p. 306; Arvanitakis (1911); Hume (1928); Sieberg (1929); (1932b), p. 189.

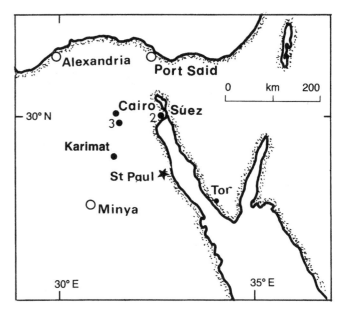

Figure 2.37. 1911 August 22, Egypt.

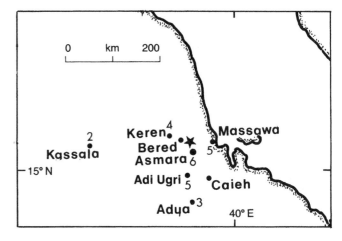

Figure 2.38. 1913 February 27, Ethiopia.

1913 February 27   Ethiopia

A damaging earthquake in the vicinity of Asmara in Ethiopia was felt throughout Eritrea and Tigre, and as far as Kassala in the Sudan (see Figure 2.38). It was followed by many aftershocks, one of which was almost as large as the main shock.[1] The data are insufficiently precise to distinguish intensities in the main shock from the cumulative effects of the series.

1 Gouin (1979), pp. 49–51; *Cairo Scientific Journal*, **7** (1913), p. 80.

1914 May 18   **Libya**

This earthquake demolished the small fort of Gheddahia. It was strongly felt and caused some damage to two other forts nearby, but was not reported from Misurata about 100 km to the north.[1] The shock was widely recorded by European stations.

1 Desio, *pers. comm.*; Lipparini (1940), p. 251.

1915 September 23   **Ethiopia**

A relatively large-magnitude earthquake in Eritrea was strongly felt in Asmara, where it caused panic and slight damage. The macroseismic details of this earthquake are not known.[1]

1 Gouin (1979), p. 51.

1920 October 1   **Egypt**

A local earthquake in the Nile Delta was felt in the region between Cairo and Faiyum. In some parts of Cairo the shock was not felt, while in others it lasted about three seconds and ruined three houses. It was perceptible at Helwan, where it was recorded, and much stronger at Faiyum where it was accompanied by noise. The seismograms from Helwan show a local shock of small magnitude of 15 seconds duration. The shock was not felt at Alexandria or Suez.[1] Sieberg places its epicentre at Faiyum where he says the shock attained an intensity of VIII. He gives an intensity of VII for Cairo and reports that the earthquake was felt in Alexandria.[2]

1 *Al-Muktataf*, **57** (1920), p. 435; *La Tunisie Française*, 4 October 1920.
2 Sieberg (1929), (1932a), p. 873, (1932b), p. 189.

1921 August 14   **Ethiopia**

An earthquake with an epicentre offshore from Massawa, followed by an aftershock of almost equal magnitude, caused considerable damage to the harbour and to the town, where a small number of people were killed. At Ras Murdur houses became uninhabitable. The shock was strongly felt as far as Asmara and Decamere. A large number of dead fish were found floating between Jizan and al-Hudaida.[1]

1 *Glasgow Herald*, 17 August 1921; Gouin (1979), pp. 108–9.

1923 December 8   **Egypt**

A local earthquake in Lower Egypt was felt by few people in Cairo, and more strongly at Beni Suef. The shock was felt and recorded at Helwan but it was not reported from Suez and Alexandria. A tremor reported about the same time from Faiyum was probably associated with this earthquake. A number of light shocks felt at Beni Suef during the early part of 1924 were perhaps aftershocks of this event.[1]

1 Helwan Station, *Monthly Bulletins* for 1924; Sieberg (1932b), p. 189, lists this event under 1 August.

1925 May 20   **Egypt**

A local shock west of Suez was sufficiently strong to wake everyone in the town (see Figure 2.39). The shock was not felt on board ships but it was widely perceptible in Cairo and at Helwan, where it was recorded. There are negative reports from Port Said, Alexandria and Beni Suef.[1]

1 *ISS* (1925), p. 83; *Bull. Volcanologique*, **7** (1926), p. 225; Sieberg (1932b), p. 189, under 9 May.

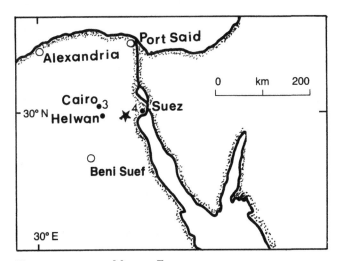

Figure 2.39. 1925 May 20, Egypt.

1926 June 26   **Hellenic Arc**

A large earthquake in the eastern part of the Hellenic Arc, most probably of intermediate depth, was felt throughout Egypt, Israel and part of Syria, in southwest Turkey, southwest Greece and Crete, and as far as Sicily and southern Italy (see Figure 2.40). The shock was perceptible in north Libya, parts of the Adriatic coast and in many parts of Italy. Along the coast of Egypt, from Alexandria to Port Said, a series of three shocks lasting for about three minutes was felt by everyone, in places causing some panic. In Alexandria five local adobe houses collapsed and about 100 suffered some damage, killing two people. It is said that the shocks caused some cracking and settling of the levees along the coast, but there was no damage to any of the engineered structures in the Nile Delta, where however, telephone communications were interrupted for some time. In the densely populated area of the Delta a few

Figure 2.40. 1926 June 26, Hellenic Arc.

local houses collapsed and some were damaged with a few casualties. In the Damanhur district, four houses fell down killing three people; another four collapsed in Minufi and a few in Tanta, killing one person and injuring five. There was no damage to better-built houses, to the railways, or to harbour and canal works. In Cairo the earthquake lasted with intermissions for about two minutes, causing considerable concern but no damage to better-built houses and public buildings. In the poorer districts of the city, six adobe houses were ruined and about 450 suffered various degrees of damage, without casualties. Another two houses fell at nearby Giza and two in the district of Faiyum, killing nine people in all. At Helwan, shortly after the earthquake, a new hot spring appeared near the town's thermal springs. The shock was reported from a number of places along the Nile as far south as Luxor where, however, the earthquake was barely perceptible. The shock was not reported from Aswan or from a number of places between Beni Suef and Qena.[1]

1 *The Times* (London), 2,5 July, 20 September 1926; *ISS* (1926), pp. 70–90; Critikos (1928); Sieberg (1932b), pp. 163–73.

1927 July 11   Dead Sea

A damaging earthquake in the Jordan Valley was strongly felt in the Nile Delta. It was slightly felt at Alexandria, Suez, Ismailia and Port Said and somewhat more strongly at Tanta, where it lasted about 20 seconds. At these places it caused no damage, and in some it was not recognised as an earthquake. At Cairo the shock was fairly severe. It was felt by many people and caused hanging objects to swing and several clocks hanging on walls to stop. The earthquake was not reported south of Beni Suef but apparently it was perceptible at Karnak near Luxor, and at Quseir on the Red Sea.[1] The earthquake was felt as far north as Tripoli in Lebanon but it was not reported from Cyprus or from northern Syria or Turkey (see Figure 2.41).[2]

1 *Egyptian Gazette*, 12–18 July 1927; *al-Muktataf*, 71 (1927), pp. 237–8.
2 Berloty (1927); Shalem (1928); Sieberg (1932b), p. 194.

1929 January 22   Ethiopia

An earthquake offshore from Djibouti, lasting six to seven seconds caused cracks in the walls of most houses in the town and the collapse of one building. Some slight damage was caused at Tadjoura. The earthquake, which was followed by many strong aftershocks till the end of September, was also felt at Assab and allegedly it was perceptible in Addis Ababa, about 560 km away (see Figure 2.42).[1]

1 Gouin (1979), pp. 150–1; *De Bilt Station Bulletin* (1929).

1930 October 24   Ethiopia

A series of earthquakes preceded by a swarm caused some damage in the region of Djibouti and Tadjoura. The largest shock of 24 October (Ms 5.8) was followed by three shocks of Ms 5.6. In Djibouti, walls of houses that had been left intact by the earthquakes of 1929 were heavily cracked, and in Tadjoura two mosques collapsed. The shocks were felt at Zaila' (Zeila), Sabieh, Diredawa and Awash, but it is not clear whether these shocks originated from the same epicentre. The series ended by the end of October. The earthquakes were not reported from Aden, Dikhil or Addis Ababa (see Figure 2.42).[1]

1 Gouin (1979), pp. 82–5; 151–2.

1931 February 18   Red Sea

Four ships steaming near Zubair Island in the south Red Sea, about 5 km from each other, experienced a severe concussion. SS *Moldavia* was badly shaken; the shock was principally felt amidships, the effect being very slight forward and only moderate aft; no disturbance of the sea was noticed. The shock was recorded by the ship's barograph. Also, SS *Rietfontein* experienced a severe shock as if the propeller had come out of the water. SS *Melbourne* was also badly shaken, the concussion felt from forward to midships, but not in the engine room. A distinct shock was felt on SS *Glenluce* which reported no disturbance or discolouring of the sea. Finally, SS *Nankin*, 25 km away, felt nothing. The shock was not recorded or felt on land.[1]

1 *The Marine Observer*, 9 (1932), p. 33.

1938 May 12   Sudan

This strong earthquake has a teleseismic location in southeast Sudan, near the coast of the Red Sea. The shock was strongly felt at Suakin where it caused some panic. Many houses in the Shata district were cracked, including the railway station. The shock was very widely felt, from nearby railway stations to Port Sudan and Atbara, as far as Khartoum, Kassala and in Nubia, the shocks lasting about 10 seconds.[1]

1 East Africa Meteorological Dept. Files, SA/ii (Khartoum); Ambraseys and Adams (1986a).

1939 January 20   Libya

This earthquake and the activity of the following three days was reported from Tripoli, al-Khums, Misurata, Buwayrat and Sirt, towns along the coast of Libya. The shock of 23 January was also felt at Beni Ulid, allegedly causing some unspecified damage at Buwayrat, but details are lacking.[1]

1 *Revue pour l'Étude des Calamités* (1941), p. 72.

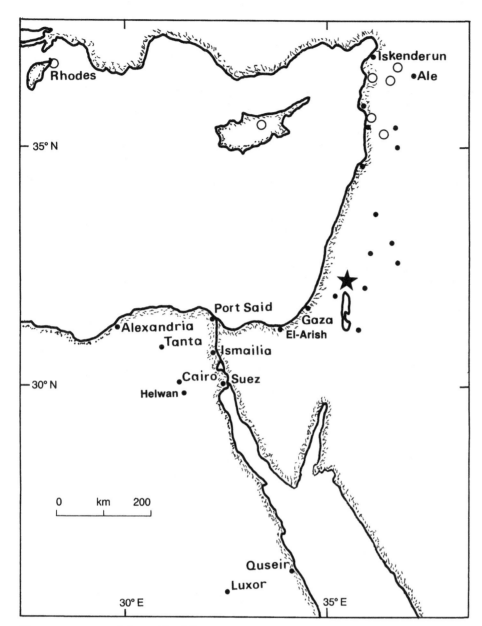

Figure 2.41. 1927 July 11, Dead Sea.

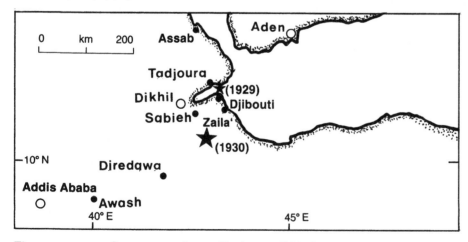

Figure 2.42. 1929 January 22 and 1930 October 24, Ethiopia.

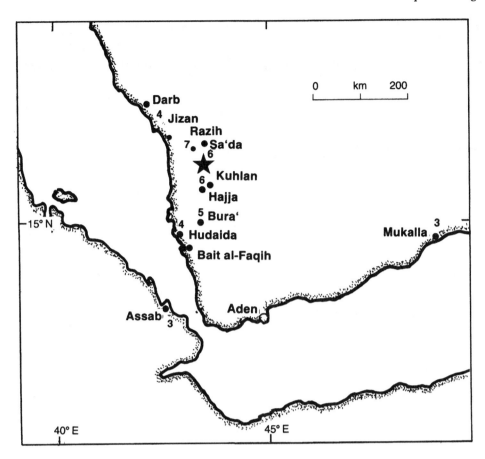

Figure 2.43. 1941 January 11, Yemen.

1941 January 11   Yemen

Following several slight foreshocks on 9 January, reported from al-Hudaida, a destructive earthquake, occurring around noon local time, affected a large area of the Yemen. The main shock was felt from al-Darb in Saudi Arabia to al-Mukalla in Hadramaut, and it was reported from Assab in Ethiopia, but not from Aden (see Figure 2.43). The region worst affected lies west of Sa'da, in the area round Razih, where a number of villages were destroyed, with small loss of life. Landslides blocked the road at the head of the Razih Valley. To the north, between Jizan and Razih, at Arida near Abu 'Arish in Saudi Arabia, the shock caused rockfalls and the drying up of spring water. In Sa'da and Rahban, and especially Wadi 'Azar, many houses were destroyed: old and new houses cracked in Sa'da, but there was no loss of life. The inhabitants abandoned their homes and camped in tents. In the district of Majz, many Jews were killed by the collapse of the roof of their synagogue. Damage also extended south to Kuhlan and Hajja, where a few houses were demolished.

Aftershocks continued daily up to the second week of March, and added to the damage, though it is not easy to distinguish the effects of individual events. Two strong aftershocks occurred at 09h 18m on 4 February and at 19h 03m on 23 February, of which the former had a magnitude of 5.2, and was reported from Haidan, Khaulan, al-Zahir and Wadi al-'Abidin near Sa'da, where landslides occurred. Damage was also reported from Abu 'Arish and Sabiya to the northwest and Harad to the southwest. The aftershock of 23 February was reported from al-Hudaida, Bait al-Faqih, al-Sa'id and Bura', and is said to have been damaging. It is reported that altogether 1200 people were killed and 200 injured in these earthquakes, which totally destroyed 300 houses, ruined 400 beyond repair and slightly damaged about 1000.

The relocation of the main shock and of the aftershock of 4 February gives 16.75° N, 43.33° E and 16.90° N, 43.90° E respectively. For neither solution are there any observations from stations between azimuth 90° and 270°, which probably has the effect of dragging the epicentres northwards by possibly as much as half a degree, although these positions are quite consistent with the rather patchy macroseismic data available.[1]

1 BRGM (1979); Ambraseys and Melville (1983); al-Maneefi, *pers. comm.*, utilising additional British archives, the Hodeida Political Records in I.O.R., R/20/C/2530, from 14 January to 3 April 1941.

1945 October 28   **Ethiopia**

A strong earthquake in the region of Djibouti was felt throughout Somalia. It is not known whether it caused any damage.[1]

1 Gouin (1979), pp. 154–5.

1951 January 30   **Mediterranean Sea**

This earthquake had an offshore epicentre north of the Nile Delta in the Mediterranean Sea (see Figure 2.44). It was felt rather strongly at Port Said, wakening sleepers on land and on board ships, and it was perceptible in Cairo and in the Negev. The shock was felt throughout Israel, in some places causing people to run outdoors, but caused no damage. It was barely perceptible in south Lebanon and was not reported from Alexandria or Jordan.[1] Existing teleseismic solutions suggest a focus deeper than normal.

1 Kallner-Amiran (1951), p. 239.

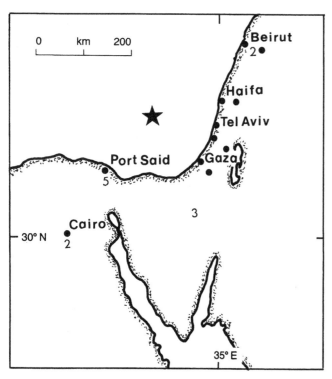

Figure 2.44. 1951 January 30, Mediterranean Sea.

1954 September 13   **Dead Sea**

This earthquake, with an epicentre just south of the Dead Sea, was felt by few people in Cairo and it was perceptible at Suez, but not in Alexandria or Port Said. It was felt as far north as Haifa and in Lebanon, but not in Beirut. Shalem, on the basis of macroseismic evidence, places the epicentre of this earthquake east of the Dead Sea in the Wadi Sirham, near 31.2° N, 37.2° E. However, we could find no macroseismic evidence to support Shalem's location, which is in a totally uninhabited region.[1]

1 *Egyptian Gazette*, 15–17 September 1954; Shalem (1955); Striem (1986).

1955 September 12   **Mediterranean Sea**

An earthquake with an epicentre offshore, northwest of Alexandria, caused more excitement than damage in Egypt (see Figure 2.45). Maximum effects were reported from around Lakes Idku and Maryut. In Alexandria many people left their houses but damage to buildings was negligible. Only seven adobe houses were ruined in the port area and eight schoolgirls, whose crowding of a staircase caused the balustrade to collapse, were injured. A few people in the old part of the city were injured by falling debris, but better-built houses and public buildings suffered no damage. Around Lake Idku about 300 adobe houses were badly damaged, including two schools at Abu Hummus and a police station at Damanhur, where some houses were totally ruined, but with no casualties. At Rosetta the shock is said to have lasted 40 seconds, and caused the collapse of a few free-standing walls, and at Idku water pipes were damaged. At Disuq there was no damage except to the high-voltage overhead lines from which a cable fell. At Mahmudiyya quite a few old houses were badly fissured, and at Kafr al-Shaikh a few collapsed causing injuries. At Kafr al-Dauwar some houses were slightly damaged, and at Tanta a house collapsed killing two people. A few people were injured by falling debris at Kaum Hamada and a few walls were badly cracked at Quweisna. Minor damage extended throughout the districts of Gharbiyya, Minufiyya and Buhaira as far as Mansura, exclusively to local types of construction. In Cairo a few old houses were badly cracked, and 24 houses and 15 school buildings collapsed in part in the districts of Hanafiyya, Hilmiyya and Imamain. The shock had no effect on public buildings. South of Cairo there was no damage. In all 18 people were killed, 89 injured, 40 houses collapsed completely and about 420 were ruined. The shock was felt as far south as Asyut, and it was perceptible at Luxor. It was also felt along the Red Sea coast as far south as Safaga, and on the Mediterranean coast up to Matruh. It was perceptible at Sur (Tyre) and Damascus, and at Athens.[1]

1 *Egyptian Mail*, 13–15 September 1955; *al-Ahram* (Cairo), 18 April 1964, 27 October 1968, 1 April 1969. See also Maamoun (1979), Striem (1986).

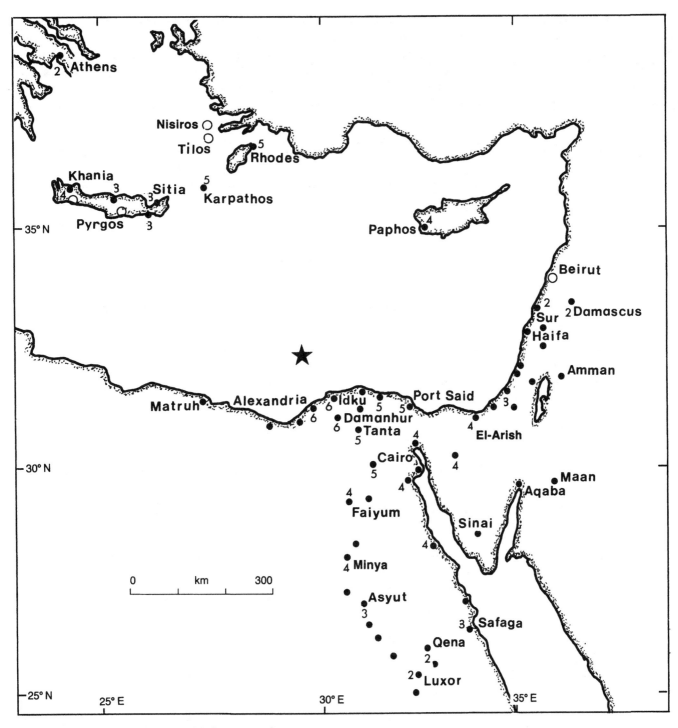

Figure 2.45. 1955 September 12, Mediterranean Sea.

1955 October 17   **Yemen**

A relatively small earthquake in the Yemen is known from its teleseismic location, which is rather poor, placing the event in the sparsely populated region northeast of Sa'da.[1] The shock is reported to have been strong in Sa'da, where buildings were severely shaken and doors and windows burst open, but without damage.[2]

Other reports, which we have not substantiated, state that the shock was severe around al-Safra, to the south, and caused damage in Sa'da.[3]

1 Ambraseys and Melville (1983).
2 *Al-Iman* (Sa'da), 15 November 1955.
3 BRGM (1979); Arya, Srivastava and Gupta (1985).

1955 November 12   Egypt

A strong earthquake southeast of Quseir on the Red Sea coast was widely felt along the Nile Valley from Aswan to Asyut, and was perceptible in Cairo. It caused no damage and was poorly reported by the press.[1]

1 *Egyptian Gazette*, 16 November 1955.

1959 August 16   Yemen

A small earthquake is located teleseismically near Bait al-Faqih. Secondary sources say the shock was strongly felt in the region between al-Mukha, Ta'izz and Ibb, possibly causing damage.[1] Newspaper reports mention strong shocks that caused panic in Ibb on 10 May 1959, and further activity was reported in the region of Hubaish, northwest of Ibb, during November and December the following year. Extensional groundcracks developed in the villages of al-Salaf and al-Qabulain, ultimately over a distance of more than two miles. Many wells disappeared and others sprang up. The shocks caused damage to cultivated land and many dwellings, including multi-storey buildings.[2] All these shocks were below the threshold of instrumental detection at this period.

1 St Ours (1976), Ambraseys and Melville (1983), Arya *et al.* (1985).
2 *Al-Nasr* (Ta'izz), 21 May 1959; *Saba* (Ta'izz), 8 December 1960. A further shock occurred at the end of Ramadan 1380/18 March 1961, triggering landslides from Jabal Qishar, southeast of Ibb, *Saba* (Ta'izz), 13 April 1961.

1961 March 11   Ethiopia

An earthquake in the Gulf of Tadjoura caused some minor damage in Djibouti and was strongly felt around the Gulf of Tadjoura. The main shock was felt as far as Dikhil and in Ethiopia, and was followed by a large number of aftershocks, which continued till the end of the month.[1]

1 Gouin (1979), p. 161.

1961 June 1   Ethiopia

This destructive shock in Ethiopia was followed by many strong aftershocks. The village of Majete was completely destroyed and in the town of Kara Kore most houses collapsed (see Figure 2.46). Within a radius of about 50 km from this epicentre, large-scale rockfalls, slides and slumping of the ground caused serious damage to the Addis Ababa–Asmara highway along which many bridges and culverts had to be rebuilt. In Dessie and Combolcia there was no destruction but elsewhere sporadic damage extended as far as Debre Berhan and Ankober. Much of this damage was cumulative, resulting from the aftershocks, of which 3500 of M_L 3.5 or more were recorded up to September. A piedmont scarp that could be followed over 15 km, in places

showing throws of up to 2 m, was formed in unconsolidated material along the escarpment of the Borkena graben. The main shock was strong in Addis Ababa, and it was felt in Assab and Djibouti, but it was not reported from Aden.[1]

1 *Revue pour l'Étude des Calamités* (1964), p. 50; Gouin (1979), pp. 57–63.

1962 August 28   Hellenic Arc

An intermediate depth earthquake with an epicentre in southern Greece was felt as far as Cairo, 1200 km away, as well as in south Italy, particularly in the region of Bari where it caused minor damage. The earthquake was particularly strong in Alexandria and Raud al-Farag where some houses toppled.[1]

1 *Al-Ahram*, 25 November 1962, 18 July 1964 (*sic.*); *Revue pour l'Étude des Calamités* (1964), p. 23.

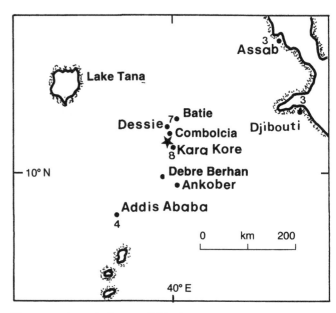

Figure 2.46. 1961 June 1, Ethiopia.

1963 February 21   Libya

The earthquake affected an area in Cyrenaica situated on an alluvial plain lying between two northeasterly trending, fault-controlled escarpments, in which Barce (Al Marj) is situated. Barce (population *c.* 13 000) and farming settlements in a strip 20 km long were badly damaged and many houses collapsed, killing about 300 and injuring 375 people. In all, 12 000 people were made homeless; maximum damage occurred in Maddalena, about 12 km northeast of Barce, where all farmhouses were demolished. The shock was not particularly large and its maximum effects were confined within an area of 6 km radius. It was perceptible in Benghazi, about 100 km to the southwest, and much of the damage was due to the very poor quality of construction. Small magnitude aftershocks continued for two days.[1]

1 Brennan (1963), Gordon and Engdahl (1963), Minami (1965), Campbell (1968), Ambraseys (1984).

1964 July 17 — Hellenic Arc

A relatively small-magnitude earthquake of intermediate focal depth in south Greece was widely felt in Alexandria, where it lasted about 40 seconds. It was also felt in Cairo by people living on upper floors, and was reported to have been presaged by animals in the Cairo Zoo.[1]

1 *Al-Ahram*, 18 July 1964.

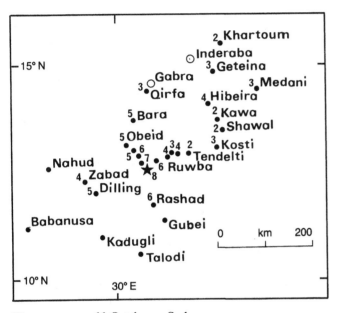

Figure 2.47. 1966 October 9, Sudan.

1966 October 9 — Sudan

This strong earthquake and its long sequence of after-shocks occurred in central Kordofan in the Sudan, where no shocks had been observed in the past (see Figure 2.47). It caused considerable damage within a radius of about 25 km from Jebel Dumbeir, mainly to poor-quality houses, all-straw huts or adobe walls and thatched roofs. The former were severely shaken only, but the latter were badly damaged with casualties but no loss of life. The shock caused the collapse of steep banks and rockfalls as well as settlements in superficial soil deposits. The earthquake was associated with 6 km of strike-slip faulting, trending N. 20° E, of about 4 cm left-lateral displacement. Felt reports were received from as far as Khartoum, Wad Ashana and from Zalin-

gei, El Fasher and Ed Daein, about 700 km west of the epicentre. Many aftershocks were felt, 20 in the first 24 hours and about 15 per day after that for two weeks. The largest aftershock had a magnitude M_s 4.9 and caused additional damage.[1]

1 Qureshi and Sadig (1967); Qureshi (1968); Ambraseys and Adams (1986a).

1967 March 13 — Red Sea

From 10 March to 11 September, an outburst of minor seismic activity was recorded from the central Red Sea. The largest shock, of magnitude 5.0, occurred on 13 March, all the activity concentrating within a zone between 19.6° N to 20.40° N and 38.50° E to 38.90° E in the axial trough with a general trend N.N.W.–S.S.E. The main shock was not felt at the ports of Port Sudan and Trikitat, and no reports were found suggesting that the earthquakes were felt at sea.[1]

1 *Al-Ahram*, 15 March–10 August 1967; Fairhead (1968).

1969 March 29 — Ethiopia

A destructive earthquake in Ethiopia, followed by a number of equally strong aftershocks, completely destroyed Sardo, the only large village in the Danakil desert, killing 24 of its 420 inhabitants and injuring 167. Heavy damage was confined within a radius of about 25 km of Sardo, where the few masonry and adobe houses were heavily damaged or destroyed, while the nomads' timber-framed huts suffered relatively little damage. In all, 40 people were killed and 160 were injured by the shocks of 29 March and 5 April which caused severe damage to road embankments and bridges along the road from Assab to Dessie in the vicinity of Sardo. These shocks triggered a few rockfalls from steep slopes up to 30 km from Sardo and they were associated with a discontinuous and incompletely mapped fault break that passes 8 km east of Sardo. This feature is about 15 km long and strikes N. 335° E, showing left-lateral displacements of about 60 cm with a throw to the northeast of 75 cm. The earthquake was strongly felt at Assab and Combolcia and it was felt by a few people at the port of Aden (see Figure 2.48). Slight shocks reported from Djibouti and al-Mukha may belong to the same event.[1]

1 *Al-Ahram*, 2–4 April 1969; Gouin (1979), pp. 128–41.

1969 March 31 — Red Sea

Preceded and followed by many strong shocks, an earthquake with an epicentre offshore, in the Gulf of Suez, affected the area of the islands of Shadwan, Tawila and Gubal (see Figure 2.49). At Shadwan the shock caused numerous rockfalls and it was strong enough to throw

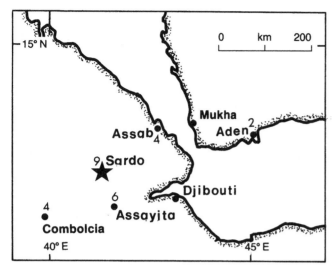

Figure 2.48. 1969 March 29, Ethiopia.

people to the ground but because of the desolate nature of the region there was no damage to property. The Shaker lighthouse at the south end of the island suffered mortar cracks near its base and mercury in a container at the top of the tower spilled out. A zone of ground deformation, probably of non-tectonic origin, was noticed extending in a north–south direction for about 1 km showing a few centimetres of right-lateral displacement. About 10 km west of this zone, in the sea, a coral reef was raised permanently above sea-level. Dead fish and some agitation of the sea were noticed after the main shock.

At Hurghada the shock caused some damage, including cracks in the brick walls of a reinforced concrete power plant, and plaster cracks in two hotels. Similar effects were reported from Gemsa. At Tor a few dilapidated adobe houses were damaged, and at the monastery of St Catherine's plaster in a building cracked and partly fell from a ceiling. At Sharm al-Shaikh the earthquake produced cracks along mortar joints in walls, articles fell from shelves and furniture was displaced. People ran outdoors and had difficulty standing. Light wooden houses remained intact, and overall damage to property was negligible. Rockfalls and talus slides triggered by the shock raised clouds of dust. There is no evidence that the shock had any adverse effect on the Balaim and other oilfields on the western and eastern sides of the Gulf of Suez.

Outside the epicentral region damage was sporadic, the shock affecting primarily a few vulnerable structures, and otherwise negligible. The earthquake was strongly

felt at the ports of Safaga and Quseir as well as further inland at Qena. At Sohag one house collapsed, and at Asyut one more and a mosque were ruined, with no casualties. At Beni Suef a few houses were slightly damaged, and a wall and a staircase of a school collapsed. There was no damage at Faiyum. In Cairo the shock caused considerable concern and some people rushed into the streets. Here one house collapsed and about ten were damaged, injuring five people. It is said that the shock was presaged by animals in the zoo. The earthquake was perceptible in the districts of Gharbiyya, Sharqiyya and Minufiyya, felt by few people at Tanta, Damanhur and Alexandria, but there are no felt reports from Kafr al-Shaikh, al-Gira, Damietta and Daqahliyya, in the northeast of the Delta. The shock was felt throughout Israel with an erratic distribution of intensity, and as far as Aswan, al-Wajh and Natanya. In all two people were killed, 15 injured, about 100 houses and 7 mosques were ruined. Aftershocks continued well into 1970.[1]

1 *Al-Ahram*, 1–7 April, 12 June 1969; *Egyptian Gazette*, 1–2 April 1969; Arieh (1969); Maamoun and El-Khashab (1978); Maamoun *et al.* (1981); Melville (1984a).

1972 January 12   Egypt
A small earthquake with an epicentre in the Gulf of Suez near Shadwan Island was strongly felt in the south Sinai region, and in the oil producing areas. At Qena some people fled their homes, and in Cairo it was felt by a few people.[1]

1 *Al-Ahram*, 13–14 January 1972.

1973 March 28   Ethiopia
Four earthquakes of almost the same magnitude in the Gulf of Tadjoura, followed by many aftershocks, caused considerable cumulative damage at Djibouti, and to a lesser extent at Tadjoura, where a number of buildings and houses were evacuated. The shocks were widely felt and they were perceptible in Diredawa.[1]

1 *Egyptian Gazette*, 4 April 1973; Gouin (1979), pp. 168–74.

1974 April 29   Egypt
An earthquake with an epicentre northeast of Cairo near Zagazig was widely felt in Lower Egypt and southwest Israel (see Figure 2.50). The shock was particularly strong in the district of Sharqiyya, as well as in Ismailia and Cairo where a few houses suffered plaster cracks, causing some panic. The shock lasted about 10 seconds and it was reported from along the Suez Canal, where it caused some concern. It was felt in Alexandria, Beersheba and Minya. This is believed to be the first instrumentally located earthquake in this part of

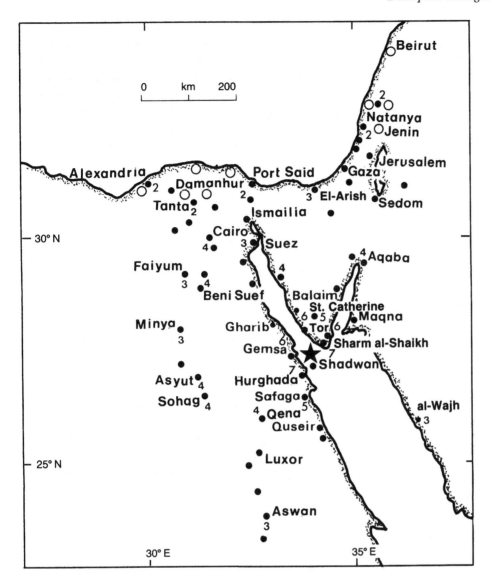

Figure 2.49. 1969 March 31, Red Sea.

Egypt. It was followed by a few weak aftershocks.[1]

1 *Al-Ahram*, 30 April, 1–4 May 1974.

1975 October   Yemen

According to local information, an earthquake in the Azal al-Radma region in the Ibb province in the Yemen caused damage to the village of Bait Badr, where 38 houses had to be evacuated. In addition in the village of Khaulan another 12 houses were rendered uninhabitable.[1] The earthquake was not recorded teleseismically.

1 St Ours (1976).

1975 December 14   Red Sea

A double shock off the Yemen coast in the Red Sea was strongly felt at al-Hudaida, where it caused some very minor cracking of walls. The shocks were felt along the coast at Salif, al-Luhaiya and Ahmadi; it was felt on board a ship and caused some concern.[1]

1 St Ours (1976).

1977 December 28   Red Sea

This rather large magnitude earthquake occurred in the Red Sea northeast of the Dahlak Islands. We could find no felt reports from the nearest ports of Massawa and Jizan, 130 km and 240 km from epicentre, respectively. There were no significant aftershocks.[1]

1 Gouin (1979), p. 146.

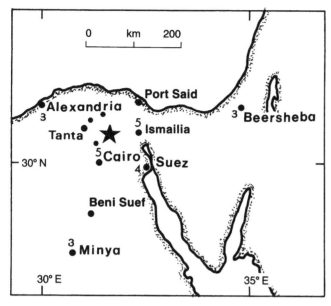

Figure 2.50. 1974 April 29, Egypt.

1978 December 9   Egypt

This earthquake occurred in the desert area of Gilf Kebir, about 650 km west of the Nile Valley in Egypt, and is the first instrumentally located event in this region. It had a surface-wave magnitude of 4.8 and a computed strike-slip motion. The shock was not reported felt. A report of a shock felt by few people in Khartoum, 1100 km away, either on the 8 or 9 December at 10h 30m local time, may belong to a separate local event.[1]

1 Maamoun, Meghahed and Allan (1980); Ambraseys and Adams (1986a).

1981 November 14   Egypt

This earthquake occurred in the Nubian Desert in the immediate vicinity of Lake Nasser at Aswan (see Figure 2.51). Within an area of about 40 km radius from the epicentre a few rubble masonry houses were badly damaged, and a few more were heavily cracked, but with no casualties. Tension features and ground cracks of doubtful tectonic origin were observed on the west side of the Lake in the aeolian sands of the desert, running in an E–W direction for more than 10 km (see Figure 2.52). This zone of cracking is close and parallel to the Kalabsha Fault, but the mechanism of the event determined teleseismically shows strike-slip faulting. Rockfalls and ground cracks were also found in the higher ground of the eastern bank of the Nile. The High Dam at Aswan itself and its appurtenant structures and buildings, 35 km from the epicentre, suffered no

damage, but the shock caused considerable panic. Minor damage was observed in a few places as far as Idfu. The shock was reported felt from as far north as Asyut, from the Red Sea ports of Gemsa to Quseir and from the railway station at Kareima. It was perceptible by some people in an upper storey in Khartoum but it was not felt in Cairo. The main shock was preceded by two foreshocks and followed by many aftershocks, two of the largest in the series (M_s 4.0) occurring in close succession on 2 January 1982.[1]

1 *Al-Ahram*, 16–20 November 1981; Adams (1983a); Kebeasy, Maamoun and Ibrahim (1981).

1982 December 13   Yemen

This was a shallow, multiple earthquake that occurred in a densely inhabited region about 70 km south of San'a in the Yemen (see Figure 2.53). Damage due to the main shock and its aftershocks extended in a 10 km wide zone between Ma'bar and Dhamar. Within this zone local types of houses, some of them more than two storeys high, built of rubble masonry or adobe, suffered most, their collapse causing the death of about 2500 people and the injury of about 1500 (see Figures 2.54–2.56). Damage was also serious to houses built adjacent to steep slopes; rockfalls and slides added to the destruction of villages built below cliffs. Loss of life was more serious in old village centres with high density habitation, than in modern centres where better-built houses and engineering structures suffered relatively little. The main shock and its aftershocks were associated with tension cracking of the ground, about 1 cm wide, along a zone 15 km long by 10 km wide, trending 350°, presumably the result of dip-slip along normal faults responsible for the earthquake. Damage and intensity of shaking decreased very rapidly away from the epicentral area, particularly to the north and south. The shock was perceptible as far as Jizan, Najran and Ta'izz, an average distance of 230 km, but it was not reported from Aden or from Yemeni and Ethiopian ports on the Red Sea. At al-Hudaida the shock was not strong and the port area and its facilities were unaffected.[1]

1 Despeyroux and Rouhban (1983); Landry (1983); Kopp (1983); Arya *et al.* (1985); Choy and Kind (1987); Langer, Bollinger and Merghelani (1987); Plafker *et al.* (1987); Witkam (1989).

1983 June 12   Gulf of Suez

A relatively small magnitude earthquake, located teleseismically in the Gulf of Suez, was felt in the region between Eilat and Beersheba, south Jordan, as well as in Cairo.[1]

1 International Seismological Centre (ISC), *Bulletin* (1983).

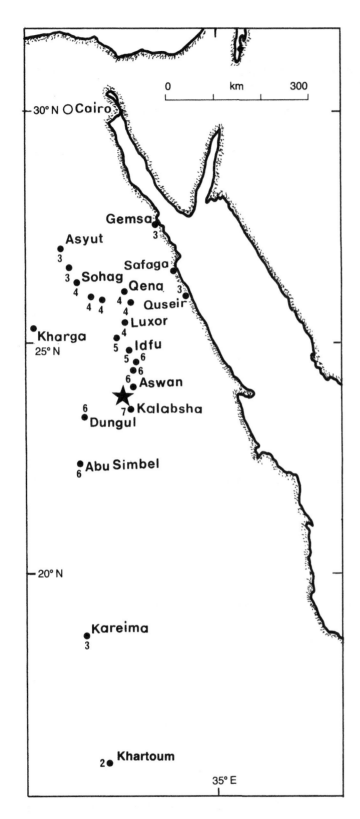

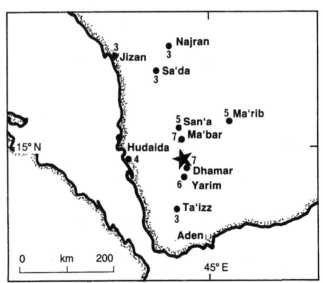

Figure 2.52. Ground cracking in area of the Kalabsha Fault, associated with the earthquake of 1981 November 14.

Figure 2.51. 1981 November 14, Egypt.

Figure 2.53. 1982 December 13, Yemen.

Figure 2.54. 1982 December 13, total destruction of traditional buildings in the Dhamar area (photo: Adnan Nowruzi).

1991 November 22   **Yemen**

An earthquake at around 3.40 a.m. local time caused moderate damage in the districts of al-'Udain and Hazm al-'Udain, about 30 km west of Ibb. It was a small event of surface-wave magnitude 4.5 and affected a mountainous region where it triggered rockfalls and slides from steep slopes. Out of 17 000 houses in these two districts, 70% were damaged beyond repair and a small number collapsed completely. The worst damage was concentrated in the sub-districts of Jabal Bahri and Bani Zuhair; 17 houses were totally destroyed in Jabal Bahri, with the death of 11 inhabitants and more than 30 wounded. In

Figure 2.55. 1982 December 13, damage to traditional buildings (photo: Adnan Nowruzi).

Figure 2.56. 1982 December 13, damage to traditional buildings (photo: Adnan Nowruzi).

97

Figure 2.57. Mosque of Sultan al-Ghauri, damaged in the earthquake of 12 October 1992; the walls were already badly cracked before the earthquake (EERI, 1992) (photo: M. Sobaih).

Bani Zuhair, more than 37 houses suffered heavy damage, mainly the older constructions. In Hazm al-'Udain, more than 50 houses were damaged, and five people injured.[1]

The very few reinforced concrete structures in the epicentral area suffered no structural damage, indicating the high vulnerability of the traditional buildings, especially those constructed on slopes and without foundations. The earthquake was preceded and followed by many small shocks, culminating in a larger earthquake on 19 May 1992.

1 *Al-Wahdawi* (San'a), 25 November 1991; *al-Sahwa* (San'a), 28 November 1991; al-Maneefi, *pers. comm.*

1992 October 12   Egypt

A relatively small-magnitude earthquake, with an epicentre about 10 km south of Old Cairo, caused a disproportionate amount of damage (estimated at £200 million), and the loss of many lives. The shock was strongly felt and caused sporadic damage and loss of

life in the Delta, around Zagazig. Damage extended throughout the Faiyum and as far south as Beni Suef and Minya. The area mostly affected was Cairo (pop. 14 million), in particular Old Cairo, Bulaq and the region to the south, along the west bank of the Nile to Gerza (Jirza) and El Rauda. In all, 350 buildings collapsed completely and 9000 were irreparably damaged, killing 545 people and injuring 6512. Many of the casualties in Cairo were victims of panic-stricken stampedes of people rushing from buildings. About 350 schools and 216 mosques were ruined, and about 50 000 people were made homeless.

Figure 2.58. Damage to the mosque of al-Tashtushi in the Cairo earthquake of 12 October 1992 (photo: M. Sobaih).

Figure 2.59. Damage to al-Fakahani mosque, Cairo (photo: M. Sobaih).

Much of the destruction was due to the high vulnerability of the ageing urban and rural building stock in the region, and to the low standards of construction of modern houses. Properly designed and built engineering structures, high-rise buildings, bridges and dams suffered very little or no damage. Terminal walls at Cairo airport cracked and windows were smashed, but flights were unaffected. Telephone and power lines were cut and water mains were wrecked in many regions of Egypt.

Many historical monuments suffered different degrees of damage, but none collapsed completely. Thus the upper part of one of the minarets of al-Azhar mosque collapsed and the shaft of the minaret structure was damaged. The arch and abutments of the Qausun Gate at al-Megherbelin were badly cracked and the wall under the dome of the Jani Bek mosque in the same district collapsed. Here also part of the minaret of the mosque of al-Hadi al-Yusufi fell, and the wall of the structure was fissured. The shock caused the separation of the minaret from the structure of the mosque of Sultan al-Ghauri, the walls of which were badly cracked (Figure 2.57). The walls of the Qayitbey mosque were slightly damaged and the first floor of the mosque of the Sabil of Muhammad 'Ali collapsed. The roofs and part of the walls of the mosques of al-Tashtushi (Figure 2.58) and al-Fakahani (Figure 2.59) collapsed. The outer walls and internal arches of the mosque of 'Amr b. al-'As in Fustat were badly shattered. Some of the walls of the 'Abdin palace were also badly damaged, as were various buildings in the citadel. Ornate finials fell from the upper parts of many minarets, causing secondary damage to adjacent houses. Of the pre-Islamic monuments, a large block was dislodged from the Great Pyramid at Giza and rolled to the ground, while some other monuments in Saqqara suffered minor cracking.

At a few places near the Nile, at Manshiyat Fadil and Dahshur, the shock caused some localised liquefaction and triggered small rockfalls from Mt Muqattam.

The earthquake was felt in Alexandria, Port Said and

southern Israel, and as far south as Asyut. It was followed by many small aftershocks, some of which caused additional damage and loss of life.[1]

1 *The Times* (London), 13–16 October 1992; *Arab News* (Jidda), 13, 15 October 1992 and Arabic press reports in *al-Ahram* and *al-Sharq al-awsat*. We are grateful to Ahmed El Ghazouli for information derived from a field visit to Cairo. The only technical report currently available is EERI (1992), which provides a preliminary isoseismal map. See also the recent publication by Degg (1993).

2.2 Catalogue of earthquakes (184 BC–AD 1899)

Table 2.1 lists the earthquakes up to the end of the nineteenth century that were discussed in the previous section. These events are plotted on Figure 4.5.

Dates are given according to the Julian calendar (Old Style) up to AD 1582 and thereafter in the Gregorian (New Style) unless indicated otherwise. Uncertain or doubtful dates are marked with a '?'.

Times are local; where precise times are not known, the following abbreviations are used: 0000 (signifies between midnight and 6.00 a.m.), 0600 (6.00 a.m. to midday), 1200 (midday to 6.00 p.m.) and 1800 (6.00 p.m. to midnight).

Epicentres are macroseismic and their accuracy depends on the availability of observational data. In most cases epicentral estimates are very approximate, and they refer to the centre of the area most heavily affected. Locations are expressed in degrees North and East.

Q (Quality) characterises the epicentre as follows.

A – indicates a location based on reports from more than one locality, either as the centre of the area affected, or where maximum intensities were observed.

B – indicates the location of events reported from one place only.

C – indicates the location of events reported from an ill-defined region without the mention of place names, or where the places named are either unidentified, or evidently not in the epicentral area.

X – indicates the approximate epicentral region of events nucleating outside the study area, but whose felt effects extended into the region under study. For details, see Section 2.1. Maps showing effects outside the study area are preliminary.

Intensities (*I*) are classified as follows (see Section 1.3, pp. 15–18).

F – Felt to strong (up to VI MSK).
S – Strong to damaging (approximately VI to VII).
D – Damaging to destructive, with casualties (approximately VII–IX).

Intensities shown on maps are on the MSK scale and have been assigned by two observers. Intensities assigned to events with epicentres outside the study area refer to maximum ratings deduced for sites within the study area.

Code symbols are assigned as follows.
A – numerous and/or strong aftershocks reported
F – associated with faulting or other surface deformations
l – probably a local shock
L – long-period effects reported
R – triggered rockfalls or landslides
S – part of a swarm sequence or a long series of shocks
s – preceded and/or followed by one or two shocks
T – event associated with tsunami or abnormal fluctuation of sea level
U – widely felt in unspecified regions
V – volcanic eruption or tremors associated with volcanism
W – effects noted in groundwaters, springs or wells
X – event probably not of seismic origin (meteor, landslide etc.).

M_F (Magnitude) denotes a macroseismic magnitude determined from felt data (see Section 3.2 for empirical relationships).

The choice of which equation to use depended on the location of the earthquake in question. The magnitude of historical events with epicentral areas in regions where twentieth-century events show high or low attenuation characteristics were calculated using equations 4 or 5 respectively (see pp. 116–17). For historical earthquakes in regions such as Sinai and the Hejaz, for which we have no twentieth-century macroseismic information, it is not possible to chose the appropriate formula with certainty. Also, these formulae have been derived from intensities not greater than V (MSK), and their application for the determination of surface-wave magnitude from a single intensity value much larger than V would tend to stretch their predictability beyond permissable limits.

However, errors arising from the choice of the appropriate formula are of about the same order as those implicit in the determination of magnitude in terms of

(continues on p. 105)

Table 2.1 *Catalogue of earthquakes up to 1899*

Date	Time	Epicentre °N – °E	Q	I	Code	M_F	Loc. (map)	I/r
184 BC		31.0 – 31.0	C				LE	
c. 95 BC		30.7 – 32.5	B	F			GS	
c. 112 AD		31.0 – 35.0	A	D		6.2	DS (2.3)	6/50
c. 200		15.6 – 44.1			V		YE	
262		32.7 – 21.6	B	D			LY	
320		31.5 – 30.0	B	S	U		MD	
365 Jul. 21	0700	36.0 – 23.0	X	D	ALT		HA (2.4)	
c. 500		15.6 – 44.1			V		YE	
520 Oct. 14		31.0 – 31.0	C	S	AlU		LE ?	
551 Jul. 9		32.0 – 36.0	X	F	T		DS (2.5)	
c. 626		24.6 – 39.6	B	F	l		HJ	
641		24.6 – 39.6	B	S	V?		HJ	
742		15.3 – 45.5	C	D	RU		YE	
747 Jan. 18	1000	31.8 – 35.7	X	F			DS (2.6)	
796 Apr.	2000	36.0 – 26.0	X	F	L		HA	
827		14.0 – 44.5	A	D		6.1	YE	5/90
857 Apr.		28.0 – 31.0	C	D	U		EG	
859		15.4 – 44.3	C		W?X		YE	
860 Jan.		37.0 – 38.0	X	F			EA	
873 Sep. ?		27.0 – 39.0	C	D	U		HJ	
879 Feb. 9		24.3 – 56.7	B	F			GO	
881		36.0 – 27.0	X	F	U		HA	
885 Nov.		30.0 – 31.2	B	D	l		LE	
912		30.0 – 31.0	C	F	lU		EG	
935 Oct. 4		30.0 – 31.2	B	D	l		LE	
950 Jul. 25		30.0 – 31.2	B	D	Al		LE	
951 Sep. 15	1800	32.0 – 30.0	B	D	AlUW		LE ?	
956 Jan. 5	1200	34.0 – 32.0	X	S	L		MD	
963 May 12	1800	35.0 – 26.0	X	F			HA ?	
997		26.0 – 34.0	B		X		UE	
1033 Dec. 5	1800	32.4 – 35.5	X	F	AT		DS (2.7)	
1068 Mar. 18	0830	28.5 – 36.7	A	D	AFLTW	7.0	HJ (2.8)	7/200, 6/340, 5/510, 4/600, 3/900
1072		14.4 – 43.7	A	D		6.0	YE (2.9)	5/80
1091 Feb. 12 ?	1800	28.5 – 34.0	B	S	ls		SI	
1111 Aug. 31	0900	31.0 – 31.0	B	D	lU		LE	
1121		23.5 – 37.0	A	S	LU	6.8	RS	4/280
1138 Oct. 15	1600	36.5 – 37.0	X	F			NS	
1145 Sep. 15		14.0 – 44.0	C	S	U		YE	
1154 Sep. 11		14.1 – 44.1	A	D	l	5.9	YE (2.10)	6/45, 4/100
1170 Jun. 29	0700	35.0 – 36.5	X	F			NS	
1195 Apr. 4		16.0 – 49.0	C	F	s		SA	
1202 May 20	0700	33.5 – 36.0	X	S	AL		DS (2.11)	
1203 Oct. 23	1800	13.5 – 41.9			V		ET	
1212 May 1	0500	30.0 – 35.2	A	D	AR	6.7	DS (2.12)	7/50, 6/100, 3/420
1222 May 11	0600	34.5 – 32.5	X	F	T		MD	
1256 Jun. 30	0800	24.6 – 39.8			V		HJ	
1259 Dec. 10		15.4 – 43.7	C	D	RsU		YE	
1264 Feb. 20		29.0 – 31.0	C	D			LE ?	
1265		15.4 – 44.2	B	F	l		YE	
1269 Oct. 29		21.1 – 40.5	B	F			HJ	
1293 Jan.		31.0 – 35.6	A	D	L	6.6	DS (2.13)	7/40, 5/130, 3/430
1293		24.5 – 39.8			V		HJ	
1299 Jan. 8		29.0 – 31.0	C	F	ls		EG	
1303 Aug. 8	0600	34.0 – 28.0	X	D	ALRT		HA (2.14)	

Table 2.1 *Catalogue of earthquakes up to 1899* cont.

Date	Time	Epicentre °N – °E	Q	I	Code	M_F	Loc. (map)	I/r
1307 Aug. 10	0000	30.0 – 31.2	C	F	l		LE	
1313 Feb. 27	1200	30.0 – 31.2	C	F	l		LE	
1335 May 29	1700	30.0 – 31.2	B	F	ls		LE	
1336		14.2 – 43.2	C	F			YE	
1347 Dec. 8		30.0 – 31.2	B	F	ls		LE	
1353 Oct. 16	2200	35.0 – 28.0	X	F			HA	
1359	1200	14.2 – 43.9	A	D	s	6.3	YE	5/120
1373 Oct. 19		30.0 – 31.0	B	F			LE	
1381		15.5 – 48.5	A	S	F?s		SA	
1385 Sep. 19	0000	30.0 – 31.2	B	F	ls		LE	
1386 Jul. 17	1000	30.0 – 31.2	B	F	l		LE	
1387 Sep. 5		13.3 – 44.8	A	D	S		YE	
1394 Mar.		13.3 – 43.5	B	F	lS		YE	
1400 Feb. 22	1200	14.0 – 44.0	C		X		YE	
1400 Apr. 6	0800	14.1 – 43.3	B		X		YE	
1400 Jul.		13.5 – 41.9			V		ET	
1408 ?		21.5 – 39.8	B	F	l		HJ	
1413 Dec.		14.1 – 44.2	C	D	S		YE	
1422 Jun. 28		30.0 – 31.2	B	F	l		LE	
1425 Jun. 23	0600	29.0 – 33.0	C	S	Ls		GS ?	
1426 Nov.		27.0 – 56.0	B	D	F?		PG	
1427 ?		14.0 – 44.0	A	D	A	5.7	YE	4/100
1432 May 20	0800	14.0 – 42.5	A	F	U	6.3	RS	3/300
1432 Dec.		15.0 – 43.0	C	F		5.7	YE	4/100
1433 Apr.		15.1 – 42.2			V		RS (2.15)	
1433 Dec. 14	1800	30.0 – 31.2	B	F	l		LE	
1434 Sep.		15.1 – 42.2			V		RS (2.15)	
1434 Nov. 6	1145	30.0 – 31.2	B	S	s		LE	
1438 Feb. 25	1600	35.0 – 28.0	X	F			HA	
1444 Jan. 20		15.1 – 42.2			V		RS	
1455 Mar. 5		30.0 – 31.2	B	F	ls		LE	
1458 Nov. 12		31.0 – 35.5	A	D		6.5	DS (2.16)	7/30, 6/70, 3/430
1463		14.2 – 43.3	B	D	lS		YE	
1466 Dec. 15	0500	14.2 – 43.3	B	S	ls		YE	
1467 Dec. 15	1800	30.0 – 31.2	B	F	l		LE	
1476 Oct.	1800	30.0 – 31.2	B	F	l		LE	
1481 Mar. 18	1700	35.0 – 30.0	X	S	LU		MD	
1483 Jun. 7	1800	14.2 – 43.3	B		X		YE	
1483 Jun. 15	1800	30.0 – 31.2	B	F	l		LE	
1484 May 9	1200	14.2 – 43.3	B	S	Al		YE	
1485 Mar.		14.2 – 43.5	C	S	US	5.2	YE	5/30
1486 Oct. 11	1200	30.0 – 31.2	B	F	l		LE	
1491 Apr. 24	1900	35.0 – 32.0	X	F	s		MD	
1498 Sep.		30.0 – 31.2	B	F	l		LE	
1500 Jul. 24	1900	36.0 – 23.0	X	F			HA	
1501 Jan. 26	0500	14.2 – 43.3	B	F			YE	
1502 Nov. 13	0000	14.2 – 43.5	B	F	lS		YE	
1502 Nov. 17		30.0 – 31.2	B	S	l		LE	
1504 Aug. 30	0000	12.5 – 43.5	A	S	s	6.6	RS (2.17)	5/140
1508 May 29	1800	35.0 – 27.0	X	F			HA	
1509 Apr.		35.0 – 27.0	X	F			HA	
1509 May 7	0500	14.2 – 43.3	B	F	l		YE	
1509 Sep. 10	2000	40.7 – 28.0	X	F			TR	
1511 Jan. 20	0600	14.2 – 43.3	B	F	S		YE	
1511 Jan. 29	0000	13.6 – 43.5	A	D	SW	5.6	YE (2.17)	5/50

Table 2.1 *Catalogue of earthquakes up to 1899* cont.

Date	Time	Epicentre °N – °E	Q	I	Code	M_F	Loc. (map)	I/r
1511 Feb. 27	1600	13.6 – 43.5	A	F	S	5.6	YE (2.17)	5/50
1511 Mar. 7	1700	40.0 – 25.0	X	F			HA	
1511 Jun. 11	0400	14.2 – 43.3	B	F	l		YE	
1513 Mar. 28	0600	30.0 – 31.2	B	F	ls		LE	
1518		13.0 – 40.0	C	F	S		ET	
1523 Apr. 4	2200	30.0 – 31.2	B	F	l		LE	
1525 Mar. 9	1800	30.0 – 31.2	B	F	l		LE	
1527 Jul. 14	0500	30.0 – 31.2	B	F	l		LE	
1529 Nov. 12	0400	30.0 – 31.2	B	F			LE	
1532 Jul. 10	0000	30.0 – 31.2	B	F			LE	
1534 Mar. 23	0600	30.0 – 31.2	B	F			LE	
1537 Jan. 8	1800	32.0 – 24.0	X	F			MD ?	
1537		32.0 – 32.0	B	F	S		MD ?	
1546 Jan. 14	1600	32.0 – 35.1	A	F?	ATW	6.0	DS	4/200
1554		13.0 – 40.0	C	F			ET	
1573 Feb. 4		36.5 – 25.5	X	F			HA ?	
1576 Apr. 1	1800	30.0 – 31.2	B	F	s		LE	
1583 ?		15.6 – 44.1			V		YE	
1588 Jan. 4	1300	29.0 – 36.0	A	D	LR	6.7	HJ (2.18)	7/200, 4/500
1588 Apr. 7	0600	29.4 – 31.6	A	F	lW	4.7	LE (2.18)	4/75
1592 May		37.0 – 21.0	X	F	s		HA	
1593 Sep.		13.0 – 39.5	C	F	s		ET	
1608 Dec. 23	1700	11.4 – 41.7			V		ET	
1609		35.0 – 28.0	X	F	L		HA	
1613		12.6 – 45.0	B	S	L?		GA ?	
1613 Jun.		35.0 – 27.0	X	F			HA ?	
1619 Jul.	1200	16.4 – 44.0	A	S	A	5.8	YE (2.19)	5/65
1631 Feb. 10	0000	11.3 – 41.7			V		ET	
1631 Mar.		15.4 – 44.2	B	F			YE	
1632 Jun.		12.4 – 37.5	C	F			ET	
1633 Nov. 5	1300	37.0 – 21.0	X	F	L		HA	
1637		15.7 – 43.8			V		YE	
1644 Sep. 22		15.4 – 44.0	A		X		YE	
1646 Apr. 30		15.5 – 43.9	B	F	s		YE	
1656		13.1 – 32.8	B	D			LY	
1664 Nov. 20		35.5 – 25.0	X	F	L		HA	
1666 Nov.	0500	15.4 – 44.2	B	F	ls		YE	
1667 Mar. 14	2200	15.4 – 44.2	B	S	U		YE	
1667 Oct. 22		12.5 – 37.5	C	F			ET	
1668 Aug. 17		40.5 – 35.0	X	F	L		TR	
1674 Aug.		14.8 – 44.2	B	F	S		YE	
1675 Dec. 21	0500	15.0 – 44.2	A	S	RS	5.3	YE	5/35
1679 Sep. 21		14.0 – 42.7			V		RS	
1694 Dec. 21	0800	29.0 – 31.0	C	S	A		EG	
1698 Oct. 2	0830	32.0 – 30.0	A	F		6.0	MD ?	4/250
1710 Aug. 27	0800	29.0 – 33.0	C	F	L	5.4	GS ?	4/200
1733 Nov. 30	0000	15.7 – 39.0	A	D	AR		ET	
1741 Jan. 31	0015	35.0 – 28.0	X	S	L		HA	
1754 Oct.		29.6 – 32.2	A	D		6.6	GS ?	7/100, 6/200
1756 Feb. 13	2100	36.0 – 23.0	X	F	AL		HA	
1775 Jun. 3	1900	13.9 – 43.5	B	F	s		YE	
1778 Jun. 22	1800	26.3 – 32.1	A	F	A	4.8	UE	4/80
1786 Jul. 28	0200	30.5 – 47.8	B	F			IQ	
1788 Nov.		15.0 – 42.0	A	S	SV?	6.2	RS (2.20)	4/180

Table 2.1 *Catalogue of earthquakes up to 1899* cont.

Date	Time	Epicentre °N – °E	Q	I	Code	M_F	Loc. (map)	I/r
1789		12.5 – 44.0	A	S	ST	5.9	RS (2.20)	4/130
1790 May 26	0000	35.0 – 25.0	X	F			HA ?	
1799 Aug. 18	0400	12.5 – 37.5	C	F			ET	
1801 Oct. 10	2100	30.0 – 31.2	B	F	l		LE	
1802 Jun. 30		12.5 – 37.5	C	F			ET	
1805 Jul. 3	0500	36.0 – 24.0	X	F	L		HA	
1809 Feb. 26		12.5 – 37.5	B	F	s		ET	
1810 Feb. 17	0100	36.0 – 23.0	X	S	LsT		HA	
c.1810		17.0 – 42.8			V		YE	
1814 Jun. 27	2200	29.0 – 33.0	A	S	LRs	5.4	GS	4/200
1818 Jun. 30		14.2 – 38.8	B	S			ET	
1824		15.1 – 42.2			V		RS	
1825 Jun. 21	2100	30.0 – 31.2	B	F	ls		LE	
1832		15.5 – 41.8			V		RS	
1832 May 4		15.0 – 39.5	B	F			ET	
c.1832		25.4 – 49.6	B	S			SA	
1838 Feb. 25		15.6 – 39.5	A	F	s	4.4	ET	4/25
c.1839		28.5 – 34.0	B	S			SI	
1842 Dec. 8		9.7 – 39.8	A	D	AR		ET	
1844 Oct. 23		15.5 – 39.5	B	F			ET	
1845 Feb. 12	1200	12.3 – 39.0	A	S	RS	6.5	ET (2.21)	5/150, 2/500
1845 Oct. 31		15.5 – 41.8			V		RS	
1846 Mar. 28	1500	36.0 – 25.0	X	F	L		HA	
1846 Jun. 15		30.0 – 31.2	B	F	l		LE	
1846 Jul. 14	1000	15.1 – 42.2			V		RS	
1847 Aug. 7	0815	29.7 – 30.8	A	D		5.8	LE (2.23)	6/70, 5/180, 3/480
1848 Aug. 1	2000	15.6 – 39.4	B	S			ET	
1849 Jul. 23	0300	30.0 – 31.2	B	F	l		LE	
1850 Jul. 17		15.4 – 44.2	B	D	l		YE	
1850 Oct. 27	0930	27.3 – 31.0	B	S	Fl		UE	
1851 Apr. 3	1630	36.0 – 28.0	X	F			HA	
1853 Aug. 5		25.9 – 14.2	B	F	A		LY	
1854 Jan. 3	2330	15.6 – 32.6	B	F			SU	
1854 Feb. 21	0300	12.8 – 39.0	A	S	FRSW	6.6	ET (2.25)	5/200
1856 Oct. 12	0045	35.5 – 26.0	X	S	ALT		HA (2.26)	
1857 Apr.		14.5 – 39.0	C	S	S		ET	
1858 Jun. 13	0500	29.6 – 50.5	A	S		5.9	PG	5/70
1858 Dec.		30.0 – 31.2	B	F	l		LE	
1859 Jan.		13.1 – 44.1	A	F	lS	5.2	YE	5/65
1860 Aug. 6	0030	18.1 – 34.0	B	S	ls		SU	
1860 Dec. 22		33.5 – 10.0	B	F			LY	
1861 May 7		13.5 – 41.8			V		ET (2.27)	
1863 Apr. 22	2230	36.5 – 28.0	X	F			HA (2.28)	
1864 Sep. 14	1115	15.6 – 39.5	B	F	S		ET	
1865 Apr. 11	0615	31.1 – 30.0	B	F	l		LE	
1868 Feb. 20	0315	32.0 – 33.0	A	S	s	5.4	MD	3/280
1870 Jun. 24	1825	35.0 – 29.0	X	S	LT		MD (2.29)	
1873 Jan. 12	1349	32.5 – 33.5	X	F	U		MD	
1873 Oct. 9	2100	14.2 – 43.3	B	F			YE	
1875 Nov. 2	1100	16.5 – 38.0	A	D	FR	6.1	ET (2.30)	7/9, 5/250, 4/430, 3/610
1878 Jun.		14.5 – 44.4	A	D	S		YE	
1878 Nov.		13.0 – 40.0	C	F	S		ET	
1879 Jul. 11	1800	29.0 – 33.0	A	S	LT	5.9	GS	5/200
1880 Jun. 30		14.1 – 38.7	C	F			ET	
1881 Jun. 10	0800	33.7 – 10.2	A	S	RS		LY	

Table 2.1 *Catalogue of earthquakes up to 1899* cont.

Date	Time	Epicentre °N – °E	Q	I	Code	M_F	Loc. (map)	I/r
1881 Jun.		16.9 – 43.8	B	F	s		YE	
1883 Mar.		11.7 – 42.0	A	F		5.6	ET	3/160
1883 Aug.		30.2 – 9.5	B	F		6.0	LY	4/350
1883 Oct. 13	1500	22.9 – 57.5	A	D	Fs	5.1	SA	4/140
1884 May 19	1800	26.9 – 56.0	A	D	AF	5.4	PG	7/12, 3/120
1884 Jul. 20	0930	15.7 – 39.6	A	D	AT	6.2	ET	7/25, 5/100
1886 May 8	1730	15.6 – 39.5	B	F	S		ET	
1886 Aug. 27	2130	37.0 – 21.3	X	F			HA (2.31)	
1886 Nov. 17	1630	30.0 – 31.2	B	F	l		LE	
1887 Apr. 1		12.8 – 45.0	B	F	s		YE	
1887 Jul. 17	0745	36.0 – 26.0	X	S	L		HA (2.32)	
1889 Oct. 4	1800	15.6 – 38.5	B	F	s		ET	
1891 Feb. 12	1240	15.6 – 39.5	B	F	s		ET	
1892 Nov. 23		15.6 – 39.5	B	F			ET	
1894 May 13	0645	15.6 – 38.5	B	F	s		ET	
1894 Sep. 12	1250	15.6 – 39.5	B	F			ET	
1894 Nov. 3	2300	15.0 – 39.4	B	F	s		ET	
1894 Nov. 8	0600	15.6 – 38.5	B	F			ET	
1895 Aug. 2	1119	13.4 – 43.6	A	F	s	5.4	YE	5/40
1895 Dec. 7	0240	30.6 – 31.2	A	F	ls	4.9	LE	4/100
1896 Jun. 29	2330	34.3 – 33.0	X	F			HA (2.33)	
1896 Jun.	2230	15.4 – 44.2	B	F			YE	
1896 Dec. 11	0220	15.6 – 39.5	B	F			YE	
1897 Feb. 20	1520	13.0 – 43.0	A	F	S	4.6	RS	3/60
1897 Sep. 30	2114	15.6 – 39.5	B	F			ET	
1898 Sep.		15.4 – 43.7	C		X		YE	
1899 Feb. 11		11.5 – 43.0	B	S	s		ET	

Those interested in obtaining machine-readable files of the basic origin parameters of the events in Tables 2.1 and 3.1 should make enquiries to the International Seismological Centre.

felt radius (*r*) and intensity (*I*) from these formulae. The combined error should not be greater than 0.5 magnitude units. Thus the magnitudes given in Table 2.1 are only of an index nature.

Location: the following symbols identify the probable location of the event (see also Figure 1.1). Doubtful locations are marked with a '?'.

AS – Arabian Sea
DS – Dead Sea system
EA – Eastern Anatolia
EG – Egypt
ET – Ethiopia and Djibouti
GA – Gulf of Aden
GO – Gulf of Oman
GS – Gulf of Suez
HA – Hellenic Arc
HJ – Hejaz
IQ – Iraq
IR – Iran

LE – Lower Egypt
LY – Libya (includes the borders with Tunisia)
MD – Mediterranean
NS – Northern Syria
PG – Persian Gulf
RS – Red Sea
SA – Arabian Peninsula
SI – Sinai
SU – Sudan
TR – Turkey
UE – Upper Egypt
YE – Yemen

Numbers appearing (in brackets) in this column refer to the Figures in Section 2.1 for the earthquakes that have been mapped.

I/r denotes the average distance in kilometres at which the shock was felt at intensity *I* (see Section 1.3 p. 18).

2.3 Unidentified events

The following section contains a handful of earthquakes that we have been unable to identify fully, either the date or the location (or both) being uncertain. They are not listed on Table 2.1, and it is very possible that some of these events were not genuinely of seismic origin.

Pre-Islamic period   Qift

An earthquake is said to have destroyed a tower in Qift, between Qena and Qus in Upper Egypt, from which it had been possible to see the Red Sea. The tower was built by the great grandson of Ham, son of Noah.[1] There are no indications of when this earthquake took place, and the information is legendary.

1 Al-Suyuti, *Kashf*, p. 60/42, saw this recorded in 'a certain history.' He mentions it at the end of his catalogue, but the shock clearly occurred much earlier than 1500. Qift b. Misr is accounted a Pharaoh by Muslim authors; for a discussion of the legends concerning the foundation of Qift and Qus, see Garçin (1976), p. 12 ff.

871 June 15   *22 Rajab 257*   Egypt ?

A strong shock of earthquake following the appearance of fire in the sky, moving from east to west, occurred around dawn.[1] The details associated with this event suggest the appearance of a comet, or the fall of a meteorite. The shock may not have been caused by an earthquake.

1 Al-Ya'qubi, **II**, 621; the author was exactly contemporary (d. 879); he correctly gives the month as Haziran in the Syriac calendar.

Late tenth/early eleventh century   Egypt

Some lines of poetry, quoted by different authors, have been variously associated with earthquake(s) in Egypt either in the reign of Kafur the Ikhshid (355–7/966–8),[1] or of the Fatimid caliph al-Hakim (386–411/996–1021).[2] It is impossible to say to which specific events the verses might refer, but there are several candidates among the series of earthquakes affecting Egypt particularly in the period before 352/963 (see catalogue).

1 Al-Suyuti, p. 30/16. Taher (1979), p. 38/230, cites these verses in connection with the earthquake of 340/951, which is certainly too early.
2 Al-Suyuti, pp. 31–2/17–18, with slight alterations, citing al-'Umari, *Masalik al-absar*. The verses are attributed to the same poet, Muhammad b. al-Qasim, who is described as al-Hakim's poet. It was not unusual for apt verses to be quoted and reused on different occasions. Taher (1979), pp. 50–1/232 connects the verses to an alleged earthquake in Egypt in 407/1016.

1240s   *640s H*   Egypt

A marginal note incorporated into Erpenius's edition of al-Makin says that an event similar to the earthquake of October 935 in Egypt occurred during the reign of al-Malik al-Salih Najm al-Din Ayyub b. al-Malik al-Kamil.[1] Al-Salih Ayyub reigned in Egypt from 637/1240 to 647/1249. No account of an earthquake in Egypt has been discovered in the sources available for his reign. The event of 935 was associated with meteorite falls or shooting stars, and it is possibly to this, rather than an earthquake, that the marginal note refers.

1 Al-Makin, ed. Erpenius, p. 208/trans. Vattier, p. 217.

1333 ?   *733 H*   Unidentified

A great earthquake occurred in Buhaira (?) this year; the ground rippled like waves and cracked open. A great many people perished and villages and castles for 10 farsakhs (*c.* 60 km) around were destroyed. A mountain there was split and smoke emerged; half the mountain disintegrated and the ground became level.[1]

This late account presents considerable problems of identification, for the best-known locality called Buhaira is the province in Lower Egypt, west of the Delta, which is not mountainous. It is always written al-Buhaira in the Arabic sources. The event appears to have been associated with large-scale landslides. In view of al-'Umari's knowledge of events in the Yemen, and on the grounds of topography, it is possible that this is an unidentified Yemen earthquake or volcanic eruption.

1 Al-'Umari, fol. 129vo; also in BL Ms. Or.6300, fol. 152. The event is not reported by the numerous contemporary Egyptian sources, nor is an earthquake this year recorded anywhere in the Islamic world in the sources we have investigated.

1365   *766 H*   Unidentified

Al-Suyuti records a great earthquake this year, which he had seen noted on the back of a book, but the place where the shock occurred was not mentioned.[1] It is possible that Egypt is involved, for we have found no reference to an earthquake elsewhere in the Muslim world this year, though there was an earthquake in Herat (Afghanistan) in 765/1364.[2]

1 Al-Suyuti, p. 56/39.
2 Ambraseys and Melville (1982), p. 44. Also, a destructive earthquake in Denizli (western Anatolia) in 767 H is mentioned in the *Tarihi Takvimler*, p. 72. Ibn al-'Imad, **VI**, 210, reports a fearful earthquake in Safar 768 (October 1366), which Taher (1979), p. 197/88, locates in Safad.

1398   *800 H*   Italy ?

About 45 earthquake shocks are reported to have affected 'Rumiyyat al-kubra.' Houses were destroyed and more than 400 people were killed in the ruins. Three churches and about 200 monasteries were ruined.[1]

The location of this event is problematical. The term Rumiyyat al-sughra (Lesser Rum) refers to the region of Amasya and Sivas in northern Anatolia, so it might be supposed that Rumiyyat al-kubra (Greater Rum) means a larger area of Anatolia. The term is not, however, used elsewhere. The details of the damage suggest that it was Christian rather than Muslim territory that was affected, and it seems likely the reference must be to Greater Rome, i.e. Italy. This suggests that the earthquake of 26 December 1397 in Bergamo is intended.[2] If so, it is interesting that al-'Umari reports the event.

1 Al-'Umari, fol. 148ro; BL Ms. fol. 169. The year 800 H began on 24 September 1397.
2 Baratta (1901), p. 59; cf. Bonito (1691), pp. 589–90.

c. 1575 **Beersheba?**

According to a pamphlet describing the activity of a French missionary on pilgrimage to Jerusalem, a destructive earthquake occurred in 'Sabee', ruining the capital 'Sabe', and five other villages, with heavy loss of life. The villages are named as 'Beem', 'Fratres', 'Lexico', 'Schillen' and 'Solim', and they were situated in a mountainous area 'between the Persian Sea and the Red Sea', a detour of 50 miles from the pilgrims' route (to Jerusalem?), though the rest of their itinerary is not clear. Some of the inhabitants of the places affected are said to have understood Latin, but they were Muslims, as is shown by the fact that they had celebrated the Birth of the Prophet (12 Rabi' I) ten days before the earthquake. As a result of the pilgrims' visit, some of the inhabitants are alleged to have adopted Christianity.[1]

It has not proved possible to identify the places mentioned; the area affected may have been around Beersheba (Be'et Sheva), between the Dead Sea and the Gulf of Aqaba, and just off the pilgrim route between Jerusalem and Sinai. Neither is the date clear; the French translation is from an undated Latin original. The whole episode might well be a pious fiction.

1 Anon. (1580). None of the places mentioned have been found on contemporary maps, for which see Tibbetts (1978).

2.4 False and mislocated events

Existing lists of earthquakes naturally form the basis for any new catalogue, by providing the dates and locations of events already noted by previous writers. As mentioned in Section 1.2, however, many existing catalogues have grave defects. For those who follow in their footsteps, the absence of evidence or support for their statements is the most frustrating shortcoming. Indeed, many authors do not name their sources, even when these are simply earlier catalogues, and it is this failure that actively encourages later writers to regurgitate their lists, since they have no certain grounds for rejecting them.

In the course of our researches, we have noted many false entries and dubious earthquakes in existing catalogues. We consider it to be an important part of producing a definitive list, to explain the sources of error wherever possible, and to demonstrate why such events have been excluded. It is not enough simply to ignore them, for the chances are that later writers will simply reinstate them.

The following list of false dates (or locations) does not claim to be comprehensive nor to be the result of a systematic study of all the erroneous earthquakes identified, but explains why these events (and others like them) are not to be found in the catalogue presented above. Many of these false events have been noted in passing in the descriptive catalogue (Section 2.1), and a few other cases are described individually below. The list underlines the kinds of problems that can arise from an exclusive reliance on secondary works.

The catalogues surveyed are those of Lyons (1907) [= L]; Sieberg (1932a, 1932b) [= Sa, Sb]; Ben-Menahem (1979) [= BM] and Poirier and Taher (1980) [= PT]. Not all these are equally blameworthy: the latter, in particular, refer almost exclusively to primary sources, but not always accurately. Sieberg, in contrast, does not give his sources, which sometimes makes it impossible to identify the event in question. For these, at least, we have been unable to find any primary evidence. Only those events within, or wrongly located within, our area of interest are covered here, and the list is taken up to the year 1900. For a list of spurious and mislocated instrumental events, see Section 3.4.

Correct identifications are given when possible. Minor inaccuracies in dates arising from conversions from the Muslim calendar, concerning only a day or two, do not always appear on the list.

2200 BC	Zagazig, Nile Delta (Sa, b)	*Not confirmed*
1210 BC	Abu Simbel (Sa, b)	*Not confirmed*
600 BC	Upper Egypt (Sb)	*Not confirmed*
27 BC	Egypt (L, Sa, b)	*See below*
20 BC	Alexandria (Sb)	*See below*
93 AD	Egypt (L, Sb)	*Anatolia, see below*
262 AD	Siwa oasis (Sb)	*Libya*
312 AD	Alexandria (L)	*320*
358 Aug. 24	Alexandria, Egypt (L)	*N. Anatolia*
396	Alexandria (L)	*Constantinople, i.e. Alexandria Troas*
497	Arabia (Sa)	*Not confirmed*
553	Alexandria (Sb)	*551 Jul. 9*
631	Arabia (Sa)	*641 ?*

645	Hejaz, Yemen (PT)	*641, Hejaz*
704	Libya and Egypt (Sb)	*742, Yemen*
742	Libya and Egypt (Sa, b)	*Yemen*
859	Bilbais (L, Sb)	*860 Jan.*
867	Mecca (Sa)	*873 Sep.*
874	Arabia (PT)	*873 Sep.*
887 May 6	Egypt (Sb)	*881, Hellenic Arc*
933	Egypt (PT)	*935 Oct. 4*
934	Egypt (L, Sb)	*935 Oct. 4*
954	Egypt (L, Sb)	*950 Jul. 25*
956 Jan. 1	Alexandria (PT)	*956 Jan. 5*
967	Upper Egypt, Karnak (Sa, b)	*Not confirmed*
969 Jul. 1	Egypt (PT)	*Not an earthquake*
1040	Egypt (Sb)	*1033 Dec. 5, Dead Sea?*
1047	Ramla (PT)	*1068 Mar. 18*
1067 Apr. 20	Eilat (BM)	*1068 Mar. 18*
1070 Feb. 25	Ramla and Cairo (BM)	*1068 Mar. 18*
1111 Mar.	Egypt (L)	*1111 Aug. 31*
1111 May 26	Egypt (Sb)	*1111 Aug. 31*
1122	Medina (PT)	*1121*
1166	Medina (Sa)	*1256 eruption?*
1203	Egypt and elsewhere (PT)	*1202 May 20*
1204	Egypt and elsewhere (L)	*1202 May 20*
1253	Aden (Sa)	*1256 Hejaz?*
1260 May 28	Egypt (L)	*1259 Jun. 6*
1261 Nov. 1–19	San'a (PT)	*1259 Nov. 22–Dec. 10*
1262	Cairo (PT)	*1303 Aug. 8*
1263 Feb. 21	Egypt (L, Sb)	*1264 Feb. 20*
1292	Ramla, Karak (PT)	*1293 Jan.*
1302	Alexandria and Egypt (L)	*1303 Aug. 8*
1303 Jul. 30	Cairo, Alexandria, Qus (PT)	*1303 Aug. 8*
1303 Dec.	Crete, Alexandria, Egypt (BM)	*1303 Aug. 8*
1312 May 1	St Catherine's, Sinai (BM)	*1212 May 1*
1326	Alexandria (Sb)	*1303 Aug. 8, see below*
1341 May	Alexandria (PT)	*1303 Aug. 8*
1343 Jan. 1	Manbij, Damascus (PT)	*1344 Jan. 3*
1375	Alexandria (PT)	*796 Apr.*
1400 Feb. 20	Yemen (Sa)	*1400 Feb. 22 meteorite*
1437 Nov. 7	Cairo (PT)	*1434 Nov. 6*
1459	Karak (PT)	*1458 Nov. 12*
1481 Mar. 18	Mecca (PT)	*Rhodes; fire in Mecca*
1481 Jul.	Cairo (PT)	*1491 Apr. 24*
1481 Oct.	Rhodes (PT)	*1481 Dec. 18*
1490 May 6	Egypt (PT)	*1491 Apr. 24*
1491 May 1	Lower Egypt (BM)	*Cyprus, aftershock of 1491 Apr. 24*
1502	Aden (PT)	*Fire*
1512 Apr. 7	Cairo (PT)	*1513 Mar. 28*
1525 Apr. 9	Cairo (PT)	*1525 Mar. 9*
1526 Jul. 14	Cairo (PT)	*1527 Jul. 14*
1576 Apr. 21	Cairo (PT)	*1576 Apr. 1*
1588 Jan. 14	Cairo, Eilat (PT)	*1588 Jan. 4*
1605 Jan. 8	St Catherine's, Sinai (BM)	*Palestine*
1608 Dec. 14	St Catherine's, Sinai (BM)	*1605, Palestine, see below*
1631 Jul.	Mecca (Sa)	*1630 Apr. 3 flood*
1672	Sinai (BM)	*Not confirmed*
1687 Mar.	Alexandria (Sb)	*Alexandretta, N. Syria*
1754 Sept.	Cairo (Sa, b, BM)	*1754 Oct.*
1803	Tripoli (Sa)	*1903 Nov. 22*
1811	Siwa Oasis (Sb)	*1810 Feb. 16, see below*

1825 Aug. 21	Cairo (Sb)	*1825 Jun. 21*
1846 Jun. 5	Cairo (Sb)	*1846 Jun. 15*
1846 Aug. 14	Red Sea (Sa)	*1846 Jul. 14*
1854 Jan. 3	Upper Egypt (L, Sa, b)	*Sudan only*
1872 Apr. 28	Alexandria (Schmidt, 1879)	*1872 Apr. 3, Alexandretta (aftershock)*
1873 Feb. 14	Cairo (Sb)	*1873 Jan. 12*
1880 Nov. 11	Massawa–Jidda (Sb, BM)	*1875 Nov. 2*
1884 Mar.	Muscat, Nejd (Sa)	*1883 Oct. 13*
1884 Jul. 23	Ethiopia, Cairo, Arabia (Sa, BM)	*1884 Jul. 20, Ethiopia only*
1884 Jul. 25	Ethiopia, Cairo, Arabia (Sb)	*1884 Jul. 20, Ethiopia only*
1886 Jul.	Nile Delta (Sb)	*1886 Aug. 27*
1899 Oct.	Karnak (Sa, b)	*1899 Jul. 14, Alaska; see below*

Some of the cases not discussed in the course of Section 2.1 are examined below.

27 BC   **Upper Egypt**

An allusion to an earthquake in Thebes is made by Strabo, who says that 'here (at the Memnonion) are two colossi which are near one another and are each made of a single stone; one of them is preserved, but the upper parts of the other, from the seat up, fell when an earthquake took place, so it is said' (Strabo, II, 217–19; xvii.1.46). Letronne (1833) and others after him, including Sieberg (1932b, p. 187), attribute the collapse of the statue to a severe earthquake in Upper Egypt. However, Quatremère (1845, II/2, 218–19) among others shows that the most likely cause of the damage was deliberate mutilation by the Persians.

As a matter of fact, another passage in Strabo refutes the possibility of a severe earthquake, saying that 'above the Memnonion . . . among the tombs, on a (standing) obelisk, are inscriptions . . .'. It is unlikely that an earthquake severe enough to break in two a monolithic colossus had no effect on nearby, less-stable structures, nor that any such effect would have escaped Strabo's notice while at Memnonion in 25–24 BC.

Later authors do mention the destruction of Thebes about this time (27 BC), but they do not attribute this to an earthquake. Most probably the city was ruined by the revolt of the inhabitants referred to by Eusebius, when the towns of Thebes, Busiris and Coptos, which revolted against Rome, were destroyed completely. It is surprising how many modern writers have misread Eusebius's text to imply earthquake damage (Eusebius, pp. 557–61/pp. 154, 362). See also Maspero (1914, p. 93).

c. 20 BC   **Lower Egypt**

Strabo says that when he was residing in Alexandria (up to *c.* 20 BC), 'the sea about Pelusium and Mt. Casius rose and flooded the country and made an island of the mountain, so that the road by Mt. Casius into Phoenice became navigable' (Strabo, i.3.17; xvi.2.26). He does not mention the occurrence of an earthquake and he probably refers to the effects of a sea-wave generated by submarine slumping off the north coast of Egypt, perhaps triggered by an earthquake.

Contrary to the assertion of Ben-Menahem, Nur and Vered (1976), no earthquake in that period is mentioned by contemporary historians. Large-scale flooding of the coast is considered by Strabo to be a rare event that has been observed, however, both in *c.* 20 and 139 BC and in earlier times, on the Egyptian and Syrian coasts. He does not associate these events with earthquakes, as implied by Clédat (1920, p. 115; 1923, p. 65). See also, Sieberg (1932b, p. 187) and Ben-Menahem (1979, p. 289).

AD 93   Egypt

Sieberg (1932b) places a destructive earthquake in Egypt in AD 93. His source of information is most probably Schmidt (1879) who paraphrases Philostratus (*Vit. Apoll.*, **VI**, 41). This author says that in AD 93 'the cities on the left side of the Dardanelles were visited by earthquakes, and the Egyptian (merchants) there went begging . . .'. This event clearly took place not in Egypt, but in western Anatolia.

1326   Egypt

Sieberg (1932b, p. 188) reports an earthquake in Alexandria in 1326, without naming a source. The origin of this mistake appears to be with Clédat (1923, p. 66), who refers to the earthquake of 702 H, wrongly stating this to be equivalent to 1324 (*sic.*) 'de notre ère'. Clédat is followed by later writers, such as Daressy (1929, p. 49) and Goby (1955, p. 35). As it happens, the celebrated traveller, Ibn Battuta, visited Alexandria in 1326 and makes no reference to an earthquake, though he does note the generally dilapidated state of lighthouse (see Ibn Battuta, I, 18–19). The lighthouse was taken down in the middle of the fourteenth century.

1605   Sinai

Ben-Menahem (1979, p. 258) mentions earthquakes in Sinai on 8 January 1605 and 14 December 1608, on the authority of the Codex 213 of the Monastery of Laura St Saba 'in Sinai'. In the first place, however, this monastery is located about 15 km S.E. of Jerusalem, not in Sinai. Secondly, the Codex actually reports earthquakes on 3 or 8 January and 14 December *the same*

year, 1605, with no reference to an event in 1608 (see Lampros, 1910, pp. 184, 274).

1811   Egypt

A large earthquake in 1811 is alleged to have caused the collapse of part of the Temple of Amon (Umm Baida) in the Siwa Oasis. Belzoni (1822, **II**, 160), noted the damage to the Temple in 1819, and a few months later Cailliaud (1826, **I**, 86, 104, 108, 123), was told an earthquake was responsible. Bayle St John (1849, p. 92), records a local tradition (in 1847) that shocks in the Siwa oasis occasionally affected the yield of spring water near Gorah; see also Forbes (1921). On the strength of this, Sieberg (1932a, p. 872), followed by later writers, lists a relatively large-magnitude earthquake in Siwa in the Libyan desert.

No information has been found to corroborate what Cailliaud was told in Siwa, and there is no contemporary evidence of damage to the villages in the oasis, where the houses were built of adobe, four to seven storeys high (Steindorf, 1904, p. 94). Nor are there any reports that a shock was felt elsewhere in Egypt in 1811. It is therefore likely either that a very small shock in Siwa caused the collapse of ruins already on the verge of giving way, or, more probably, that the earthquake in question was the large shock of 17 February 1810, about 500 km north of Siwa, which was strongly felt throughout Lower Egypt. The 1811 Siwa earthquake is, in all probability, a spurious event (cf. Ambraseys, 1991).

1899 October   Upper Egypt

Some earthquake catalogues give an early instrumentally located event on 14 July 1899 at 23° N – 33° E, near the present position of Lake Aswan. Earthquakes were recorded at about the same time on widely separated seismographs throughout the world, and there is an alternative solution for the event, in Alaska, see Milne (1911), Wood (1966).

Sieberg (1929; 1932a, p. 873; 1932b, p. 189), however, lists this earthquake as the strongest in recent times in Upper Egypt and postdates it by two months, to make it coincide with the date on which the temple at Karnak sustained considerable damage due to the collapse of a number of its columns. It can be shown, however, that the damage at Karnak was due to the rapid draw-down of the Nile, which induced the foundation failure of a number of the columns, which collapsed at about 9 a.m. on 3 October 1899 (three days before the seismological station at Abbasiyya in Cairo came into operation). Neither of the alleged earthquakes of 14

July and October 1899 was reported to have been felt anywhere within the region of interest. Examination of available station bulletins shows no evidence that such an event occurred in the eastern Mediterranean or North Africa, see Ambraseys and Adams (1986a); Ambraseys (1991).

CHAPTER 3

Instrumental information

In compiling definitive earthquake catalogues, there must be a critical reassessment of all available information, and particularly a reconciliation between instrumental and macroseismic data. This is particularly important during the early part of the instrumental era, when misinterpretation of the scanty recorded information then available could lead to gross errors of location.

Work on the present catalogue was carried out systematically in the following stages. First, we made a macroseismic study of all the events for which information was available. This is presented in the previous chapter (Section 2.1). This was followed by the recalculation, confirmation and assessment of the position of all significant events. Spurious, mislocated and unsubstantied events were removed (see Section 3.4). A uniform recalculation was made of surface-wave magnitudes for all teleseismically recorded events, and calibration functions were derived from felt reports, to allow the determination of magnitude for earthquakes with no teleseismic information.

The information available to us, both instrumental and macroseismic, is not uniform in time, and falls into three distinct periods. In the early part of the century, up to the First World War, there are full station reports of instrumental recording, and also many sources of felt information (see above, Section 2.1). From about 1916 onwards, the availability of information decreases until the rapid increase in seismological recording and reporting in the 1960s.

3.1 Instrumental coverage

Since the mid-1980s there has been an increasing number of seismographs in our area, but previously the coverage was extremely sparse. Milne instruments were installed at Helwan in Egypt in 1899 and at Ksara in Lebanon in 1904. Some stations in southern Europe, such as Athens, operated from the early 1900s, as did Milne stations in India, such as Bombay, Madras and Kodaikanal. Later a station was established at Tiflis in Georgia. Stations in Istanbul, Asmara, Massawa and Dar es-Salaam had a very short life and ceased to function before the First World War. To the south, initially the only stations were at Cape of Good Hope and Mauritius. It was not until 1925 that a station was set up at Entebbe in Uganda, which operated for a few years. Some stations were established in the Belgian Congo in the 1950s, but recording has continued there and in present-day Zaire only intermittently. Thus in the period up to the establishment of the World-Wide Standard Seismograph Network (WWSSN) in the 1960s, the coverage of stations in and around our area was extremely poor. As part of the WWSSN, new stations were installed at Jerusalem, Shiraz and Mashhad in Iran, and Addis Ababa and Nairobi in East Africa, but there still remained no stations in Arabia, and coverage in northern Africa was very sparse.

A close network was established in Djibouti from 1972, and at about this time, three new stations were established in Egypt (at Marsa Matruh, Aswan and Abu Simbel), but these were not effective in reporting earthquakes. There was no major improvement in Egypt until the early 1980s, when a multi-station telemetered network was established around Aswan, following the major induced earthquake in November 1981 (see Figure 3.1). Until recently, these stations have only seldom reported regional events. An extensive network has grown up in Israel, and since September 1983 a telemetered network has been operating in Jordan. Initially, this network comprised about seven stations and in the early years teleseisms and regional events were often interpreted as locals, resulting in spurious earthquakes appearing in Jordanian bulletins. The network has now grown to about thirty stations, extending to the eastern parts of the country, and contributes to reliable locations in this area. A station was opened at Baghdad in 1979

Figure 3.1. Simple solar-powered telemetric seismograph station Gebel Marawa (code: AGMR) installed as part of the network to monitor the Aswan region of Egypt following the earthquake near Lake Nasser in November 1981. This station transmits its information by radio to be recorded at the base station at Aswan, and is continuing to provide information on local, regional and global earthquakes into the 1990s.

and later four additional stations were installed in Iraq (Mosul, Sulaimaniya, Rutba and Basra).[1] Some of these appear to be sited on poor ground, and there is sometimes difficulty in interpreting arrivals from these stations for regional events. Recent conflict in the region has interrupted the regular operation of some of these stations.

A close network of five stations has been set up in Bahrain, and from the mid-1980s an extensive network has been progressively installed in Saudi Arabia, including stations in the central regions, near the Gulf of Aqaba, and near the Yemeni border. There are plans to install a network centred in Dhamar in Yemen in the early 1990s.

Figure 3.2 shows the location of seismograph stations listed as operating in 1993. Unfortunately, there are still serious gaps: in Libya, Sudan, Chad and also in Syria, no stations are operating. In Egypt there are plans for network expansion, but at present the only stations are the Aswan network and those immediately south of Cairo, where a new station, Kottamia, has been installed at a quieter site to the east of Helwan. The two outstations in Ethiopia, at Asmara and Alemaya, are not currently operating due to conflict in that country.

Thus the seismographic coverage of our area has been extremely uneven in time. For almost all the period of our interest there has been inadequate local control, and only large events located teleseismically have been detected. Since the mid-1980s, however, detection capability has improved greatly, particularly in the Levant and in Saudi Arabia.[2] Caution should be taken, however, in accepting as genuine all small events listed as local in Jordan.

3.2 Reassessment techniques and derivation of the catalogue

Location

For the first part of the century, the instrumental locations were controlled mainly by the global network of Milne stations, the reports of which were published in the Shide Circulars under the auspices of the Association for the Advancement of Science (BAAS, 1899–1912) and the International Seismological Summary (ISS, from 1913). The undamped Milne seismographs had magnifications from 10 to 20 and periods between 10 s and 20 s, and responded mainly to the surface-wave

[1] Fahmi, Ayar and Al Salim (1987).
[2] For a detailed review of the situation in the early 1980s, see Adams (1983b).

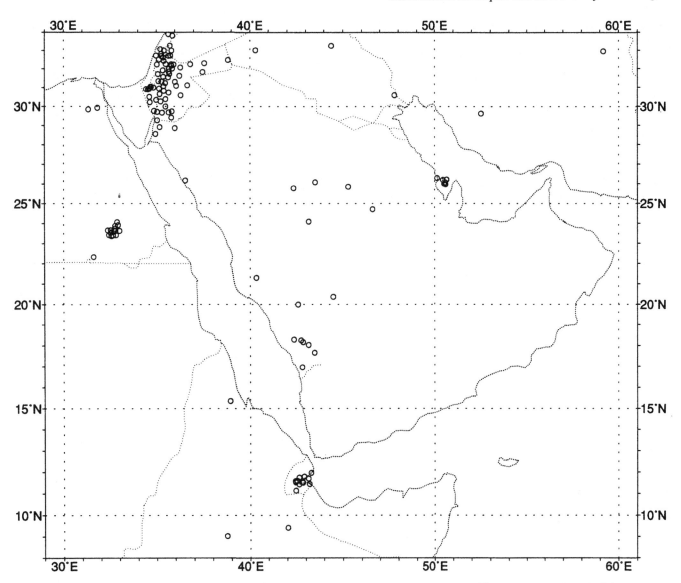

Figure 3.2. Internationally registered seismograph stations operating in our region in 1993. There are no stations in Libya or Sudan.

train. Body-wave, P and S, arrivals were only poorly recorded and only for the largest events, and cannot be relied upon for location purposes. In addition, it was difficult to maintain accurate timing, with clocks having to be adjusted by astronomical observations. Milne's determinations were also made without the benefit of adequate knowledge of seismic propagation or earth models. In our reassessments of instrumental readings, we therefore employ a technique previously used,[3] in which we assume that the maximum reported phase corresponds to an Airy phase of surface-wave propagation, and travels with a velocity of 2.8 to 3.0 km/s.

As noted above, most stations recording events in our region were to the north, in Europe, which makes location more difficult, for by choosing different origin times, an event's distance from the station can vary significantly in a North–South direction. We rely heavily on the two closest Milne stations, at Helwan and Ksara, to provide the main local control of instrumental locations in our area in the early part of the century. Readings were seldom reported from southern stations such as Mauritius or Cape of Good Hope.

Our reassessment has confirmed the positions given for many early events, but some major adjustments have been found necessary, including the relocation elsewhere of events previously listed as occurring in our area. The earthquake of **6 August 1901** (18h) was

---

[3] Ambraseys and Adams (1986a, 1986b).

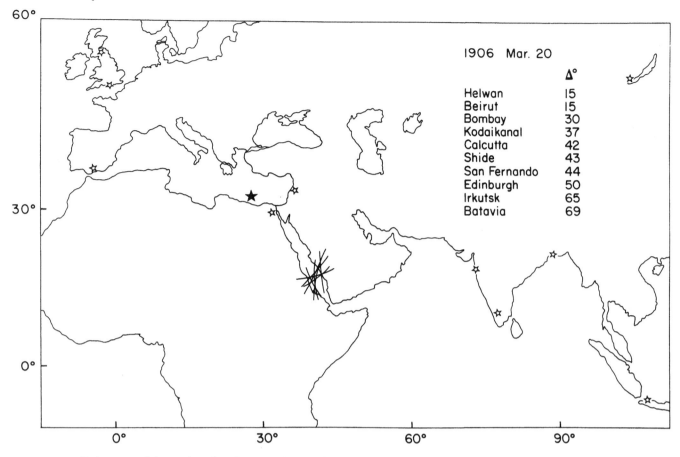

Figure 3.3. Relocation of the earthquake of 20 March 1906 from its original position in the Mediterranean Sea (filled star). Positions of seismograph stations used in relocation are shown by open stars, with arcs of re-determined distances intersecting near the revised position in the southern Red Sea.

originally located by BAAS at 20° N–55° E, in the eastern Arabian Peninsula. Reports are available from Milne stations in India and Europe, and also from Mauritius and North America. The times of maximum reported arrivals determine a position further north, in the known active area of southern Iran, near 28° N–56° E, although there is some discrepancy in the origin time determined from reports of body phases at European stations. The error of determination is about ± 5°, but the instrumental data do not allow an epicentre as far south as the original position. The lack of felt reports from Iran is surprising in view of the widespread recording of the earthquake.

A good example of our relocation is given in Figure 3.3, for an earthquake on **20 March 1906** (03h). BAAS assigned this an unusual position in the Mediterranean Sea, about 180 km north of Matruh, Egypt. There was only meagre reporting of P and S phases, but a set of consistent times for maximum phases (M) was found from Milne instruments at a wide range of distances and azi-

muths. These gave distances as listed in Figure 3.3. This shows that arcs corresponding to these distances intersect near 17° N, 41° E, close to the Farasan Islands in the southern Red Sea. The slow velocity of these M phases (about 3 km/s) means that a timing uncertainty of 0.1 minute corresponds to a distance of only about 20 km, and phases given to the nearest minute can still be used to provide useful control. In this instance the estimated error of position is about ± 3°, which is small compared with the shift in location of about 20°. Although the times of P and S phases are not good enough to determine the position of the earthquake by themselves, reports from some European stations, such as Trieste, Göttingen, Strasbourg, Jena and Uccle are adequate to confirm the revised position of the earthquake. It is not clear whether statements in some station bulletins (e.g. de Bilt) that the earthquake affected Eritrea, about 270 km from the new position, represent felt reports, or simply a proposed location for the shock.

Many early events are best located by their felt effects,

even if some instrumental observations are available. Such is the case with the Gulf of Suez earthquake of **6 March 1900**. The closest instrument, at Helwan, was thrown off scale, and it was recorded at distances out to 63°. These readings confirm the macroseismic epicentre, based on reports of intensity up to VI MSK (see Figure 2.34). The use of combining instrumental with macroseismic information is also illustrated by an earthquake on **25 January 1908**. This earthquake is very poorly recorded and was originally given a position in the Sudan near 15° N–35° E. About this time, however, there were reports of earthquakes from Yemen and a solution here, equally consistent with the instrumental data, was preferred.[4]

With the development of better instrumentation from the 1920s, earthquake locations became more reliable, and in particular more use could be made of P and S phase arrivals. Because of the growing diversity of instruments, however, it is not always easy to compare readings at different stations, and uncertainty due to misinterpretation of phases becomes more of a problem.

A continuing difficulty is that almost all stations are to the north, and particularly in Europe, and locations may be moved significantly according to the interpretation of later phases, and the consequent variation in origin time. International agencies place many earthquakes in the Mediterranean Sea, north of Libya and Egypt, that are on or just inside the northern boundary of our area. Local determinations by Athens Observatory, however, suggest that these events are further north.

The earthquake of **1 May 1931** (09h) is a good example of these difficulties in interpretation. It was recorded by only seven stations, and was originally placed by ISS in a position about 50 km west of the Red Sea, in the Sudan (18° N–37.75° E). Gouin (1979, pp. 109–10) carried out a computer relocation using an early program, which gave weight to only four stations, and placed the event near Jidda on the other side of the Red Sea (21.2° N–39.2° E). There were, however, no felt reports. In an earlier re-appraisal (Ambraseys and Adams, 1986b), we mistakenly found a position much further south in Zaire (3° N–27° E), by giving a different interpretation to S phases. By re-interpreting these phases, we now place the earthquake in a more likely position, in the central Red Sea (19° N–39° E, ± 1°). A late S phase at Helwan fits the phase S*, consistent with travel along oceanic-type crust in the median trough of the Red Sea, and a similar phase at Ksara fits

Sg, consistent with a path of continental type through the Arabian shield.

Even in modern times, gross errors of location have been found. The large, spurious event of **7 September 1953**, located near Iraq's border with Saudi Arabia, was included in regional and global seismic catalogues and used in the assessment of earthquake hazard for the design of a number of engineering projects in Saudi Arabia, until noticed by the authors in 1987. The month and day have been transposed, and ° W mistaken for ° E: the earthquake in fact occurred on 9 July 1953, on the Mid Atlantic Ridge.

In recent years, the number, location and sensitivity of recording stations has improved considerably, particularly in the Levant (see Figure 3.2). The combination of the additional local stations, with more comprehensive reporting from regional and distant stations, has resulted in greatly improved earthquake detection capability in the area.

Our investigations have not only removed some spurious instrumental events from the area, but have revealed the existence of an unknown earthquake on **18 February 1991**. This event occurred about 350 km southwest of Cairo in an area with no previously located instrumental earthquakes. Its location was not determined by NEIC or any local agency, but it was found by the 'search' procedure at ISC, which seeks previously undetected events. It is well located from 18 stations, ranging in distance from Helwan at 3.05° to Norsar at 35.0°, and including stations in Egypt, Saudi Arabia, Israel, Jordan, Cyprus, Turkey and Norway. The amplitude and period reported from Norsar give M_b 4.0, and local magnitudes of 4.2 and 4.1 respectively are reported by Helwan and Kottamia (Egypt). The earthquake is located near the Bahariya oasis, but we are unaware of any felt reports.

It is noteworthy that the advent of more stations in Saudi Arabia has not resulted in the detection of earthquakes in the interior of the Arabian shield, which supports the belief that the aseismicity of this area is genuine, and not the result of inadequate station coverage.

Magnitude determination and attenuation relationships

For the recent period, magnitudes of most recorded earthquakes can be determined instrumentally. For surface-wave magnitude M_s, we use the 'Prague' formula (Vanek *et al.*, 1962), and for body-wave magnitude m,

---

[4] See Section 2.1, under 1908 April.

the standard procedure of Gutenberg and Richter (1956).

For early events, those recorded poorly and those with only macroseismic observations, we have to employ other means. For early shocks recorded on Milne seismographs (up to about 1916), an equivalent M_s is determined with a formula developed by Ambraseys and Melville (1982). Other techniques are discussed below.

In Table 3.1 (see Section 3.3) values of M_s and m before 1965 are generally recalculated by us from measurements of amplitudes and periods read from available seismograms.

MAGNITUDE DERIVED FROM NUMBER OF RECORDING STATIONS   A semi-empirical estimate of a magnitude M_N, equivalent to surface-wave magnitude M_s, can be obtained from the number of stations recording an earthquake (Ambraseys and Melville, 1982). The constants in a relationship $M_N = a + b \log (N)$ may be estimated from known values of magnitudes and the corresponding number of recording stations for events in a given region and for a period during which there is no drastic change in the number or sensitivity of stations. For small to medium sized earthquakes the value of N depends on the regional rather than the global distribution of stations.

Table 3.1 shows the recalculated values of M_s and values of N extracted from world-wide and regional bulletins. N is the number of stations that recorded any phase and is often larger than the number of stations quoted by agencies such as BAAS, BCIS, USCGS, ISS, ISC or in special studies. For a period in the 1950s and early 1960s ISS published data only for larger events, and N is incompletely known for many smaller events; in Table 3.1 these values of N are followed by a plus (+) sign.

The analysis falls into three periods. For the early period, 1921 to 1949, consistent results give
$$M_N = 3.91 + 1.06 \log (N) \tag{1}$$
for $4.0 < M_N < 6.0$. For the period 1966 to 1987, after the establishment of the World-Wide Standard Seismograph Network, another consistent relationship was found for $4.0 < M_N < 5.8$
$$M_N = 2.05 + 1.40 \log (N). \tag{2}$$
For the period 1950 to 1965, the number and sensivity of stations varied considerably and only an approximate relationship
$$M_N = 3.86 + 0.74 \log (N) \tag{3}$$
can be found for the narrow range $4.3 < M_N < 5.5$.

For all three relationships the standard deviation of M_N is between 0.15 and 0.2, comparable with that with

which M_s can be determined from individual station readings. Equation 2, for the most recent period, is similar to that found for the same period in Iran (Ambraseys and Melville, 1982). Equations 1 to 3 were used to estimate magnitudes, mainly for events with $M_N < 5.0$ for which there was not enough instrumental information to determine M_s. These values appear bracketed in the M_s column of Table 3.1.

MAGNITUDE DERIVED FROM FELT EFFECTS   From information given in Table 3.1, relations can be established between the recalculated values of surface-wave magnitudes, M_s, and the various intensities I_i and their corresponding radii, r_i, to form an attenuation equation of form $M_s = f(I_i, r_i)$.

Figure 3.4 shows such a plot for events for which sufficient data exist. The plot suggests relationships between r_i and M_s, as well as between r_i and I_i, but there is considerable scatter, likely to be due to different attenuation rates and source efficiencies, probably of a regional character. There is not enough detailed information available to carry out a full regional analysis, but the information we have is just adequate for the events in Table 3.1 (and Figure 3.4) to be separated into two groups with different attenuation characteristics. This can be done by trial and error, and optimising the goodness of fit.

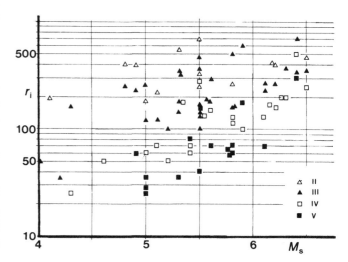

Figure 3.4. Plot of average distances r_i at which earthquakes of magnitude M_s were felt with intensity I_i, shown by different symbols, for the entire region.

The first group, shown in Figure 3.5, satisfies the relationship
$$M_s = 0.30 + 0.55\,(I_i) + 1.4 \times 10^{-3}\,(r_i) + 1.14 \log (r_i) \tag{4}$$
with a goodness of fit 0.84.

The second group, shown in Figure 3.6, satisfies
$$M_s = -0.38 + 0.56\ (I_i) + 1.6 \times 10^{-3}\ (r_i)$$
$$+ 1.14 \log\ (r_i) \tag{5}$$
with a goodness of fit 0.91.

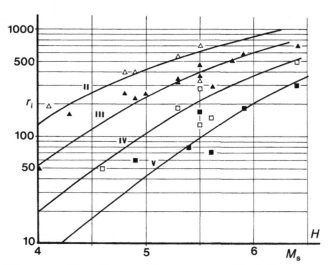

Figure 3.5. Plot of average distances r_i at which shallow earthquakes of high efficiency (or slow attenuation) of magnitude M_s were felt with intensity I_i, shown by different symbols. The location of this class of earthquakes is shown on Figure 3.7.

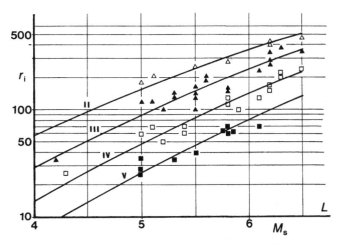

Figure 3.6. Plot of average distances r_i at which shallow earthquakes of low efficiency (or fast attenuation) of magnitude M_s were felt with intensity I_i, shown by different symbols. The location of this class of earthquakes is shown on Figure 3.7.

These equations are valid for shallow earthquakes in the magnitude range 4.0 to 6.5 and are restricted to intensities between II and V (MSK). The lack of near-field data of intensities greater than V arises from effects of low source efficiency, including focal depth, rather than from absorption along the path. Equation 4 is associated with more efficient sources, with slower

attenuation of intensity with distance, whereas equation 5 applies to less efficient sources, and also more rapid attenuation.

These equations can be used to estimate a felt magnitude, M_F, equivalent to surface-wave magnitude M_s for those earthquakes for which there is macroseismic information, but no instrumental determination of magnitude. For each event, an estimate of magnitude can be derived from each pair of values of intensity and radius and the results averaged. Such magnitude determinations appear in the M_s column of Table 3.1, with the value followed by an apostrophe ('). Equations 4 and 5 have also been used to determine magnitudes for historical earthquakes, see Table 2.1 (Section 2.2), appearing in the M_F column.

It is necessary to apply the appropriate relationship to each event, and Figure 3.7 shows the geographical distribution of events giving rise to the two types of attenuation.

Earthquakes showing low efficiency, that is, affecting a smaller area with more rapidly attenuating intensities, are found along the Zagros in Iran, in the Afar and in regions of Tertiary and Quaternary volcanism in Ethiopia and Yemen, as well as along the Red Sea and Gulf of Suez. On the other hand, the few earthquakes for which we have macroseismic data in the interior of Africa and southern Saudi Arabia, right up to the margin of the Red Sea and the Mediterranean, demonstrate high efficiency. Thus, earthquakes originating in the main trough of the Red Sea, which is on average about 150 km from land, need to be of magnitude greater than 6.0 in order to be felt with noticeable intensity, say, IV. In contrast, earthquakes of the same magnitude occurring in the interior will be felt with an intensity IV or greater out to about 380 km. Even where there is a low density of population, therefore, the chances of identifying historical events are better for those in shield areas than for events that are offshore, or in the volcanic areas of Ethiopia or Yemen. This is an important consideration not only in assessing hazard but also in the retrieval of historical information based on felt reports from coastal areas of the Gulf of Suez and the Red Sea. Even with a reliable reporting system from coastal areas, offshore earthquakes in these areas need to be of a relatively high magnitude before they will be reported.

FELT EFFECTS IN EGYPT FROM OFFSHORE EARTHQUAKES Most of the historical earthquakes felt in the densely populated region of the Nile Delta and Lower Egypt, and many of the recent ones, originate offshore in the Mediterranean Sea and the Hellenic Arc.

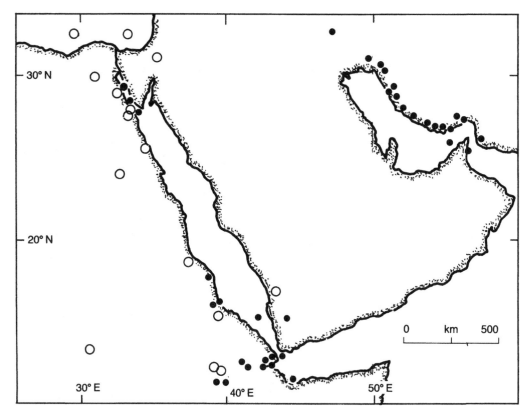

Figure 3.7. Location map of earthquakes of low (solid) and high (open) efficiency, deduced from their attenuation of intensity.

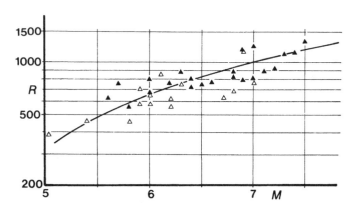

Figure 3.8. Maximum felt distance R (in km measured in the direction of Egypt) of shock originating in the Mediterranean, of intermediate (solid) and shallow (open) depth, plotted against magnitude M_s.

From Table 3.1, we see that the large distances at which twentieth-century shallow shocks from these areas are felt in Egypt are consistent with equation 4, for intensity II or III. Both shallow and intermediate shocks from these areas are very efficient, and intensities attenuate only slowly south of the Egyptian coastline. Felt reporting is enhanced by the extremely high population density along the Nile Valley, particularly in the region of Cairo, and by the effects of the unconsolidated alluvium.

Figure 3.8 shows the maximum felt distance in the direction of Egypt (R) against the magnitude of the earthquake, taken as M_s for shallow events, and m_b for intermediate events. The relationship may be approximated by

$$M = 2.48 + 2.4 \times 10^{-3}\,(R) + 0.7 \log(R). \qquad (6)$$

3.3 Catalogue of earthquakes (1899–1992)

Table 3.1 lists all the known events recorded in our study area, from the beginning of the instrumental era up to the end of 1992. The catalogue combines instrumental data with the macroseismic information presented in Section 2.1, although in many cases no macroseismic information has been found. This is true of the first instrumentally located earthquake in the area (1899 August 17), which was reported by the BAAS, without felt effects. Conversely, the location and magnitude of many earthquakes in the early part of the century depend entirely on macroseismic data, as is indicated symbolically on Table 3.1. Most, but not all, of these events are described in Chapter 2; those excluded are either located in Ethiopia, Iran or the Hellenic Arc, or so little macroseismic information was available that it did not justify an account of them. In later periods, instrumental determinations predominate. The list is the result of a systematic retrieval of data from worldwide and regional catalogues, as well as from special studies; spurious and unsubstantiated entries have been removed (see Section 3.4). The data in Table 3.1 are plotted on Figure 4.7.

The coverage is not uniform in time or area. For the period up to the end of 1987, all events originally located in Libya, Egypt, the Sudan, the Red Sea and the Arabian Peninsula have been reexamined and in most cases reassessed, using both teleseismic and macroseismic information. We claim to have a reasonable completeness of earthquakes of magnitude 4 and above for these regions. In other parts of the study area, only events of doubtful origin were reexamined. We have also attempted to list events outside the area that could have been felt within it, or whose errors of location are such that they could have occurred inside.

To the south, we have included major earthquakes in Ethiopia, the Gulf of Aden and the Arabian Sea, without claiming completeness below magnitude 5, and have deliberately not considered the area lying south of 10° N and east of 40° E. There is no attempt at completeness south of 3° N. To the north, in the active areas of Iraq and Iran, only earthquakes above magnitude 5 occurring immediately north of the Persian Gulf coast have been systematically included. North of 33° N in the Mediterranean and 31° N in the Levant, we have included only earthquakes greater than magnitude 5 and those events reported felt in North Africa. For reasons given earlier in this chapter (see p. 115), some small events in the Mediterranean Sea near the border of our area have been omitted. These geographical limits are shown in Figure 4.7, with broken lines denoting diffuse boundaries.

For the last five years, 1988 to 1992 inclusive, the catalogue is less complete. We have included all known events of magnitude 4 or greater in Libya, Egypt, the northern Sudan and the Yemen, and of magnitude 5 or greater for the remaining areas to the south. The list is dominated by sequences in Ethiopia (1989) and Sudan (1990). We have not included any additional events in southern Iran for this period. For earthquakes after mid-1991, for which ISC analysis is not yet available, we have relied mainly on the results of the Preliminary Determination of Epicenters service of NEIC. There is no easy way of searching these, and there may be a few omissions. Magnitudes are as given by the main source, not re-evaluated by us. The number of recording stations is not given from PDE solutions, as these would not be compatible with the numbers given for earlier events.

Aftershocks, except when of comparable magnitude with the main shock, have been excluded from Table 3.1.

Instrumental coverage is not adequate to allow precise determination of focal depth, and values are not given in Table 3.1. There is no evidence to suggest that the events studied were other than normal depth, though some earthquakes of intermediate depth, nucleating in the Hellenic Arc to the north of our area, are included in the catalogue.

Table 3.1. *Catalogue of earthquakes, 1899 to 1992*

(A key to this table appears on p. 138.)

Date	Time	Epicentre	M_s	m	N	I_o	Notes
1899							
Aug. 17	2038	16.0–56.0C	6.6	—	16+	—	GA
1900							
Jan. 18	0529*	29.0–33.0A	4.1*	—	10	III	GS 2/190
Mar. 6	1758	29.0–33.0R	6.2*	—	12	VI	GS 2/410; 3/270; 4/160; (2.34)
Apr. 2	1910	15.3–39.2*	4.6′	—	—	III	ET 3/55
Oct. 10	0302	16.0–60.0R	6.5*	—	9+	—	AS
—	—	31.8–47.2*	—	—	—	VIII	IQ
1901							
May	—	15.8–38.8*	—	—	—	VI	ET
Jun.	—	3.9–31.7*	—	—	—	VI+	SU
Nov. 11	—	15.3–39.3*	4.6′	—	—	IV	ET sr; 3/60
1902							
Jul. 9	0338	27.0–56.4*	6.4*	—	20+	VII	IR 3/340; af
1903							
May 6	02	30.1–31.2*	—	—	—	F	EG
Jun. 4	1458	2.0–35.0R	5.8	—	—		UG
Jul. 19	1808	(35.0–30.0)R	5.8	6.2	—	F	MD R450
Aug. 11	043335	(36.0–23.0)	6.6	7.2	—	F	HA R900; (2.35)
Sep. 9	—	30.4–48.2*	—	—	—	F	IR
Sep. 26	0542	15.5–39.0*	—	—	—	F	ET
Nov. 22	0945	32.5–13.5*	5.0	—	3	F	LY
1904							
Oct. 3	0307	12.0–58.0R	7.1*	7.1	31+	F	AS
1905							
Mar. 9	1710	15.2–38.9*	4.3′	—	—	F	ET 3/40
Apr. 25	1401	27.0–56.0*	5.8	—	4+	VII*	IR 3/160; 5/60; af
May 13	—	23.6–58.6*	—	—	—	F	GO af
Dec. 23	—	14.3–39.0*	4.0′	—	—	F	ET 3/30
1906							
Feb. 28	1145	18.4–31.8*	—	—	—	F	SU
Mar. 20	0348	17.0–41.0R	6.1	5.9	23+	F	RS 3/270
Aug. 15	1840	19.5–30.3*	—	—	—	F	SU
Dec. 26	1343	27.2–33.4*	4.9	—	3	V	EG 3/230; 5/60; af; (2.36)
1907							
Feb. 2	0904	33.0–21.0R	5.8	6.1	35	—	LY
Feb. 4	2105	26.3–34.8*	—	—	—	F	RS
Jul. 4	0921	27.0–55.0A	5.7	6.0	34	V	IR
Jul. 20	0337	12.8–45.0*	—	—	—	F	GA
1908							
Jan. 25	2004	15.0–44.0R	5.6	—	7+	VII	YE af
Apr. 2	0553	3.0–26.0R	6.1	—	—	—	SU
Dec. 18	153630	13.5–54.0R	6.9	6.7	38+	—	AS
Dec. 28	14	15.7–38.5*	—	—	—	F	ET
1909							
Feb. 19	03	11.6–43.1*	—	—	—	F	ET
Feb. 26	2057	27.6–33.6*	—	—	—	V	EG
1910							
Jan. 6	1920	29.9–33.8*	—	—	—	F	SI
Feb. 18	050918	(35.7–24.0)	5.7	6.8	—	F	HA R830
Mar. 6	1855	26.3–34.8*	4.5*	—	2+	F	RS
May 30	1230	10.0–27.0R	5.5	—	7	—	SU

Table 3.1. *Catalogue of earthquakes, 1899 to 1992* cont.

Date	Time	Epicentre	M_s	m	N	I_o	Notes
1911							
Jan. 5	015906	30.0–31.3*	—	—	1	IV	EG
Jan. 20	0405	29.3–51.2*	5.2	—	5+	IV	IR
Jan. 26	173536	28.0–33.7A	4.3*	—	1	IV	GS 4/25
May 13	—	14.1–43.3*	—	—	—	VI	YE
Aug. 22	2023	28.8–32.6*	4.3*	—	1	V	EG 3/160; af; (2.37)
Nov. 12	1930	28.2–33.6*	—	—	—	F	SI af
1912							
May 12	1545	11.7–43.2*	—	—	—	F	ET sr
Jul. 9	0818	3.0–33.0R	6.8	—	11	—	UG
1913							
Feb. 27	162254	17.2–38.8A	5.8*	—	27	VII+	ET 2/270; 4/130; 5/60; af; (2.38)
Mar. 24	103411	26.8–53.7*	5.8	—	22	VIII	IR 4/120
Mar. 27	1313*	16.0–39.0A	5.7*	—	29	—	ET
1914							
May 18	1046	31.3–15.3*	5.0	6.0	—	D	LY
1915							
Mar. 18	210030	14.0–42.0R	5.7	6.5	26+	—	RS
May 21	0419	4.4–31.9*	6.3	6.8	—	VIII	SU
Sep. 23	081443	16.0–39.0C	5.8	6.2	31	VI	ET
1916							
Jun. 19	0350	16.0–57.0I	5.4	5.6	16	—	AS
Jun. 21	0059	15.7–52.1A	4.8?	—	5+	—	GA
Jun. 24	0403	14.0–59.0A	5.3	—	12	—	AS
1917							
Feb. 22	102210	28.2–33.6*	—	—	—	F	GS
1918							
Sep. 23	021320	(36.0–28.0)	5.1	5.6	—	F	HA R620
Oct. 14	140605	33.0–22.0I	5.4	5.9	10	S	LY
1920							
Oct. 1	0210	29.5–31.3*	4.2′	—	1	V	EG 3/50
1921							
Aug. 14	131528	15.6–39.6C	5.9	—	41	VIII	ET 5/100; af
Sep. 21	110131*	15.6–39.7C	5.8	—	39	—	ET
1922							
Apr. 20	102200	(35.8–26.5)	5.5	6.4	—	F	HA R760
Jun. 18	193857	28.8–32.0A	(4.2)	—	2	—	EG
Jul. 29	204820	30.3–31.2A	4.0′	—	2	F	EG 3/55
Aug. 2	05	15.6–32.6*	—	—	—	F	SU
Aug. 11	081936	(36.0–28.0)	6.2	6.4	—	F	HA R620
Aug. 13	000953	(36.0–28.0)	6.7	7.1	—	F	HA R620
Sep. 7	010746	28.0–34.0A	(4.2)	—	2	—	SI
1923							
Jun. 24		4.3–33.7*	4.5	—	—	VI+	SU
Aug. 1	081630	(35.0–25.0)	7.1	6.9	—	F	HA R800
Aug. 5	2315	19.2–30.5*	—	—	—	F	SU
Dec. 8	133251	29.2–31.4*	4.3′	—	1	F	EG 3/90
Dec. 10	235328	13.5–50.0I	5.2	—	18	—	GA
1924							
Mar. 13	124740	16.0–38.5A	5.4	—	16	—	ET
Apr. 20	142654	14.5–52.0I	6.1	6.9	52	—	GA
Oct. 4	064105	15.0–44.0R	5.2	—	6+	—	YE
Oct. 19	—	11.3–43.5*	—	—	—	F	ET sr
Dec. 11	230100	25.2–56.3R	5.1	—	11	F	GO 2/220

Table 3.1. *Catalogue of earthquakes, 1899 to 1992* cont.

Date	Time	Epicentre	M_s	m	N	I_o	Notes
1925							
May 20	232733	30.0–32.0A	4.4'	—	1	IV	EG 3/80; 4/50; (2.39)
Jul. 30	184316	28.5–51.8R	5.1	—	16	F	IR
Sep. 24	043839	25.6–55.4R	5.5	6.1	38	F	PG 3/130
Dec. 18	055338	28.8–51.3*	5.4	—	19	VII	IR 4/60; af
1926							
Mar. 18	140609	(36.0–29.5)	6.8	7.3	—	F	HA R680
Apr. 23	013131	26.9–56.4R	5.3	—	24	V	IR 5/35
May 19	211405	26.2–58.8R	5.1	—	14	—	IR
Jun. 26	194634	(36.5–27.5)	7.1	7.5	—	VIII	HA R1300; (2.40)
Aug. 30	113812	(36.7–23.3)	7.4	7.4	—	F	HA R1100
Oct. 30	013810	11.8–43.5R	5.4	—	17	F	ET af
Oct.	—	25.7–32.6*	—	—	—	F	EG
1927							
Feb. 26	235424	13.5–50.0A	4.4	5.5	6+	—	GA
May 2	062024	32.5–31.0I	4.6	4.9	10+	—	MD
Jul. 1	081904	(36.7–22.7)	7.3	7.3	—	F	HA R1100
Jul. 11	130355	(32.0–35.5)	6.0	6.3	—	V+	DG R640; (2.41)
Sep. 24	002750	29.0–35.0R	4.9	—	5+	F	SA 2/400
Oct. 18	003923	28.7–31.7A	—	—	2+	F	EG
Nov. 12	144603	32.5–47.4*	5.6	6.0	32	VII	IR 3/180; af
1928							
Mar. 19	100206	14.5–53.5C	5.6	—	26	—	GA
Apr. 15	100928	27.5–52.1R	5.0	—	9+	—	IR
Apr. 27	131820	(35.0–30.0)	5.1	—	10	F	MD R560
Sep. 18	195237	14.0–52.0C	5.9	—	78	—	GA
Sep. 20	145900	25.2–56.8A	4.6	—	4	—	GO
1929							
Jan. 22	144305	11.7–43.2A	5.8	—	45	VII	ET 3/160; 5/70; af; (2.42)
Feb. 5	015713	31.5–47.5I	4.8	—	8	—	IR
Mar. 16	123040	14.2–53.8I	(4.9)	—	9	—	GA af
Apr. 28	045832	14.5–54.0I	(5.2)	—	15	—	GA
May 1	193644	(34.0–28.0)	5.1	—	—	F	MD R540
May 18	010216	12.1–41.4R	5.6	—	35	F	ET
Jul. 16	194315	28.7–51.9A	5.1	—	11	—	IR
Nov. 11	073615	(36.8–26.5)	5.9	—	—	F	HA R870
1930							
Feb. 14	183820	(35.7–24.8)	7.2	6.8	—	F	HA R88C
Feb. 15	190705	28.7–51.8R	5.2	—	24	IV	IR 3/100
May 30	065852	29.4–34.7A	(4.2)	—	2	—	SI
Jun. 24	214750	27.3–30.9*	—	—	1	F	EG
Aug. 5	2030	30.0–48.5*	—	—	—	F	IQ
Oct. 9	213030	21.0–60.0I	5.2	—	20	—	AS
Oct. 24	104706	10.4–42.8C	5.8	—	28+	VII	ET af; sr; (2.42)
Nov. 16	204630	(34.1–32.2)R	5.4	—	—	F	MD R460
1931							
Feb. 18	0212	14.8–42.2*	—	—	—	F	RS
Apr. 24	152037	31.1–19.9I	4.2	—	8	F	LY
May 1	094829	19.0–39.0R	5.5	—	8+	—	RS
May 5	064221	26.5–54.3R	5.1	—	26	—	IR af
Jun. 23	121253	13.5–52.0A	5.2	—	6+	—	GA
Jun. 24	234704	15.0–59.7I	5.4	—	28	—	AS
Oct. 1	054449	30.2–31.7A	3.7'	—	1	F	EG 3/30

Table 3.1. *Catalogue of earthquakes, 1899 to 1992* cont.

Date	Time	Epicentre	M_s	m	N	I_o	Notes
1932							
Feb. 1	073824	10.5–42.6C	(5.0)	—	11	—	ET
Jun. 11	083256	15.0–53.5C	5.4	—	38	—	GA
Aug. 14	123608	13.5–56.0C	5.2	—	18+	—	AS
Dec. 7	075546	(36.5–27.8)	5.4	—	—	F	FA R660
1933							
Jul. 24	0344	29.9–31.3A	3.7′	—	1	IV	EG 3/30; 4/10
1934							
Feb. 16	075950	25.9–55.4R	(4.9)	—	9	—	PG
Mar. 18	224431	26.1–53.6R	5.0	—	15	F	PG af
Mar. 19	032828	27.3–52.7R	5.2	—	21	—	IR
May 12	164621	29.1–34.9A	(4.2)	—	2	—	SA
Sep. 5	022103	14.0–49.0I	5.3	—	10+	—	GA
Nov. 9	134056	(36.7–25.7)	6.6	6.3	—	F	HA R900
1935							
Jan. 18	020837	14.5–56.0C	5.5	—	25	—	AS
Feb. 25	025137	(35.7–25.0)	6.7	7.1	—	F	HA R890
Mar. 18	084041	(35.5–27.0)	6.4	6.5	—	F	HA R740
Apr. 19	152324	31.4–15.4R	7.0	7.6	164	D	LY
Apr. 19	161841	31.2–15.4R	4.7	—	33	—	LY
Apr. 19	173654	31.2–15.4R	—	—	12	—	LY
Apr. 19	1750	31.2–15.4R	—	—	14	—	LY
Apr. 19	175747	30.8–15.5I	5.3	5.9	63	—	LY
Apr. 19	203140	31.6–15.4R	5.6	6.3	79	—	LY
Apr. 20	051051	31.1–15.6R	6.1	6.4	117	—	LY
Apr. 29	1927	30.8–15.5I	4.0	—	15	—	LY
May 31	131641	31.9–47.9R	5.2	—	42	F	IQ
Jun. 14	190214	29.9–31.3*	4.1′	—	1	F	EG 3/65
Jul. 2	152459	26.5–55.0R	5.1	—	20	—	IR
Oct. 15	170245	28.9–51.3*	5.2	—	14	VI	IR 4/50; af
Oct. 20	045130	18.0–60.0C	5.3	—	22	—	AS
1936							
Jan. 8	123438	26.9–52.9R	4.9	—	20	—	IR
Apr. 21	021438	26.3–55.3R	5.5	6.2	31	—	IR
Jun. 13	003234	32.4–22.5R	4.8	—	54	—	LY
Jun. 17	084053	25.9–54.7R	4.6	—	6	—	PG
Aug. 16	213701	26.0–54.4R	5.1	—	27	—	IR
Aug. 20	020847	30.4–51.5R	5.3	—	22	—	IR
Oct. 20	124726	13.5–52.0A	5.1	—	6+	—	GA
1937							
May 29	152240	(35.5–32.0)	5.0	—	—	F	HA R610
Jul. 27	2044	30.6–15.4*	—	—	1	S	LY
—	—	14.4–44.5*	—	—	—	F	YE V
1938							
Mar. 12	130440	10.5–44.5I	5.6	—	18+	F	ET 3/190; af
May 12	213135	18.3–37.6R	5.8	—	41	VI	SU 3/500
Sep. 18	003900*	10.4–40.8C	5.6	—	27	—	ET af
Sep. 27	023149	10.8–40.6R	6.1	6.0	94	—	ET af
Sep. 28	215605	32.5–33.5A	—	—	2	F	MD
Oct. 20	131505*	10.1–40.2C	5.7	—	45	—	ET
Oct. 23	022854*	10.0–40.3C	5.8	—	57	—	ET
Nov. 6	153618	30.2–31.8A	4.0′	—	1	F	EG 3/50
1939							
Jan. 20	012412	31.0–15.8R	5.4	—	42	S	LY
Jan. 20	142226	31.1–15.9R	5.1	—	48	F	LY

Table 3.1. *Catalogue of earthquakes, 1899 to 1992* cont.

Date	Time	Epicentre	M_s	m	N	I_o	Notes
Jan. 23	022251	31.7–16.1R	5.6	—	56	F	LY
Jan. 25	110225	29.3–51.0*	5.4	—	31	VII*	IR af
Feb. 2	235044	31.7–16.1R	5.0	—	30	—	LY
Feb. 25	050508	21.0–60.0I	5.3	—	20	—	AS
Aug. 18	225235	26.8–54.5R	5.0	—	16	—	IR
Sep. 29	2015	4.8–31.6*	—	—	—	F	SU
1940							
Aug. 13	052011	13.5–51.5I	5.2	—	9+	—	GA af
Aug. 14	084922	11.0–46.0I	5.4	—	16	—	ET
1941							
Jan. 11	083259	16.4–43.5R	5.9	6.5	93	VIII	YE 3/600; 5/180; af; (2.43)
Feb. 4	091744*	16.0–43.0A	5.2	—	17	VI+	YE
Mar. 4	234510	30.8–15.8C	4.8	—	21	—	LY
Mar. 19	013150	10.9–44.0R	5.3′	—	12	F	ET 3/120
Apr. 28	092128	32.4–32.0A	(4.2)	—	2	—	MD
Jun. 15	123855	26.4–53.3R	5.2	—	20	—	IR
Jul. 17	233430	14.0–52.0I	5.4	—	22+	—	GA
Sep. 25	034547	14.5–53.5C	5.0	—	10+	—	GA
Nov. 13	0611	3.0–30.9*	—	—	—	V	UG
1942							
Jun. 21	043843	(36.6–27.2)	6.1	6.4	—	F	HA R820
Sep. 21	212902	29.3–31.0A	4.1′	—	1	F	EG 3/60
Nov. 18	120120	11.3–39.2C	5.4	—	23	F	ET 5/80
Dec. 21	212130	14.0–52.0A	5.2	—	19	F	GA
1943							
Jun. 10	185151	27.8–33.8A	—	—	1+	—	GS
Jul. 16	015358	32.3–20.7R	4.8	—	18	S	LY
Oct. 16	130846	(36.4–28.4)	5.5	6.6	—	F	HA R760
Nov. 22	22	11.4–34.6*	—	—	—	F	SU
1944							
Dec. 3	1050	15.6–32.6*	—	—	—	F	SU
1945							
Jan. 15	172133	27.0–54.8R	5.6	—	25	—	IR af
Jan. 27	031336	32.0–32.0A	—	—	1	—	MD
Mar. 12	013845	31.1–47.1R	5.4	—	23	—	IQ
Mar. 31	220748	14.9–39.5C	5.3	—	34	—	ET 3/340
Apr. 1	1020	15.5–36.4*	—	—	—	F	SU
Sep. 2	115357	(33.7–28.5)	5.9	6.4	—	F	MD R490+
Oct. 28	001710	11.2–42.7I	5.5	—	36+	VI+	ET af
Dec. 1	0850	19.2–30.5*	4.5	—	1	VI+	SU
1946							
Jun. 16	100455	14.0–52.0A	(5.3)	—	21	—	GA
Aug. 24	002924	13.5–51.5A	5.5	—	17+	—	GA
Sep. 29	202207	13.0–48.0C	(5.6)	—	38	—	GA
Oct. 23	080205	30.0–47.5A	4.9	—	10	—	IQ
1947							
Jan. 2	141108	28.5–51.9R	(5.0)	—	10	—	IR
Jan. 16	112906	13.7–55.4I	5.4	—	22	—	AS
Feb. 26	1550	11.9–34.4*	—	—	—	F	SU
May 4	223408	26.5–55.3R	5.1	—	26	—	IR af
May 24	001023	12.1–48.7I	5.5	—	45	—	GA
May 24	151140*	12.1–48.7A	5.3	—	15	—	GA
Jun. 16	001806	13.9–56.7I	5.1	—	34	—	AS
Jun. 27	150816	10.0–47.0I	(5.0)	—	11	—	ET
Jul. 3	033028	28.9–33.1A	(4.2)	—	2	—	GS

Table 3.1. *Catalogue of earthquakes, 1899 to 1992* cont.

Date	Time	Epicentre	M_s	m	N	I_o	Notes
Oct. 3	061350	25.9–57.4R	5.8	5.8	74	V	IR 5/60
Oct. 6	195534	(37.0–22.0)	6.9	—	—	F	HA R1150
Dec. 10	191801	27.5–34.0A	(4.2)	—	2	—	RS
1948							
Feb. 1	233828	26.9–57.8R	5.3	—	21	—	IR
Feb. 17	221106	13.7–39.8C	(5.5)	—	35	—	ET
Apr. 30	145042	(35.9–31.1)	5.8	6.0	—	F	MD R650
May 12	0215	3.0–30.9*	—	—	—	VI	UG
Jul. 24	060350	(34.4–24.5)	7.1	7.0	—	F	HA R800
Nov. 27	1132	4.9–31.6*	—	—	—	F	SU
1949							
Jan. 6	021801	32.0–33.0A	(4.4)	—	3	—	MD
Apr. 24	042216	27.0–56.5R	6.3	6.5	122	VII+	IR 4/200; 7/20; af
May 22	130150	29.9–31.3*	—	—	1	F	EG
Jun. 16	175809	11.6–42.6A	5.3	—	63	F	ET af
Jun. 17	042056	(34.4–28.5)	4.9	5.8	—	F	MD R550
Nov. 5	234714	27.4–34.0A	(4.2)	—	2	—	RS
Dec. 1	103158	29.0–33.0A	—	—	1	—	GS
1950							
Jan. 19	172716	27.3–52.8*	5.5	5.8	78	VIII*	IR 3/140; 7/12; af
Feb. 2	154600	29.0–49.0C	4.7	—	12	—	PG
Feb. 2	224513	25.5–54.0I	4.7	—	12	—	PG
Feb. 27	035512	26.0–52.0C	4.5	—	13	—	PG
Mar. 2	101935	28.0–34.0A	—	—	1	—	GS
Mar. 26	165330	14.3–39.3R	5.2	—	47	—	ET
May 9	061030	12.0–47.3I	5.5	5.9	95	—	GA
Aug. 2	135002	14.5–39.7C	5.5	6.3	191	—	ET
Sep. 18	003930	13.7–42.2B	—	—	—	—	RS
1951							
Jan. 30	230723	32.4–33.4I	5.5	6.0	125	F	MD 2/330; 4/130; (2.44)
May 15	115147	12.1–46.7I	4.8	—	35	—	GA
May 28	141621	31.8–27.0B	(4.8)	5.3	20	—	MD
Jul. 3	052346	12.2–45.8A	5.5	5.7	79	—	GA af
Jul. 3	181601*	12.2–45.8A	5.3	5.2	73	—	GA
Jul. 21	032257	13.9–56.7A	5.3	5.7	69	—	AS af
Jul. 23	164003*	13.9–56.7A	5.2	5.6	43	—	AS
1952							
Feb. 25	224805	29.6–32.3A	4.0'	—	1	F	EG 2/120
Mar. 9	011544	31.5–13.0B	—	—	—	—	LY
Mar. 22	045233	27.2–34.5B	5.0	—	20	—	RS
Sep. 10	090617	14.5–40.2C	5.3	—	34	—	ET
Oct. 10	1306	16.0–56.0B	—	—	—	—	AS
Dec. 23	2230	30.9–49.4*	5.1'	—	—	V	IR 3/100
1953							
Feb. 1	195236	(33.5–32.0)	5.0	5.5	—	F	MD R390
Feb. 14	084313	(35.5–26.5)	5.1	6.2	—	F	HA R750
Mar. 30	044935	13.9–51.7B	—	—	—	—	GA
Apr. 14	111224	27.0–31.5A	—	—	1	—	EG
May 14	032145	27.0–31.5A	—	—	1	—	EG
Sep. 10	040603	(34.9–32.2)	6.2	6.3	—	F	MD R540
Sep. 25	2303	13.0–50.5B	—	—	—	—	GA
Nov. 12	012032	11.5–43.0B	4.8'	—	—	V+	ET 5/20
Nov. 18	183929	13.1–57.4I	5.3	5.8	33	—	AS
Dec. 21	0140	4.8–31.6*	—	—	—	V	SU
Dec. 30	040933	31.8–46.5R	4.8	5.2	11	—	IQ

Table 3.1. *Catalogue of earthquakes, 1899 to 1992* cont.

Date	Time	Epicentre	M_s	m	N	I_o	Notes
1954							
Jan. 11	224506	13.7–51.2B	5.4	—	—	—	GA
Feb. 28	212343	27.0–56.0B	4.7	—	—	—	IR
Mar. 31	182545	12.4–57.9I	7.0	7.4	275	—	AS
Apr. 6	081248	32.0–30.0A	—	—	1	—	MD
Jun. 28	182926	27.5–34.0A	—	—	1+	—	RS af
Jul. 3	090001*	27.5–34.0A	—	—	1+	—	RS af
Jul. 24	005240	31.8–30.6R	(4.8)	—	18	F	MD 3/70
Aug. 20	153030	27.8–52.1*	5.0	5.6	14+	VII+	IR 2/180; 7/10
Sep. 13	151332	28.7–34.8A	—	—	2+	—	DS
Sep. 13	214630	30.9–35.3*	4.8	—	17	VI	DS 2/400; 3/250
Sep. 19	041637	25.0–53.0A	(4.4)	—	6	—	PG
Oct. 28	133901	32.4–31.4B	(4.6)	—	11	—	MD
1955							
Jan. 4	2226	4.8–29.5*	—	—	—	VI+	SU
Jan. 17	153513	12.3–46.0C	—	—	5	—	GA
Jan. 28	074159	33.9–23.6I	—	5.4	43	—	LY
Feb. 22	132031	28.7–34.8A	—	—	1+	—	DS
Mar. 3	004340	16.5–41.2C	(4.5)	5.0	5	—	RS
Apr. 18	191610	27.7–52.3*	5.0	5.4	—	VII	IR 6/15
Apr. 26	013720	14.5–56.5R	—	6.0	7+	—	AS
Sep. 4	2213	1.7–30.9C	5.2	—	25	F	UG
Sep. 12	060924	32.2–29.6I	6.4	6.5	268	VI	MD 3/700; 4/500; 5/300; af; (2.45)
Oct. 17	200853	17.2–43.6C	4.8	5.7	14	F	YE
Nov. 12	053214	25.3–34.6C	5.3	6.2	92	F	EG 2/550; 3/350; 4/180; af
1956							
Feb. 15	154925	27.8–53.1*	5.3	5.4	76	VII	IR af
Jun. 25	201027	20.7–38.0C	(4.7)	—	12	—	RS
Nov. 10	081855	25.9–54.7B	5.0	—	10+	—	IR
1957							
Mar. 14	001139	14.9–40.2C	4.5	—	13	—	ET
Mar. 26	044927	27.7–52.5R	—	—	24	—	IR
Apr. 12	155844	11.5–43.1I	5.0	—	12+	VI	ET af; 5/25
Apr. 25	022542	(36.5–28.8)	7.0	—	—	F	HA R750
May 2	02	—	—	—	—	F	LY
Aug. 8	011215	32.4–25.2I	(5.1)	5.6	54	—	MD
Sep. 3	204530	31.0–47.5B	—	—	—	—	IQ
Oct. 2	130908	26.7–54.8R	5.2	5.5	17+	—	IR
Oct. 14	212011	30.5–31.5A	—	—	2	F	EG
Dec. 19	155635	14.5–53.4C	(4.6)	—	11	—	AS
1958							
Jan. 9	075627	17.7–40.1C	(4.5)	—	7	—	RS
Feb. 13	102334	14.3–41.9C	4.8	—	20	—	RS
Apr. 9	043634	29.1–52.0R	5.0	5.8	81	—	IR
Apr. 13	120203	29.0–34.8A	4.0	—	4+	F	DS
May 17	052533	31.8–11.3I	4.6	—	38	—	LY
May 24	235338	12.1–43.6C	5.2	5.6	44	—	GA af
May 25	025348*	12.1–43.4C	4.9	5.7	37	—	GA
Jun. 28	170516	11.9–45.4C	4.6	—	9	—	GA
Jun. 30	084244	(36.6–27.3)	5.2	6.0	—	F	HA R810
Sep. 16	185227	33.0–31.5A	—	—	3	—	MD?
Oct. 12	063653	32.5–33.5A	—	—	3	—	MD
Nov. 4	050612	14.1–53.6C	(4.6)	—	11	—	GA
Nov. 13	231922	15.0–53.8C	—	—	14+	—	GA af
Dec. 4	102548	13.8–51.7C	(4.9)	—	26	—	GA
Dec. 25	183328	26.9–54.1R	5.0	—	30	—	IR af

Table 3.1. *Catalogue of earthquakes, 1899 to 1992* cont.

Date	Time	Epicentre	M_s	m	N	I_o	Notes
1959							
Jan. 4	035815	15.6–53.8I	5.4	5.8	42	—	AS
Jan. 5	081715*	13.7–51.6C	(4.9)	5.8	25	—	GA
Jan. 7	051309	27.0–54.2R	(5.2)	5.3	76	—	IR
Jan. 21	135730	13.6–51.8C	5.2	5.8	29+	—	GA af
Apr. 1	164729	27.5–34.0A	—	—	2+	—	RS
Apr. 29	002345	27.0–54.8R	5.0	5.4	34	—	IR
Jun. 7	090348	14.5–53.7C	(4.7)	—	13	—	AS
Aug. 16	133115	14.5–43.1C	(4.6)	—	9	—	YE
1960							
Jan. 4	061631	11.6–42.8C	5.4	—	56	VI	ET af
Jan. 25	213424	27.5–51.0B	—	—	—	—	PG
Feb. 13	020138	27.5–34.0A	—	—	2	—	RS
Mar. 25	094540	12.1–46.4C	5.2	—	21+	—	GA
May 20	—	4.9–31.6*	—	—	—	F	SU
May 25	003029	33.0–33.0A	—	3.5	3	—	MD
May 31	002351	14.7–54.8R	(5.1)	—	50	—	GA
Jun. 10	134921	26.5–53.0C	4.5	5.0	—	—	PG
Jul. 9	041012	33.0–25.0B	(4.7)	—	15	—	MD
Jul. 10	225610	26.5–53.0B	4.5	5.0	6	—	PG
Aug. 8	122808	12.1–44.5I	5.3	—	44	—	GA
Aug. 13	222814	15.1–40.1C	4.5	—	12	—	ET af
Aug. 23	042836	13.0–52.0B	—	—	5	—	GA
Sep. 12	031344	11.8–46.6C	—	—	6+	—	GA
Oct. 23	192108	17.5–40.1C	(4.5)	—	8	—	RS
Nov. 4	165200	27.0–54.0B	—	5.8	6	—	IR
Dec. 16	164915	14.8–42.5C	5.3	—	30+	—	RS
1961							
Jan. 13	152906	33.0–27.0A	—	—	4	—	MD
Feb. 7	025753	14.6–53.9C	5.0	—	26	—	GA
Mar. 11	084104	11.6–43.0I	5.5	6.0	85	VI	ET 3/100; af
May 11	050241	27.5–34.0A	—	—	3	—	RS
May 23	024518	(36.5–28.6)	6.3	6.5	—	F	HA R750
Jun. 1	232919	10.6–39.8C	6.3	6.5	147	VIII	ET 3/370; 4/200; af; (2.46)
Jun. 2	045111*	10.4–39.9C	6.1	6.2	142	—	ET af
Jun. 20	032129	12.2–44.3C	5.3	—	77	—	GA
Jul. 15	061121	27.5–34.0A	—	—	2	—	RS af
Aug. 3	004131	14.5–52.2C	4.7	5.5	29	—	GA
Sep. 15	014610	(34.9–33.8)	5.7	6.0	—	F	MD R590
Oct. 25	162412	14.2–56.4C	5.3	—	61	—	AS
Nov. 10	135233	13.2–51.7C	(4.7)	—	14	—	GA
Dec. 8	104038	13.4–50.2C	(4.7)	—	13	—	GA
1962							
Feb. 8	201015	31.0–49.0C	4.4	—	6	—	IR
Mar. 8	2138	3.7–29.0R	5.1	—	76	—	
May 24	082812	13.2–48.0C	4.7	—	—	—	GA
Jun. 23	050501	29.7–49.3R	5.4	5.6	43	—	IR
Jun. 24	150824	13.3–48.9C	4.9	—	—	—	GA
Jul. 15	215220	14.1–53.5C	4.8	5.2	37	—	GA
Aug. 25	005417	17.1–40.1C	4.7	—	12	—	RS
Aug. 28	105955	(37.7–22.3)	6.0	6.9	—	F	HA R1200
Sep. 1	003813	12.7–48.1C	4.7	—	23	—	GA
Sep. 9	—	15.3–37.5*	—	—	—	F	ET
Sep. 16	07	29.7–31.5*	3.7'	—	1+	F	EG 3/30
Nov. 11	151534	17.2–40.6C	5.3	6.2	95	—	RS

Table 3.1. *Catalogue of earthquakes, 1899 to 1992* cont.

Date	Time	Epicentre	M_s	m	N	I_o	Notes
Nov. 24	0208	29.4–30.7A	4.0′	—	1+	F	EG 3/50
Dec. 21	174726	13.9–51.6C	5.3	—	44	—	GA
1963							
Feb. 7	164506	14.5–53.5C	4.6	—	31	—	GA
Feb. 21	091712	12.0–43.0B	(4.6)	4.8	11	—	ET
Feb. 21	171431	32.6–21.0*	5.4	5.3	174	D	LY
Feb. 21	1724	32.6–21.0IU	—	4.0	7	—	LY
Feb. 21	1833	32.6–20.9R	—	4.6	35	—	LY
Feb. 21	183619	32.6–21.0IU	—	4.0	9	—	LY
Feb. 21	202640	32.6–21.0R	4.4	4.8	70	—	LY
Feb. 22	024715	32.6–21.0IU	—	4.3	23	—	LY
May 20	101920	25.7–56.5B	(4.8)	—	16	—	GO
Jun. 28	0811	33.5–13.0B	—	—	—	—	LY
Jul. 2	191224	28.8–33.3B	—	—	9	—	GS
Jul. 14	171810	15.6–39.0B	(4.9)	—	29	IV+	ET 4/30; af
Jul. 21	060148	14.1–56.5I	5.1	—	83	—	AS
Aug. 19	20	4.1–30.7*	—	—	—	F	SU
Sep. 18	2249	26.0–53.0C	—	—	2+	—	PG? af
Oct. 4	132944	18.0–60.0B	4.7	5.4	48	—	AS
Oct. 5	145743	11.5–42.8I	5.7	5.7	125	V	ET af
Nov. 1	070235	14.2–53.6B	5.0	—	40	—	GA
1964							
Jan. 19	091353	26.8–54.9*	5.3	5.6	131	VII*	IR 3/140; af
Feb. 9	060730	25.7–36.5C	(4.9)	—	26	—	RS
Mar. 19	094236	14.4–56.4I	5.4	5.8	84	—	AS af
Mar. 31	213832	32.6–20.9B	—	4.3	13	—	LY
Jun. 7	0510	4.8–31.6*	—	—	—	F	SU
Jul. 17	023427	(38.0–23.6)	5.7	—	—	F	HA R1100
Jul. 20	133112	14.1–53.7I	(4.8)	—	18	—	GA
Aug. 3	104509	17.5–39.1B	(4.9)	—	25	—	ET
Aug. 12	192627	30.9–49.7I	(5.4)	5.1	109	VII	IR 3/130
Aug. 25	092942	32.9–27.5I	(4.6)	—	11	—	MD
Sep. 7	112713	15.1–53.4I	5.2	5.1	72	—	GA
Sep. 11	102850	12.5–48.5A	—	—	—	—	GA
Oct. 14	172603	14.5–53.7I	4.8	5.3	59	—	GA af
Oct. 18	173127	31.5–32.5A	(4.2)	—	3	—	MD
Nov. 15	093348	30.0–50.9R	5.0	5.1	48	F	IR
Dec. 9	182144	32.0–29.5I	(4.2)	—	3	—	MD
Dec. 31	161802	(35.8–25.5)	5.0	—	—	F	HA R830
1965							
Jan. 25	1125	31.0–31.5C	—	—	2+	—	EG
Mar. 7	073236*	12.2–46.3I	5.1	5.2	68	—	GA
Mar. 7	074232	12.1–46.3I	5.2	5.4	94	—	GA af
Mar. 31	094726	(38.1–21.6)	6.6	7.0	—	F	HA R1250
Apr. 9	235702	(35.1–24.3)	6.1	6.3	—	F	HA R850
May 16	004557	11.0–45.6I	—	—	8	—	GA
May 18	102714	13.3–49.7I	(4.7)	4.5	12	—	GA
Jun. 7	134358	11.5–41.5I	(5.0)	4.9	41	IV	ET 3/50; af
Jun. 15	164116	14.1–51.7I	5.0	5.2	75	—	GA
Jun. 20	163120	13.3–50.3I	4.8	5.0	50	—	GA
Jun. 29	220433	15.5–39.0*	—	—	1	F	ET
Jul. 19	154935	11.8–42.5C	—	4.7	7	—	ET
Sep. 11	015321	12.7–50.4U	—	—	7	—	GA?
Sep. 21	062603	32.8–21.4U	—	4.2	13	—	LY
Oct. 23	100417	27.5–34.0A	—	—	3	—	RS
Nov. 9	100121	27.8–34.0A	—	—	3	—	RS

Table 3.1. *Catalogue of earthquakes, 1899 to 1992* cont.

Date	Time	Epicentre	M_s	m	N	I_o	Notes
Nov. 10	100334	27.0–54.6R	5.0	4.3	36	—	IR
Nov. 28	052605	(36.1–27.4)	5.4	5.7	—	F	HA R760
Dec. 29	203546	12.7–48.1U	—	—	5	—	GA?
Dec. 30	085414	18.9–39.7C	(4.6)	4.1	10	—	RS
1966							
Jan. 21	123946	12.1–43.6C	—	4.7	11	F	GA sr
Jan. 25	112442	30.2–31.5A	3.0'	—	3	F	EG 2/25
Feb. 5	042010	15.5–39.0*	—	3.8	1	F	ET
Mar. 6	022346	33.0–34.0A	—	—	2+	—	MD
Apr. 9	191112	14.5–40.7C	—	4.8	16	—	ET
Apr. 18	081422	13.0–48.4I	4.8	5.4	128	—	GA
May 9	061905	14.2–52.0U	(4.2)	—	32	—	GA
Sep. 2	104126*	13.4–50.9I	4.1	4.6	39	—	GA
Sep. 9	204207	14.7–52.3I	4.4	4.7	77	—	GA
Oct. 9	064839	12.6–30.7I	5.6	5.1	139	VIII*	SU 3/290; 4/150; 5/70; af; F; (2.47)
Oct. 9	102828	12.7–30.9I	4.8	4.1	38	VI	SU
Oct. 15	065421	13.1–50.5I	4.1	4.6	33	—	GA
Dec. 25	054250	14.3–53.5I	4.7	5.0	45	—	GA af
1967							
Jan. 2	081937	32.4–22.6I	4.6	4.6	114	—	LY
Jan. 29	075640	26.5–55.2I	5.5	5.1	166	VI	IR 3/170; af
Mar. 13	192216	19.8–38.8C	5.0	5.6	181	—	RS af
Mar. 14	093105	31.5–31.5A	—	—	3	—	EG
May 8	135827	4.3–35.7C	4.4	—	—	VI	KE
May 17	175040*	19.8–38.8C	4.7	5.2	134	—	RS
May 19	155239	14.7–40.2I	(4.6)	5.0	69	F	ET af
Jun. 7	212233	33.8–23.4I	—	—	7	—	LY
Jul. 6	185840	13.3–50.8I	5.0	4.9	81	—	GA af
Sep. 9	223559	3.9–32.6C	4.1	4.7	8	F	SU
Sep. 18	020302	15.8–38.9I	4.0'	4.8	33	F	ET 3/50; af
Sep. 21	183627	18.1–40.1C	—	4.4	16	—	RS
Nov. 16	022203	15.2–39.5C	—	5.1	7+	F	ET
Nov. 23	083555	14.5–52.0I	6.6	6.7	273+	—	GA af
Nov. 25	090625	30.6–30.2A	—	—	1	—	EG
Dec. 11	182725	27.5–33.8A	—	—	1+	—	EG
Dec. 11	223021	13.6–51.6I	5.2	5.5	139	—	GA
Dec. 14	022026	14.2–53.7I	5.0	5.2	105	—	GA af
1968							
Feb. 8	122826	14.6–54.1I	5.5	5.4	165	—	AS
Mar. 30	051325	33.1–24.8I	—	—	8	—	LY
May 23	233608	14.9–39.9I	(4.1)	4.8	28	F	ET 2/90
Jul. 16	081218	18.0–33.2A	—	—	1+	—	SU 4.0L
Aug. 6	083440	14.0–51.5I	4.9	5.0	96	—	GA
Sep. 26	1730	29.0–31.3A	—	—	1+	—	EG
Dec. 12	173030	12.1–46.1I	(4.5)	4.5	53	—	GA
1969							
Jan. 14	231206	(36.1–29.2)	6.3	5.9	—	F	HA R700
Feb. 8	232334	29.8–50.9C	5.0	5.1	98	—	IR
Mar. 5	2130	31.2–29.9*	—	—	—	F	EG
Mar. 29	091554	11.9–41.2I	6.2	6.0	204	IX*	ET 2/400; 4/170; af; F; (2.48)
Mar. 29	110452*	11.9–41.4I	5.8	5.5	166+	—	ET
Mar. 29	110745*	12.0–41.1I	5.7	5.3	55+	—	ET
Mar. 31	071554	27.6–33.9I	6.6	6.1	295	VII	RS 2/460; 3/350; 4/240; af; (2.49)
Apr. 5	021830*	12.0–41.3I	6.1	5.8	175	—	ET
Apr. 5	152349	4.5–31.6I	—	—	8	—	SU

Table 3.1. *Catalogue of earthquakes, 1899 to 1992* cont.

Date	Time	Epicentre	M_s	m	N	I_o	Notes
Apr. 22	223440	12.8–58.2I	5.3	5.6	162	—	AS
Jun. 12	151331	(34.4–25.0)	6.0	5.8	—	F	MD R570
Jun. 13	125507	33.9–24.8I	—	—	8	—	LY
Jun. 24	1947	31.1–29.9*	—	—	—	F	EG
Jul. 15	1633	3.5–31.4C	—	—	—	—	SU Strasbourg only
Sep. 20	232115	32.3–28.0I	—	—	7	—	MD
Sep. 26	045438	16.4–41.0I	5.0	5.0	94	—	RS
Oct. 5	084438	33.0–19.5A	—	—	3	—	LY
Oct. 13	1754	31.2–29.9*	—	—	—	F	EG
Oct. 24	101243	11.9–45.0I	(4.0)	4.6	25	—	GA
Nov. 15	235850	26.7–53.6I	(4.4)	4.9	46	—	PG
Dec. 5	113839	14.4–53.3I	(4.3)	4.7	42	—	GA
Dec. 14	235357	12.4–48.2I	(4.0)	4.4	23	—	GA
1970							
Jan. 26	0634	30.0–32.0A	—	—	1	—	EG?
Jan. 27	212117	31.0–30.0*	3.7'	—	1+	F	EG 3/30
Feb. 14	154839	5.5–31.8I	—	—	8	—	SU
Apr. 11	124518	30.2–31.0A	—	—	1+	F	EG
Apr. 20	034458	21.2–38.9A	—	—	3	—	RS?
Jun. 7	074706	13.3–51.3I	(4.5)	4.8	52	—	GA af
Jul. 30	045846	14.4–51.7I	6.6	6.2	275	—	GA af
Sep. 5	113847	14.5–53.7I	5.4	5.3	135	—	GA
Oct. 12	20	4.5–31.5*	—	—	—	V+	SU
Dec. 19	121536*	27.4–33.9I	(4.3)	4.8	39	F	RS
Dec. 19	224412	27.5–33.9I	(4.6)	4.5	67	F	RS
1971							
Mar. 3	021558	22.1–59.4I	(4.7)	5.0	75	—	SA
Mar. 27	1018	30.4–31.4*	3.3'	—	1+	F	EG 3/15
Apr. 15	185727	12.8–48.6I	4.5	4.8	76	—	GA
Apr. 25	174215	11.8–43.8I	(4.2)	4.3	35	—	GA
Jul. 8	234056	27.5–33.8I	(4.7)	4.8	75	—	RS
Sep. 2	182447	30.3–50.7*	5.0	5.1	158	VII*	IR 3/120; 5/25; af
Nov. 9	001656	26.9–54.5I	4.7	4.7	67	VI	IR 5/35; af
Nov. 13	154744	10.5–39.5I	5.0	5.6	124	V	ET 4/60; af
1972							
Jan. 12	081544	27.5–33.6R	(4.8)	5.1	87	VI	EG 2/370; 4/180
Feb. 4	163319	13.2–49.5I	(4.2)	5.2	33	—	GA
Feb. 28	184454	29.8–50.6*	5.1	4.7	22+	VI	IR af
Mar. 3	073053	11.7–45.7I	—	5.1	17	—	GA
Mar. 18	085819	15.2–54.6I	—	—	14	—	AS
Mar. 27	133330	32.2–21.8I	—	4.6	41	—	LY
Jun. 28	094935	27.7–33.8I	5.5	5.7	233	VI	GS 3/370; af
Jul. 2	125606	30.0–50.9*	5.3	5.4	253	VIII*	IR 3/189; af
Sep. 2	145348	31.4–16.1I	5.4	5.1	215		LY
Nov. 14	084530	30.0–32.5A	—	—	1+	F	GS
Nov. 28	101926	14.7–53.7I	(4.4)	5.1	50	—	GA
1973							
Mar. 5	235950	27.7–33.4I	(4.3)	4.6	39	F	EG
Mar. 14	011644	29.2–49.0I	—	4.6	6	—	PG
Mar. 28	134211	11.7–42.7C	5.5	5.5	154	VI	ET 2/250; 5/40; af
Mar. 28	141855*	11.7–42.9I	5.3	5.5	151	—	ET
Mar. 28	145907*	11.8–42.8I	5.4	5.4	126	—	ET
Apr. 1	071241*	11.6–43.0I	5.5	5.4	199	—	ET
Apr. 5	015909	12.1–46.4I	(4.7)	5.0	77	—	GA
Apr. 7	173643*	11.6–43.0I	5.0	5.0	83	—	ET

Table 3.1. *Catalogue of earthquakes, 1899 to 1992* cont.

Date	Time	Epicentre	M_s	m	N	I_o	Notes
Apr. 22	220341	4.1–31.3I	—	4.8	17	—	SU
May 3	074424	28.1–52.0C	(4.6)	4.9	68	—	IR
May 13	111937	14.7–55.6I	(5.0)	5.2	128	—	AS
May 29	174640	14.0–53.6I	(4.2)	4.6	33	—	GA
Jul. 16	140801	32.6–23.6I	—	—	7	—	LY
Aug. 24	020603	27.9–52.8I	5.0	5.2	177	—	IR af
Sep. 28	113226	13.2–50.7I	4.8	5.5	137	—	GA af
Oct. 27	154155	14.3–53.4I	(4.6)	4.6	62	—	GA
Nov. 19	062824	4.3–31.3I	—	—	18	—	LY
Dec. 6	075345	14.3–53.6I	(4.5)	4.6	57	—	GA
Dec. 22	023518	33.9–24.7I	—	4.0	15	—	LY
1974							
Mar. 17	073126	13.3–30.9I	4.9	5.1	39	—	SU
Mar. 17	132728	14.2–51.7I	(4.1)	4.5	30	—	GA
Apr. 17	182734	17.3–40.3I	5.2	5.3	179	—	RS
Apr. 29	200437	30.6–31.6I	(5.0)	5.0	123	V+	EG 3/230; 4/120; 5/70; (2.50)
May 20	104207	13.1–50.3I	(4.7)	5.1	84	—	GA
May 30	1654	33.6–23.2A	—	—	6	—	LY 2.8L
Jun. 7	001430	27.7–50.8I	—	—	5	—	PG
Jun. 15	033249	13.5–50.9I	(4.7)	5.2	77	—	GA
Jun. 21	160356	12.6–47.0I	(4.4)	4.4	48	—	GA
Jun. 30	132626	16.0–39.6I	—	4.5	20	—	ET
Jul. 2	182847	10.9–43.9C	—	—	5	—	ET 3.3L; af
Sep. 3	2345	29.3–30.8*	—	—	1+	—	EG
Sep. 4	062914	33.1–13.5I	5.5	5.2	226	F	LY
Nov. 1	124635	14.7–52.1I	(4.2)	4.7	33	—	GA
Nov. 15	001757	13.2–50.5I	—	4.4	18	—	GA
Nov. 25	163545	13.6–51.0I	(4.8)	4.9	88	—	GA af
Dec. 9	1957	30.4–30.9*	—	—	1	F	EG
1975							
Jan. 1	070355	32.8–21.0I	—	4.1	26	S	LY
Mar. 5	110005	4.4–31.1I	—	—	15	—	SU
Mar. 29	093623	13.3–50.8I	5.6	5.3	232	—	GA
Apr. 16	025509	14.6–40.7I	(4.0)	4.3	25	—	RS
Apr. 19	134554	14.3–56.4I	5.3	5.2	193	—	AS af
Apr. 19	201548*	14.4–56.5I	5.2	5.4	179	—	AS
May 6	232921	12.9–51.2I	(4.3)	4.6	41	—	GA
May 12	031852	27.8–34.0I	—	4.1	10	—	RS
Jun. 4	163728	11.9–43.8I	—	4.8	16	—	ET
Jun. 28	181218	17.2–39.8I	—	—	9	—	RS
Jun. 29	214559	18.6–39.8I	4.6	5.0	43	—	RS af
Jul. 6	103856*	13.1–51.9I	(4.3)	4.9	38	—	GA af
Jul. 21	132747	13.7–51.6I	5.1	4.8	113	—	LY
Jul. 27	114900	33.0–13.7I	—	3.9	62	—	RS
Aug. 7	224314	15.4–40.4I	(4.4)	4.7	48	—	RS
Aug. 14	035356	28.1–31.1A	—	—	20	—	EG
Oct. 26	193750	27.0–52.9A	—	—	5	—	IR
Oct.	—	14.0–44.2*	—	—	—	VI	YE
Nov. 13	030721	33.4–22.8I	4.5	5.1	203	—	LY
Nov. 26	001810	13.7–56.7I	5.2	5.4	206	—	AS af
Nov. 27	231015	13.4–50.9I	—	—	11	—	GA
Dec. 10	152629	15.6–42.6I	—	—	5	—	RS
Dec. 14	231649*	14.6–42.3I	5.1	5.3	131	V	RS 3/120; 4/70
Dec. 14	232727	14.7–42.3I	5.1	5.4	144	—	RS af
Dec. 24	114857	26.9–55.7*	5.4	5.7	166	VII	IR 4/70; af
Dec. 24	225527	25.6–55.4A	—	—	6	—	SA?

Table 3.1. *Catalogue of earthquakes, 1899 to 1992* cont.

Date	Time	Epicentre	M_s	m	N	I_0	Notes
1976							
Jan. 2	043035	28.5–49.0I	4.7	4.5	30+	—	PG
Jan. 14	105626	13.9–51.7I	(5.0)	5.0	120	—	GA
Jan. 31	023610	19.0–39.2I	—	4.3	15	—	RS
Feb. 5	081406	14.2–53.3I	(4.5)	5.0	60	—	GA
Feb. 10	014052	12.6–48.0I	(4.3)	4.6	44	—	GA
Mar. 6	163522	32.0–26.4I	—	4.0	14	—	MD
Mar. 7	213758	26.0–54.8I	3.1	4.2	5	—	PG
Mar. 18	173940	19.7–39.0I	—	4.3	11	—	RS af
Apr. 22	162925	19.8–38.7I	—	4.5	21	—	RS
Jun. 17	233220	23.3–36.8I	—	4.3	12	—	RS
Jul. 2	081810	14.2–50.9I	—	—	7	—	GA
Jul. 24	172927	21.8–37.4I	—	—	11	—	RS
Aug. 29	200124	31.6–47.4I	—	—	18	—	IQ
Sep. 2	195307	25.0–52.7I	—	4.0	8	—	PG?
Sep. 26	001247	29.9–47.3I	—	3.4	5	—	SA?
Sep. 27	022415	28.9–48.2I	—	3.8	10	—	SA?
Nov. 7	055307*	15.9–41.4I	(4.0)	4.8	23	—	RS
Nov. 9	235118	13.3–49.5I	(4.3)	4.5	38	—	GA
Dec. 1	050341	15.8–41.8I	(4.1)	4.8	32	—	RS
Dec. 19	212153	33.0–14.0I	4.0	4.5	81	F	LY
1977							
Jan. 16	203126	28.8–48.1I	—	—	9	—	SA
Jan. 18	051748	29.3–32.7C	—	—	5	—	EG
Jan. 26	123835	13.2–51.1I	—	4.8	17	—	GA
Feb. 28	173507	14.8–55.1I	5.2	5.5	184	—	AS af
May 15	041801	17.8–37.3I	—	4.3	19	—	ET
Jun. 1	015318	14.0–51.6I	—	4.6	8	—	GA
Jun. 27	094217	15.7–39.7I	—	—	6	—	ET
Jun. 27	141321	16.2–39.8I	—	4.3	12	—	ET
Jul. 4	195008	14.9–51.6I	—	4.6	18	—	GA
Jul. 8	062303	11.1–39.6I	5.0	5.1	144	VI+	ET 3/250
Aug. 16	232202	14.7–52.5I	(4.6)	4.8	70	—	GA
Oct. 4	061858	14.3–53.8I	—	4.6	13	—	GA
Oct. 17	214644	11.8–42.9I	—	—	5	—	ET
Dec. 5	100609	11.4–40.1I	—	4.9	14	—	ET
Dec. 15	080244	13.1–50.4I	—	4.6	7	—	GA af
Dec. 17	235755	13.2–41.0I	(5.1)	5.0	142	—	ET
Dec. 28	024533	16.5–40.3I	6.4	6.3	339	—	RS af
1978							
Jan. 17	150031*	16.5–40.3I	(5.2)	5.1	169	—	RS af
Jan. 29	173209	13.2–51.3I	(4.5)	4.8	54	—	GA
Feb. 9	050043	11.6–42.5I	—	—	11	—	ET
Feb. 11	125418	13.2–51.0I	5.4	5.3	205	—	GA af
May 30	201721	11.0–57.3I	(5.5)	5.2	294	—	AS
Jun. 7	072123	25.6–15.2I	3.6	4.0	42	—	LY
Jun. 8	190445	13.2–50.3I	—	4.6	15	—	GA
Jun. 10	190741	12.1–46.5I	—	4.6	6	—	GA
Jun. 22	214106	12.6–48.1I	(4.3)	4.7	40	—	GA
Sep. 14	040558	11.7–48.1I	—	—	6	—	GA
Sep. 14	071418*	12.5–47.9I	(4.5)	4.8	65	—	GA
Sep. 15	121751	12.4–47.9I	(4.5)	4.9	57	—	GA
Oct. 16	203456	11.9–43.6I	—	—	5	—	ET
Nov. 7	002327	10.2–42.8I	—	4.2	10	—	ET
Nov. 17	203401	12.5–48.0I	(4.3)	4.7	42	—	GA
Dec. 7	020217	14.2–50.7I	—	4.3	8	—	GA

Table 3.1. *Catalogue of earthquakes, 1899 to 1992* cont.

Date	Time	Epicentre	M_s	m	N	I_o	Notes
Dec. 7	162203	12.9–49.7I	—	4.3	10	—	GA
Dec. 9	071252	24.0–26.4I	5.0	5.3	205	—	EG
Dec. 21	040354	11.5–42.9I	(5.1)	5.1	145	—	ET
1979							
Mar. 29	100511	14.3–53.6I	4.5	4.7	63	—	AS
May 13	204801	18.8–39.3I	(4.6)	4.8	68	—	RS
May 13	205548	19.6–39.1I	(4.2)	4.5	33	—	RS
May 25	171059	25.2–36.5I	(4.4)	4.6	51	—	RS
Jun. 26	134327	4.1–30.5I	—	—	8	—	SU
Jul. 6	181113	14.8–52.4I	—	4.2	8	—	GA
Jul. 8	040911	14.6–53.7I	5.3	4.8	217	—	GA
Jul. 10	231104	12.4–45.7I	—	4.6	15	—	GA
Jul. 17	170704	17.7–40.1I	(4.8)	5.1	84	—	RS
Jul. 22	185304	11.9–42.7I	—	—	5	—	ET
Aug. 6	004830	12.3–40.8I	—	4.5	15	—	ET
Aug. 15	022042	15.3–41.9I	(4.3)	4.7	38	—	RS
Sep. 10	205725	11.9–46.0I	(4.1)	4.8	28	—	GA
Sep. 19	184814	12.5–40.5I	(4.7)	4.9	79	—	ET
Sep. 24	234136	12.7–48.3I	4.6	5.0	107	—	GA
Oct. 16	155750*	11.8–43.6I	5.2	5.0	141	—	ET
Oct. 16	205449	11.8–43.8I	5.5	5.3	173	—	ET
Oct. 20	202214	13.2–51.3I	—	—	9	—	GA
Nov. 10	174245	14.9–39.7I	—	4.3	6	—	ET
Nov. 12	233429	14.5–53.8I	(4.3)	4.9	42	—	AS
Nov. 13	170137	13.2–40.0I	—	4.1	7	—	ET
Dec. 22	154334	13.8–51.6I	5.0	5.1	192	—	GA
1980							
Jan. 14	041053	16.6–40.3I	5.8	5.4	291	F	RS
Jan. 14	122822*	16.5–40.3I	5.2	5.3	224	—	RS
Apr. 4	041447	14.4–53.4I	(4.8)	4.9	98	—	AS
Apr. 7	164538	17.6–40.2I	(4.1)	4.8	31	—	RS
Apr. 20	023752	11.8–57.7I	6.3	5.7	359	—	AS
Apr. 29	185425	10.1–43.1I	—	4.6	12	—	ET
May 3	153814	29.6–32.6I	—	4.2	9	—	EG
Jun. 13	020837	33.8–23.1I	4.2	5.0	194	—	LY
Jul. 14	08	21.9–31.3*	—	—	—	F	SU af
Jul. 16	231455	17.2–40.4I	(4.8)	4.7	94	—	RS
Aug. 31	163055	11.3–57.5I	5.3	5.3	205	—	AS
Aug. 31	220533	12.1–46.0I	(4.6)	4.7	73	—	GA
Oct. 18	211443	21.0–37.8I	—	5.0	19	—	SU
Nov. 28	021133	12.5–43.0I	—	—	7	—	ET
1981							
Jan. 19	081348	12.4–48.0I	5.2	5.1	132	—	GA
Jan. 25	022623	33.6–14.3I	—	3.8	24	—	LY
Apr. 17	215411*	13.4–49.6I	(4.0)	4.6	26	—	GA
Jul. 3	205823	13.2–40.5I	5.0	4.9	156	—	GA
Jul. 17	001034	33.6–16.3I	—	4.1	13	—	LY
Sep. 28	013239	33.8–24.9I	—	4.2	29	—	LY
Nov. 14	090523	23.8–32.6I	5.5	5.1	270	VII	EG 2/700; 3/460; 4/280; 5/170; (2.51)
Nov. 18	1850	26.6–30.8*	—	—	—	V	EG 4.0L
Dec. 5	184657	14.5–58.1I	5.4	5.6	323	—	AS
Dec. 9	233123	30.5–35.1I	—	—	—	—	DS 3.2L
1982							
Jan. 6	060914	13.0–50.7I	4.2	4.7	36	—	GA
Mar. 14	165035	13.0–50.8I	—	4.1	7	—	GA
Mar. 23	104828	28.0–34.4I	4.6	4.9	81	F	SI

Table 3.1. *Catalogue of earthquakes, 1899 to 1992* cont.

Date	Time	Epicentre	M_s	m	N	I_o	Notes
May 14	144258	14.4–53.7I	5.2	5.3	208	—	AS
Jul. 11	221755	12.9–51.1I	4.1	4.5	31	—	GA
Aug. 20	125734	23.6–32.6I	(4.5)	4.7	53	F	EG
Oct. 14	134023	26.8–34.8I	(4.2)	4.8	37	—	RS
Oct. 15	083753	5.9–31.9I	—	5.1	11	—	SU
Oct. 30	043646	27.6–33.8I	(4.1)	4.6	29	—	GS
Nov. 10	054117	13.2–50.4I	4.9	5.1	128	—	GA
Dec. 8	061936	12.1–46.1I	5.6	5.8	279	—	GA
Dec. 13	091251	14.7–44.2I	6.1	6.0	419	VIII*	YE 3/230; 4/130; 5/70; af; F; (2.52)
1983							
Feb. 3	134605	29.2–34.8I	4.8	4.9	106	F	SI af
Feb. 3	233027*	29.3–34.8I	4.8	4.8	101	F	SI
Mar. 7	065203	32.4–25.8I	—	—	9	—	MD
Mar. 18	213047	11.8–43.4I	—	4.5	17	—	ET
Mar. 19	18	30.1–31.3*	—	—	—	F	EG
Apr. 8	022829	11.4–57.5I	6.4	5.8	437	—	AS
Jun. 12	120009	28.5–33.1I	4.7	5.0	217	F	GS 3/200; af
Jun. 15	221949	27.2–34.5C	—	—	5	—	RS 4.1L
Jul. 22	072023	12.1–47.5I	—	4.6	10	—	GA
Jul. 29	180359	10.4–56.9I	5.4	5.7	333	—	AS
Aug. 3	140853	12.5–48.1I	—	—	9	—	GA
Aug. 6	020720	26.7–32.4I	—	—	—	—	EG 4.2L
Aug. 7	104218	16.4–41.3I	(4.0)	4.7	24	—	RS
Sep. 30	112414	13.2–50.9I	4.5	5.3	52	—	GA
Sep. 30	185815	11.8–43.4I	5.6	5.5	249	F	ET
Oct. 30	053523	27.4–32.1I	—	—	8	—	EG 4.2L
1984							
Jan. 28	224751	14.0–51.7I	5.4	5.1	187	—	GA
Feb. 12	233643	14.7–53.9I	4.6	4.6	56	—	GA
Feb. 17	035043	30.8–35.8I	—	—	11	—	DS 3.0L; af
Feb. 18	052147	14.5–52.1I	4.7	5.1	102	—	GA
Mar. 2	195439	28.0–36.1I	—	—	8	—	SA 4.1L
Mar. 4	020317	30.8–35.3I	—	—	4	—	DS
Mar. 7	131411	29.0–34.4C	—	—	—	—	SI 3.7L; af
Mar. 15	204658	27.6–33.9C	—	—	—	—	GS 3.5L
Mar. 29	213607	30.2–32.2I	(4.8)	4.9	95	IV	EG 3/100; 4/40; af
Apr. 6	183054	30.4–33.8I	(4.0)	—	23	—	SI 4.6L
Apr. 7	034528	28.5–32.5C	—	—	—	—	EG 3.6L
Apr. 10	004449	30.5–35.2I	—	—	7	—	DS 2.8L; af
Apr. 11	201052	13.1–51.1I	—	4.8	8	—	GA
Apr. 12	103640	29.6–34.8C	—	—	—	—	DS 3.1L
Apr. 12	111910	25.4–56.0I	—	4.5	11	—	SA
Apr. 16	083921	29.8–35.2C	—	—	—	—	DS
Apr. 18	202025	28.8–33.3I	—	—	13	—	GS 4.1L
Apr. 21	033427	32.4–30.9C	—	—	—	—	MD 3.4L
Apr. 23	143415	27.9–35.0C	—	—	—	—	RS 3.7L
May 8	141152	30.5–35.1C	—	—	—	—	DS 2.1L; af
May 9	145621	31.0–34.9C	—	—	—	—	DS 3.0L
May 10	224204	29.1–34.8C	—	—	—	—	DS 3.3L; af
May 24	000153	13.9–51.7I	4.7	5.0	100	—	GA
May 25	070506	27.4–35.2C	—	—	—	—	RS
May 28	192557	28.4–34.5C	—	—	—	—	DS 3.1L; af
May 28	201820	28.2–35.5C	—	—	—	—	DS 3.5L; af
May 28	234236	31.0–35.5I	—	—	11	—	DS 2.9L; af
Jun. 21	110835	33.9–22.8I	—	—	11	—	LY 4.2L
Jun. 21	132908	33.9–22.6I	—	—	13	—	LY 4.1L

Table 3.1. *Catalogue of earthquakes, 1899 to 1992* cont.

Date	Time	Epicentre	M_s	m	N	I_o	Notes
Jun. 25	020641	13.6–44.6I	4.5	4.9	58	—	YE
Jun. 28	005134	27.3–34.0C	—	—	—	—	RS 3.6L; af
Jul. 2	014659	25.2–34.5I	5.0	5.1	243	F	EG 3/220
Jul. 22	234051	32.0–32.6C	—	—	—	—	MD
Jul. 27	120620	28.1–30.4C	—	—	—	—	EG 3.9L
Jul. 27	162510	29.3–32.1C	—	—	—	—	EG 3.5L
Jul. 27	235459	28.0–32.1C	—	—	—	—	EG 3.5L
Aug. 10	012126	28.4–33.0C	—	—	—	—	EG 3.5L; af
Aug. 16	185233	13.5–49.4I	—	4.7	20	—	GA
Sep. 7	232649	27.1–35.0C	—	—	—	—	RS 4.2L; af
Sep. 14	043840	27.7–33.8C	—	—	—	—	GS 3.7L
Sep. 23	212341	14.7–44.3I	(4.5)	4.7	53	—	YE
Sep. 26	172426	27.6–33.0C	—	—	—	—	EG 3.7L
Sep. 30	040458	26.0–35.0C	—	—	—	—	RS 4.7L
Oct. 7	145856	30.7–32.6C	—	—	—	—	GS
Oct. 12	063830	14.6–40.3I	—	4.1	9	—	ET
Oct. 30	104055	14.6–54.0I	—	4.7	9	—	AS
Nov. 22	033257	30.8–35.2I	—	—	13	—	DS 3.4L
Dec. 8	074154	28.0–34.1I	—	4.4	10	—	SI
Dec. 14	104750	30.9–35.1I	—	—	8	—	DS 3.1L; af
Dec. 18	010059	29.8–35.1C	—	—	—	—	DS 3.1L
Dec. 21	031747	29.4–32.4C	—	—	—	—	GS 4.0L
1985							
Jan. 13	150849	27.0–34.7C	—	—	—	—	RS 4.2L
Jan. 14	135253	13.2–50.2I	—	4.8	5	—	GA
Jan. 19	224006	30.9–35.5I	—	—	12	—	DS 3.4L; af
Feb. 24	191322	29.9–34.9C	—	—	—	—	DS 3.1L
Feb. 26	042710*	13.0–51.0I	4.5	4.9	48	—	GA
Feb. 26	054647	13.0–51.0I	4.8	5.0	95	—	GA
Feb. 27	032500	12.9–51.4I	—	4.8	14	—	GA
Feb. 28	165547	27.7–33.7I	4.0	4.5	28	F	GS
Mar. 16	172515	28.9–34.7C	—	—	—	—	DS 3.3L; af
Apr. 6	014333	27.9–33.6C	—	—	—	—	EG 3.4L
Apr. 11	134230	29.4–36.6C	—	—	—	—	SA 3.0L
Apr. 27	033514	28.9–34.9C	—	—	—	—	DS 3.4L
May 26	163925	26.2–34.3I	—	—	5	—	EG 4.2L
Jun. 4	024350*	11.8–43.8I	5.0	5.0	75	—	ET af
Jun. 4	035224	11.7–43.5I	5.4	4.8	150	—	ET
Jun. 13	081445	16.0–39.7I	—	4.6	7	—	ET
Jul. 20	102627	12.7–47.0I	—	4.4	7	—	GA
Jul. 20	131207	12.6–48.2I	5.0	5.0	152	—	GA
Jul. 22	142726	19.1–39.2I	4.4	4.8	32	—	RS
Jul. 22	213230	(34.4–28.3)	5.1	5.7	404	IV	MD 3/410
Aug. 20	054601	5.5–35.9I	4.8	5.3	231	—	SU
Oct. 10	101954	26.8–54.9I	5.1	5.3	268	VII	IR af
Oct. 26	010935	27.9–34.7C	—	—	—	—	RS 3.7L
Nov. 7	141446	28.9–34.7C	—	—	—	—	DS 3.5L
Nov. 13	133136	30.8–36.1I	—	—	11	—	DS 3.0L
Nov. 13	213110	29.8–32.1I	—	—	18	—	GS
Nov. 23	070649	30.5–35.1I	—	—	12	—	DS 3.4L; af
Nov. 26	090736	14.7–53.6I	(4.1)	4.6	31	—	AS
Dec. 1	153217	15.0–57.4I	—	4.4	17	—	AS
Dec. 14	181331	14.6–57.9I	5.6	5.5	346	—	AS
Dec. 18	021535	4.7–33.2I	—	4.3	7	—	SU
Dec. 31	170423	30.4–35.0I	—	—	10	IV	DS 4.3L
Dec. 31	194241	29.1–34.9I	4.6	4.8	70	V	DS 4/50; 5.1L; af

Table 3.1. *Catalogue of earthquakes, 1899 to 1992* cont.

Date	Time	Epicentre	M_s	m	N	I_o	Notes
1986							
Jan. 8	165300	29.7–31.7I	—	—	8	—	EG
Feb. 20	070728	27.4–35.3I	—	—	6	—	SA
Mar. 24	191107	13.9–51.7I	—	4.8	58	—	GA
Mar. 25	013233	26.8–54.8I	5.4	4.9	180	VI	IR
Apr. 8	084334	27.1–34.4I	—	4.4	22	—	RS
May 16	081759	11.9–43.3I	—	3.9	7	F	ET
May 23	095125	12.7–48.2I	5.4	5.5	354	—	AS
Jul. 7	162657	10.3–56.8I	6.1	6.2	565	—	AS
Aug. 20	172725	11.8–43.3I	—	4.0	8	—	ET
Sep. 17	212515*	10.4–57.0I	5.9	5.7	439	—	AS
Dec. 23	08	31.0–29.8*	—	—	—	IV	EG
1987							
Jan. 2	101446	30.5–32.2I	—	5.0	32	V	GS 5/15; 3/100
Jan. 19	002323	13.9–51.7I	—	4.7	26	—	GA
Jan. 22	042106	12.2–41.3I	—	4.5	16	—	ET
Jan. 27	003628	14.7–54.7I	—	4.9	101	—	AS
Feb. 21	201952	2.9–30.4I	—	5.1	19	—	SU
Mar. 1	031932	33.8–23.0I	—	5.0	315	—	LY
Apr. 9	030005	32.4–29.0I	—	4.6	153	—	MD
Jun. 28	005016	32.8–24.3I	4.4	5.3	483	—	LY
Jul. 7	012551	11.5–42.7I	—	—	7	—	ET
Jul. 28	195203	13.4–31.3I	—	4.6	30	F	SU
Sep. 13	235457	32.4–24.3I	—	4.2	65	—	LY
Oct. 7	222925	6.3–37.8I	4.1	5.2	91	—	ET
Oct. 9	034918	5.5–36.0I	—	4.3	11	—	SU
Oct. 18	010542	29.5–35.11	—	4.4	25	F	DS 4.4L
Oct. 25	164614	5.4–36.8I	5.9	5.6	431	S	ET Felt strongly
Oct. 28	085830	5.8–36.7I	5.6	5.4	336	D	ET Damage at Arba Minch
Nov. 1	170022	32.4–26.2I	3.2	4.0	62	—	MD
Nov. 12	103504	5.7–36.7I	—	4.4	13	—	ET
Dec. 14	215100	30.7–31.7I	—	4.1	34	VI	EG 4.1L; 5/18; 3/90
1988							
Jan. 28	154808	32.4–21.1I	4.8	4.9	338	—	LY
Jan. 28	191218	32.3–21.2I	—	4.8	76	—	LY
Apr. 20	145428	10.7–57.0I	—	5.1	105	—	AS
Jun. 5	182658	28.0–33.7I	—	4.5	52	—	EG
Jun. 9	021823	32.2–27.9I	—	4.7	222	—	MD
Jul. 10	024255	12.9–57.5I	4.5	5.1	122	—	AS
Jul. 16	084202	13.9–51.6I	5.1	5.5	254	—	GA
Aug. 6	093240	13.7–51.6I	—	5.2	130	—	GA
Aug. 21	160539	28.5–34.4I	—	—	14	—	EG 4.1L
Nov. 26	035152	22.0–31.4U	—	—	13	—	SU 4.0L
Nov. 29	170659	28.5–33.8I	—	—	29	V	EG 4.0L
Dec. 10	173319	16.3–41.1I	5.3	5.3	282	—	RS
1989							
Apr. 4	113212	26.9–34.9I	—	4.2	24	—	RS
Apr. 12	050514	13.3–39.9I	—	5.0	103	—	ET
Apr. 13	075510	13.3–39.9I	5.2	5.3	195	IV	ET
Apr. 16	062054	13.9–43.4I	—	—	10	—	YE 4.1L
May 11	205352	9.0–40.0I	4.7	5.0	146	IV	ET
Jun. 8	062412	6.8–37.9I	4.8	5.2	142	IV	ET
Aug. 20	111657	11.8–42.0I	6.1	5.7	569	IX	ET Damage and casualties
Aug. 20	111755	11.9–42.0I	6.5	5.7	72	—	ET
Aug. 20	114628	11.9–41.8I	5.8	6.0	460	—	ET
Aug. 20	115617	11.8–42.0I	6.0	5.3	233	—	ET

Table 3.1. *Catalogue of earthquakes, 1899 to 1992* cont.

Date	Time	Epicentre	M_s	m	N	I_o	Notes
Aug. 20	132527	12.0–41.8I	5.9	5.2	275	—	ET
Aug. 20	132620	11.9–41.9I	5.9	5.2	150	—	ET
Aug. 20	182733	11.8–41.7I	6.6	5.1	100	—	ET
Aug. 20	183949	11.9–41.9I	—	5.4	184	—	ET
Aug. 20	185405	11.8–41.8I	—	5.3	187	—	ET
Aug. 20	192557	11.9–41.8I	5.8	6.0	531	—	ET
Aug. 21	010907	11.9–41.9I	6.1	6.1	635	F	ET
Aug. 21	050306	11.9–41.8I	5.7	5.7	460	F	ET
Aug. 21	050546	11.9–41.7I	5.3	5.2	31	—	ET
Aug. 21	070739	11.8–41.7I	4.6	5.1	127	—	ET
Sep. 6	212728	27.6–33.9I	—	—	30	—	EG 4.0L
Sep. 9	051653	28.8–34.7I	—	—.	22	—	EG 4.1L
Dec. 18	214812	28.4–33.3I	—	4.3	38	F	EG
1990							
Jan. 24	035752	27.6–34.2I	—	—	14	—	RS 4.1L
Feb. 23	045815	27.0–32.7U	—	—	4	—	EG 4.0L
Mar. 15	112430	28.0–34.5I	—	—	13	—	EG 4.0L
Apr. 20	145131	28.9–34.6I	—	—	8	—	EG 4.2L
May 4	101206	11.7–41.0I	4.8	5.1	185	S	ET
May 18	182752	31.7–24.8I	—	—	15	—	LY 4.3L
May 20	022201	5.1–32.2I	7.2	6.5	639	D	SU
May 24	193447	5.3–31.8I	6.4	5.9	513	F	SU
May 24	200008	5.4–31.9I	7.0	6.4	521	D	SU
May 24	221604	5.5–31.9I	5.4	5.4	279	—	SU
May 25	004232	5.4–31.9I	4.9	5.3	248	—	SU
May 26	142240	5.2–31.8I	4.7	5.0	107	—	SU
May 27	185657	13.1–39.9I	—	5.1	165	—	ET
Jun. 3	162344	5.4–32.1I	4.6	5.0	109	—	SU
Jul. 9	151120	5.4–31.7I	6.4	5.9	534	—	SU
Jul. 28	164604	5.2–32.6I	5.1	5.4	271	—	SU
Sep. 7	001126	5.4–31.7I	5.3	5.2	273	—	SU
Sep. 12	152840	15.1–59.3I	4.8	5.5	330	—	AS
Sep. 13	221008	27.2–35.1I	—	4.6	101	—	RS
Sep. 14	204019	13.4–51.5I	4.9	5.4	332	—	GA
Nov. 3	112019	14.6–54.3I	4.6	5.1	204	—	AS
Dec. 4	091543	29.1–33.2I	—	—	17	—	EG 4.0L
Dec. 11	050909	5.4–32.6I	—	5.0	20	—	SU
1991							
Jan. 10	070627	5.1–31.8I	4.6	5.2	142	—	SU
Jan. 25	232252	13.6–43.9I	—	—	11	—	YE
Feb. 18	062821	28.0–28.5I	—	4.0	18	—	EG 4.2L
Mar. 27	001826	27.7–33.8I	—	—	9	—	EG 4.0L
Mar. 29	090607	5.2–32.7I	4.9	5.4	199	—	SU
May 11	152630	12.4–47.5I	4.8	5.2	263	—	GA
Nov. 22	004024	13.9–44.1U	—	4.7	—	D	YE Damage and deaths
1992							
Jan. 10	111514	11.8–42.3U	—	4.9	—	—	ET
Feb. 26	034520	11.8–57.8U	5.5	5.8	—	—	AS
Mar. 5	085506	11.5–42.8U	6.2	5.5	—	S	ET Felt strongly
Apr. 8	012852	12.0–46.0U	4.5	5.0	—	—	GA
May 19	012645	13.8–44.0U	—	4.5	—	D	YE Damage and casualties
Jun. 1	235305	25.5–36.1U	—	4.3	—	—	RS 4.5L
Sep. 25	222017	27.4–34.0U	—	3.9	—	—	EG 4.2L
Oct. 12	130956	29.9–31.2U	5.2	5.9	—	D	EG Damage and deaths
Oct. 22	173858	29.5–31.5U	3.4	4.5	—	D	EG Damage and deaths
Oct. 27	110248	28.8–33.2U	—	4.0	—	—	EG

Table 3.1. *Catalogue of earthquakes, 1899 to 1992* cont.

Date	Time	Epicentre	M_s	m	N	I_o	Notes
Nov. 5	184149	29.6–31.1U	—	4.5	—	F	EG 4.3L
Nov. 5	191647	29.7–31.1U	—	4.7	—	—	EG 4.1L
Dec. 6	014353	10.9–57.3U	5.3	5.2	—	—	AS

Those interested in obtaining machine-readable files of the basic origin parameters of the events in Tables 2.1 and 3.1 should make enquiries to the International Seismological Centre.

Key to Table 3.1.

Time. Origin time: Universal time (hours, minutes, seconds). Asterisked time indicates foreshock or aftershock.

Epicentre. I: determined by ISS/ISC.

  B: determined by BCIS Strasbourg.

  C: determined by other agencies: BAAS; CSEM; MOS; IPRG; Gutenberg and Richter (1965); Gouin (1979); Nowroozi (1971); Sykes and Landisman (1964); Fairhead (1968).

  A: adopted epicentre, low accuracy in need of authentication.

  R: relocated epicentre in this study or other special studies.

  *: macroseismic epicentre.

  U: determined by USCGS/NOAA/NEIS.

  ( ): felt event in study area originating from focus outside the region.

Magnitudes. M_s: recalculated surface-wave magnitude from amplitude–period data using the Prague formula.

  *: surface-wave magnitude calculated from Milne recorders.

  ': magnitude derived from felt effects (equations 4 and 5, see Section 3.2)

  m: body-wave magnitude, recalculated for events prior to 1965.

  ( ): estimated from number of recording stations N (equations 1 and 3)

  N: Number of stations that recorded an event.

  +: minimum number of stations.

Intensity. I_o: maximum recorded intensity (MSK).

  *: epicentral intensity.

  F: felt.

  S: strong.

  D: damaging.

Notes.

AS = Arabian Sea	IR = Iran
DS = Dead Sea	KE = Kenya
EA = Eastern Anatolia	MD = Mediterranean
EG = Egypt	PG = Persian Gulf
ET = Ethiopia and Somalia	RS = Red Sea
GA = Gulf of Aden	SA = Arabian Peninsula
GO = Gulf of Oman	SI = Sinai
GS = Gulf of Suez	SU = Sudan (includes northern parts of Uganda and Zaire)
HA = Hellenic Arc	TR = Turkey
IQ = Iraq	YE = Yemen

  Poor or unsubstantiated positions are marked with a ?

I/r: r = average distance in kilometres at which the shock was felt at intensity I. For events not mentioned in Section 2.1 or lacking radii of perceptibility, we have no macroseismic data.

RX: X = radius of perceptibility in kilometres of shocks with epicentres outside the study area.

af: event preceded and/or followed by aftershocks not listed in Table 3.1.

sr: swarm of shocks associated with event.

XL: X = local magnitude determined by others.

V: volcanic eruption.

F: faulting.

(X): X = figure number (see Section 2.1).

3.4   Spurious and mislocated events

As is usual in such studies, we have found errors and uncertainties in previously published catalogues. Details of these events are given at the end of this section. Examples of seismological uncertainties of interpretation were given in Section 3.2. These can arise from low sensitivity of instruments, poor timing and inadequate knowledge of travel-time tables and Earth models. There are, in addition, other sources of error. Macroseismic reports can be misinterpreted, and errors can arise from misreporting of dates and times. We have found examples of errors in month, year and decade. One particular case was caused by misinterpretation of the American date order, resulting in an event on 9 July being reported as 7 September (see Section 3.2).

Many errors also result from uncritical acceptance of early results from computers. In particular, locations made with the Large Aperture Seismic Array in Mon-

tana between 1966 and 1970 must be disregarded unless supported by other readings, for they often arise from misinterpretations of core phases and later arrivals from other events. There is similar uncertainty for events determined by the array-type agencies NORSAR in Norway and Hagfors in Sweden. These locations appear under the agency codes LAO, NAO and HFS. LAO determinations do not appear in ISC publications after 1970. NAO and HFS determinations are still retained by ISC, but usually as secondary determinations following more reliable estimates. The positions given by these agencies may be several degrees in error, and in particular events in Iran and Iraq are often placed at too large a distance from the station and given positions too far south, in the Arabian Peninsula. These secondary solutions are not listed below, except the few instances where origins from these agencies appear as primary estimates in ISC publications.

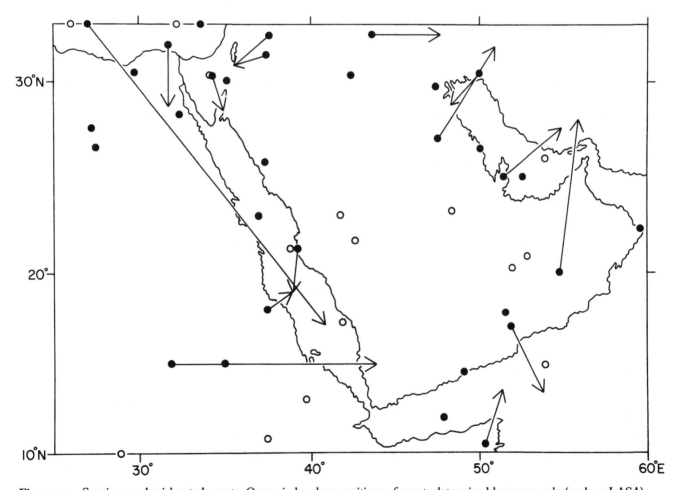

Figure 3.9. Spurious and mislocated events. Open circles show positions of events determined by arrays only (such as LASA), which cannot be substantiated. Filled circles denote other events which have been removed as unsubstantiated, relocated outside our area, or relocated to new positions within the area as shown. See Section 3.4 for details. Events listed with errors in date or timing only are not shown.

The following is a chronological list of events which have appeared in other sources, but which can either be shown to be spurious, grossly mislocated or of such doubtful validity that they should be discarded. After the date and general area, the origin time and position are given, followed by the agency or the reference in which the event appears. Finally, some brief comment on the event is given, with reference where appropriate. Figure 3.9 shows the location of these spurious events, and the revised positions of those relocated within the study area.

1899 July 14   Egypt
13h   23° N, 33° E   BAAS
Mislocation of event in Alaska (See Section 2.4).

1900 April 30   Gulf of Aden
20h 17m   12° N, 48° E   BAAS
No supporting instrumental evidence.

1900 December 7   Egypt
01h 32m   North of Cairo   Maamoun *et al.*
(1984)
Date error for event in Indian Ocean on 1903 December 6, which was recorded at Helwan.

1901 August 6   Saudi Arabia
18h 39.3m   20° N, 55° E   BAAS
Conflicting instrumental information suggests a location near 28° N, 56° E in south Iran, but there are no confirmatory felt reports (see above, Section 3.2).

1903 June 4   Red Sea
15h 12m   23° N, 37° E   BAAS
Relocated to 2° N, 35° E in Sudan. Ambraseys and Adams (1986a).

1903 July 19   Dead Sea
18h 07m 54s   30° N, 35° E   Ibrahim (1985)
Likely transpose of position 35° N, 30° E in Mediterranean Sea.

1903 December 6   Egypt
22h   'Distinct tremor   BAAS
felt locally'
(Helwan).
Time given corresponds to instrumental recording of earthquake in Indian Ocean. No evidence of local shock.

1904 August 15   Egypt
Reported strongly   Plassard (1960)
felt near Suez.
No instrumental or macroseismic evidence found.

1906 March 20   Mediterranean Sea
03h 48m   33° N, 27° E   BAAS
Relocated at 17° N, 41° E in Red Sea, with confirmatory felt reports (see above, Section 3.2).

1907 March 31   South Iraq
14h 14m   BAAS
Relocated in Turkey at 38.5° N, 42.2° E. Ambraseys and Finkel (1987).

1908 January 25   Sudan
20h 06m   15° S, 35° E   BAAS
Ambraseys and
Adams (1986a)
Reassessment of poor instrumental data, combined with felt information moves this event from Sudan to Yemen, near 15° N, 44° E (see Section 3.2).

1908 April 2   Sudan
05h 53m   18° N, 30° E   BAAS
Relocated at 3° N, 26° E, in Zaire, Ambraseys and Adams (1986a).

1908 December 18   Saudi Arabia
15h 35m   17° N, 52° E   BAAS
Relocated at 13.5° N, 54° E, Arabian Sea, (this study).

1910 May 30   Chad
12h 30m   15° N, 22° E   BAAS
Relocated at 10° N, 27° E in Sudan, Ambraseys and Adams (1986a).

1919 August 3   Libya
09h 45m   31.5° N, 19.5° E   ISS
Macroseismic position near 38.3° N, 20.5° E, in Greece.

1922 August 29   Egypt
20h 48m 20s   30.3° N, 31.2° E   Maamoun *et al.*
(1984)
One month error. Event occurred 1922 July 29.

1925 May 9   Egypt
23h 27m 33s   30.3° N, 31.2° E   Sieberg (1932b),
Maamoun *et al.*
(1984)
Date error for 1925 May 20, when event was both felt and recorded.

1925 November 11   Egypt
26.5° N, 27.5° E   Maamoun *et al.*
(1984)
No macroseismic or instrumental evidence for this event.

1925 November 19   Egypt
12h 25m 05s   26.5° N, 27.5° E   ISS
Mislocation of felt event in Albania near 40.3° N, 20.0° E.

1926 August 30   Libya
11h 38m   32° N, 21° E   Macroseismic
Local felt effects from event near 36.8° N, 23.3° E, in Greece.

1927 September 24   Sinai
  ooh 28m 14s   30.4° N, 34.1° E   Plassard (1960),
                                        Kárník (1968)
We can find no evidence that this event was felt anywhere except in Cairo, where its effects were slight. Our redetermination of instrumental data gives a range of possible positions from near Aqaba to near the mouth of the Gulf of Aqaba, depending on interpretation of crustal phases.

1927 October 15   Saudi Arabian Coast
  09h   14.5° N, 49.1° E   *Bull. Volcan.*, **273**
                                (1928), 15–18
Reported strong sea current caused vessel to drag anchor and run aground off Mukalla. No supporting seismological evidence.

1927 October 18   Egypt, Mediterranean Sea
  ooh 29m 28s   Maamoun *et al.*
                   (1984)
Re-evaluation of instrumental and felt information places this event southeast of Beni Suef, near 28.7° N, 31.7° E, rather than off Damietta (Dumyat).

1929 October 29   Qatar
  08h 57m 35s   25.0° N, 51.5° E   ISS
This appears to be a mislocated aftershock (m_b 5.0) of a larger event in Iran at 05h 53m, near 27.5° N, 55.0° E.

1930 April 3   Iraq
  12h 08m 40s   32.5° N, 43.7° E   ISS, Kárník (1968)
This event was originally located in an unusual position west of Karbala in central Iraq. On re-examination a better solution could be found in the Zagros area of Iran, near 32.5° N, 47.5° E.

1931 May 1   Saudi Arabia, Sudan, Zaire
  09h 48m 29s   21.2° N, 39.2° E   Gouin (1979)
  09h 47m 55s   18.0° N, 37.5° E   ISS
  09h 45m   3.0° N, 27.0° E   Ambraseys and
                                    Adams (1986a,
                                    1986b)
Closer examination, including reinterpretation of crustal phases, now places this event at 19.0° N, 39.0° E (± 1°) in the central Red Sea (see Section 3.2).

1940 January 2   Libya
  ooh 07m 08s   30.3° N, 22.0° E   ISS
Seismograph readings suggest position further north.

1940 April 13   Somalia
  17h 21m 42s   10.5° N, 50.5° E   ISS
More likely to be in same place as event of 1940 August 13, at 13.5° N, 51.5° E in Gulf of Aden.

1949 October 28   Mediterranean Sea
  19h 10m   33.0° N, 33.5° E   Maamoun *et al.*
                                    (1984)
Believed to be near 33.0° N, 35.5° E in Israel, where it was strongly felt at Kinneret. It does not appear to have been recorded at Helwan, and reports that it was felt in Egypt cannot be substantiated.

1953 September 7   Iraq–Saudi Arabian
                          border
  21h 23m 49s   30.2° N, 42.5° E   Jalil (1986)
Double error of day/month and E/W transposition for earthquake on 9 July at 30.2° N, 42.5° W on Mid Atlantic Ridge (see Section 3.2).

1954 January 17   Sudan
  17h 39m 38s   16.5° N, 36.0° E   Gouin (1979)
Misprint for event at 16.5° S, 36.0° E in Mozambique.

1954 August 20   Saudi Arabia
  15h 29m 30s   20° N, 52.25° E   BCIS
Poor distribution of stations. Relocated by us with combined macroseismic and instrumental data to 27.8° N, 52.1° E in southern Iran.

1954 September 13   East Jordan
  21h 46m 50s   32.2° N, 37.5° E   Kárník (1968)
                   31.2° N, 37.2° E   ISS macroseismic
                                         report
Poor locations for felt event at 30.9° N, 35.3° E near Dead Sea, Arieh *et al.*, 1983.

1955 November 11   Egypt
  18h 27m 35s   27.5° N, 27.2° E   Maamoun *et al.*
                                        (1984)
10° error for earthquake in Turkey at 37.5° N, 27.2° E.

1958 September 28   Near Bahrain
  12h 36m   26.5° N, 50.0° E   Strasbourg
Unusual position not substantiated.

1960 May 10   Saudi Arabia
  21h 51m 55s   27.0° N, 47.5° E   USGS
Relocated by Sykes and Landisman (1964) in southern Iran at 31.77° N, 50.84° E.

1960 May 31   Gulf of Aden
  ooh 24m 03s   14.7° N, 54.8° E   ISS
                   depth 127 km
Unusual depth not substantiated. Depth is likely to be normal and origin time 23m 50s.

1960 October 20   Libya
  12h 23m 01s   9.1° N, 20.8° E   USGS
Relocated by BCIS at 34° N, 26.5° E.

1963 May 11   Libya
  01h 11m 48s   32.8° N, 19.4° E   USGS
Felt in Greece, near 39° N, 22° E.

1964 May 15   Libya
  01h 12m 41s   33.1° N, 15° E   ISC
Poorly recorded, likely to be further north.

1965 October 17   Yemen
20h 08m 53s   17.2° N, 43.6° E   Rothé (1969)
Ten year error for event on 1955 October 17.

1966 December 3   Sudan
14h 24m 29s   10° N, 29° E   LAO (ISC) Array
solution

1966 December 26   Mediterranean Sea
07 06m 20s   33° N, 32° E   LAO (ISC) Array
solution

1967 March 28   Red Sea
02 41m 34s   19.8° N, 38.6° E   USGS
m_b 6.7
High magnitude not substantiated. $M_s \not> 4.0$.

1967 August 20   Mediterranean Sea
17 28m 50s   33° N, 26° E   LAO (ISC) Array
solution

1967 November 22   Southern Iran
06 15m 08s   26° N, 54° E   LAO (ISC) Array
solution

1968 February 10   Saudi Arabia
11 07m 43s   20.2° N, 52.0° E   LAO (ISC) Array
solution

1968 March 14   Saudi Arabia
09 57m 05s   17.1° N, 42.0° E   LAO (ISC) Array
solution

1968 April 20   Saudi Arabia
07 18m 25s   20.9° N, 53.0° E   LAO (ISC) Array
solution

1968 May 2   Saudi Arabia
13 51m 04s   23.0° N, 41.9° E   LAO (ISC) Array
solution

1968 May 24   Libya
17h 57m 22s   23.9° N, 11.9° E   LAO (ISC) Array
solution

1968 May 26   Libya
00h 05m 44s   26.7° N, 14.6° E   LAO (ISC) Array
solution

1969 November 19   Libya
08h 39m 30s   30.6° N, 24.7° E   LAO (ISC) Array
solution

1969 December 4   Libya
04h 55m 27s   23.1° N, 11.0° E   LAO (ISC) Array
solution

1969 December 20   Arabian Sea
06h 07m 49s   14.9° N, 54.1° E   LAO (ISC) Array
solution

1969 December 20   Ethiopia
07h 51m 33s   10.8° N, 37.6° E   LAO (ISC) Array
solution

1970 January 9   Israel
06h 16m 14s   30.2° N, 34° E   LAO (ISC) Array
solution

1970 April 20   Red Sea
03h 44m 58s   21.2° N, 38.9° E   HFS (ISC) Array
solution

1970 May 3   Libya
18h 58m 42s   31.9° N 22.8° E   LAO (ISC) Array
solution

1970 May 14   Ethiopia
18h 45m 40s   13° N, 39.8° E   HFS (ISC) Array
solution

1970 June 8   Saudi Arabia
22h 41m 06s   23.1° N, 48.2° E   HFS (ISC) Array
solution

1970 August 9   Saudi Arabia
02h 20m 34s   21.7° N, 42.8° E   LAO (ISC) Array
solution

1970 November 23   Libya
23h 00m 48s   32.8° N, 20.9° E   ISC
HFS array mislocation of event near 35.0° N, 21.2° E.

1970 December 28   Libya
03h 49m 44s   30.0° N, 21.5° E   LAO (ISC) Array
solution

1971 July 4   Egypt
02h 08m 51s   30.4° N, 29.7° E   ISK (ISC)
No supporting observations. Likely to be 9° latitude error for
event near 39° N.

1974 May 24   Libya
01h 29m 13s   31.4° N, 14.4° E   ISC
Poor location, likely to be farther north, near Sicily.

1975 August 12   Libya
14h 52m 09s   32.25° N, 24.5° E   ISC
Poor ATH solution, cannot be substantiated.

1975 September 26   Libya
22h 42m 21s   28.7° N, 17.2° E   USGS
Very poor solution, cannot be substantiated.

1975 November 26   Egypt
    01h 02m 14s   28.2° N, 32.3° E   ISC
Located by five stations only, the closest being in
Czechoslovakia. Not likely to be genuine.

1976 January 30   Saudi Arabia
    22h 59m 39s   17.88° N, 51.69° E   ISC
Very poor determination. Event probably further south in Gulf
of Aden.

1976 September 2   Saudi Arabia
    19h 53m 06s   25.0° N, 52.7° E   ISC
Very poor solution based on few readings. Event probably
further north in Zagros area.

1976 September 26   Arabian Peninsula
    00h 12m 47s   29.91° N, 47.27° E   ISC, USGS
                depth 98 km
Very poor determination, based on only five readings. No
evidence for deeper than normal depth. Event probably further
north in Zagros area.

1976 September 27   Iran
    02h 24m 15s   30.35° N, 49.92° E   ISC
                depth 152 km
No justification for deeper than normal depth. Position given
by USGS at 28.8° N, 48.2° E for event with normal depth is
likely to be more correct.

1981 May 16   Libya
    12h 01m 32s   31.7° N, 22.0° E   ISC
Likely to be closer to Crete.

1981 October 6   Libya
    21h 59m 29s   32.1° N, 24.9° E   ISC
Poor solution, likely to be further north.

1984 October 28   Arabian Peninsula
    12h 93m 50s   22.3° N, 59.86° E   ISC
                depth 33 km
Very poor determination from bad distribution of stations.
Position not substantiated. Event is possibly in Arabian Sea.

1992 June 2   Arabian Peninsula
    20h 12m 47s   25.91° N, 37.53° E   NEIC
                m_b 4.8
Position and magnitude unreliable. They are largely controlled
by five mis-associated Nepalese stations. ISC relocates 150 km
to west in Red Sea; M_s 4.2.

CHAPTER 4

Conclusions

In this chapter we review the evidence that we have assembled for the long-term seismicity of the region. The occurrence of earthquakes in time and space is not necessarily constant, but some variations in the record will undoubtedly be the product of changes in the circumstances in which events were reported, rather than of changes in seismic activity itself. The chance of an earthquake being reported in historical sources, or recorded by instruments, depends partly on where chronicles were written, or where seismographs were located, in relation to where the earthquake originated. It also partly depends on the size of the event. The coverage naturally varies at different periods and for different areas. Chroniclers tend to be subjective and may have no interest in reporting even the most destructive shocks. Documentary sources are far less sensitive than purpose-built instruments, but in their own way they can quite subtly reflect the level and rate of seismic activity affecting the society whose history they record.

In an area of high seismicity, where large and destructive earthquakes are frequent, it is likely that only the most severe will be reported in sufficient detail to indicate their size. In an area of relatively low seismicity, the threshold at which events are perceived and reported is correpondingly lower, and it is possible to distinguish small earthquakes that cause little or no damage. In such regions, we would expect the rarer, large event to be well documented, always assuming it affected a populated area. At the same time, it is also probable that the buildings in such areas will be more vulnerable to earthquakes, as local builders will not have needed to evolve more resistant construction techniques. Damage may thus be disproportionate to the size of the earthquake concerned.

In what follows, we discuss the record of earthquakes in our region in the light of these considerations. First, fluctuations in reporting earthquakes over time and in their spatial distribution are linked: both depend on the availability, provenance and scope of documentary sources. Variations largely reflect the changing historical and cultural factors that influence the production of documents. Secondly, the spatial distribution of earthquake locations in the pre-instrumental period is linked very closely to the population geography of the region. The extent to which the pattern might have been biased or distorted by this factor can be determined by comparing the macroseismic with the instrumental information, though the latter may not be all that complete either. Such comparisons allow us to identify and seek to explain gaps in the long- and short-term record of seismic activity in the region.

4.1 Completeness of the historical earthquake catalogue

Table 2.1 lists 245 events identified as having affected the study area before 1900, not all of them originating within the area and some of them of dubious seismic origin. Volcanic eruptions and earthquakes associated with them are also included in this number. For southern Iran, the data are incomplete, as a handful of events affecting the northern littoral of the Persian Gulf have been excluded. From Table 3.1, it will be seen that

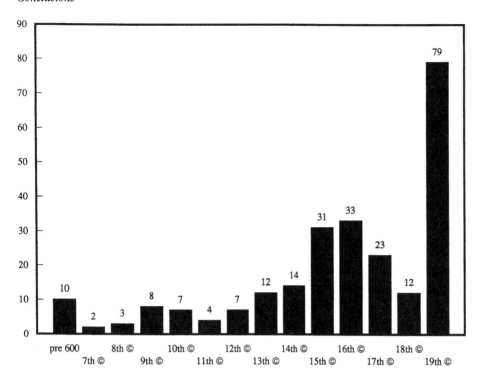

Figure 4.1. Time distribution of the earthquakes listed in Table 2.1, by century (all events).

macroseismic information has been found for a slightly larger number of events (271) since 1900 (see Figure 4.9), and the total would of course be much higher if earthquakes below magnitude 5 were included from peripheral areas such as southern Iran. A comparison between the spatial distribution and level of seismicity in the twentieth century and in historical periods is made below (Sections 4.2 and 4.3). Here, in view of the fact that the twentieth-century record comfortably exceeds that of the whole historical period, it is necessary to consider the numerical incompleteness of macroseismic information reported before 1900, and how genuinely representative it is of actual earthquake activity. The use of statistical methods for testing completeness, which assume stationarity, are of course unsuitable in the present context. For long-term observations and large areas such as ours (more than 10 million square kilometres), the problem is how to establish stationarity *a priori*, and to discriminate between random and non-random processes.

In Figure 4.1, the time distribution of the earthquake record is presented in arbitrary but convenient 100-year units, from the beginning of the Islamic period in the early seventh century AD. Only ten earthquakes are reported in the previous centuries, from 184 BC onwards. This early period is not discussed further, but the absence of earthquake reports is evidence of both relatively low seismicity and a paucity of sources com-

pared with other areas of the eastern Mediterranean world at that time. Lower Egypt, Sinai and southern Palestine enjoy the longest and most continuously documented history in our area and most of the earliest earthquakes reported affected these regions. But despite the written material preserved in inscriptions and papyri, and the evidence provided by the temples and monuments themselves, it is not possible to identify any datable earthquakes or their effects in Ancient Egypt before the Ptolemaic period (323–30 BC).

Figure 4.1 shows, as one would expect, a steady increase in the number of events recorded per century, from 2 in the seventh century to 79 in the nineteenth. The majority of these occurred after the ninth century, that is from the time that Arabic chronicles become available. Nevertheless, the increase in information is not steady; a peak in the fifteenth and sixteenth centuries is followed by a sharp decrease, particularly in the eighteenth century. We must briefly consider to what extent these variations are the product of cultural or historical factors, rather than genuine fluctuations in the level of seismic activity over time.

To do so, a further geographical subdivision is necessary (Figure 4.2). First, because the total numbers conceal important differences between various regions and periods; secondly, because it allows us to concentrate

---

[1] See the convenient maps in Guidoboni (1989), pp. 720–2.

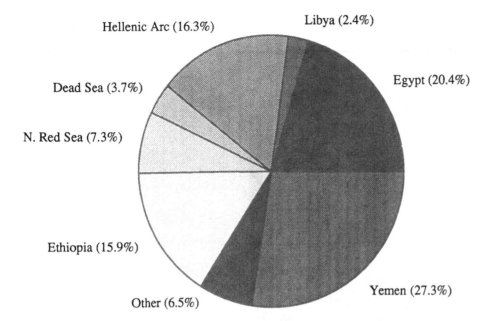

Hellenic Arc (16.3%)

Libya (2.4%)

Dead Sea (3.7%)

N. Red Sea (7.3%)

Egypt (20.4%)

Ethiopia (15.9%)

Other (6.5%)

Yemen (27.3%)

Figure 4.2. The proportion of earthquakes before 1900 reported in different regions of the study area (from Table 2.1, all events). The total for Yemen (67 events) includes those located in the Gulf of Aden and the southern Red Sea; Hellenic Arc (40 events) includes the Eastern Mediterranean; northern Red Sea (18 events) includes Sinai, the Hejaz and the Gulf of Suez; Other (16 events) includes Northern Syria, Turkey, Iraq, the Persian Gulf, the Gulf of Oman, the Arabian Peninsula (excluding Hejaz) and the Sudan.

on the central area of interest, for which we would claim the greatest degree of thoroughness in our search for data. We may note that the record for the Sudan (which only starts in 1850), and Libya (only two events before 1853) remains incomplete throughout the period. Before 1900, it is increasingly difficult to find data even for coastal areas of Libya. With the exception of Tripoli and to a lesser extent Barce (Al Marj), the region contains few towns of historical importance, and its few oases on the traditional caravan routes are separated by large tracts of barren country. Barce, like Benghazi after the fifteenth century, is rarely mentioned by Arab writers or in Ottoman documents, in contrast with the situation in neighbouring Tunisia. From the seventeenth century, European sources provide some scant information about earthquakes and other natural phenomena, such as the distant effects in Tripoli of the 18 July 1787 Mount Etna eruption, and the meteorite fall near Murzuq on 24 December 1870.[2] In general, however, the lack of macroseismic data for Libya must be attributed chiefly to a dearth of sources of local origin.

Indigenous data become available for Ethiopia in the fifteenth century, but there remain large gaps in the record for the next 400 years. Only 10 events in our area are reported between 1400 and 1799, though this is twice as many as in the northern half of the Red Sea and along its coasts. However, when compared with the 28 earthquakes reported in the nineteenth century alone (comfortably the highest for any region within the study area), the extent to which earlier evidence is deficient becomes apparent.

[2] Tully (1817), p. 156; Coumbary (1870b), p. 100.

The great majority of earthquakes listed in Table 2.1 have been located either in Egypt (50 events: 20% of the total) or the Yemen, including the southern Red Sea and Gulf of Aden (67 events: 27%), see Figure 4.2. The historical record for the Yemen is fairly continuous and provides an adequate basis for assessing regional seismicity. The time distribution of these events (Figure 4.3) shows that a small but steady number of earthquakes was reported up to the end of the fourteenth century, reaching a peak between the fifteenth and seventeenth centuries. After a decline in the eighteenth century, reported events reach their highest level in the nineteenth century. Throughout the period, Yemeni earthquakes form a substantial proportion of the total number reported each century for the whole study region (between 20 and 50%).

These fluctuations are closely related to variations in the availability and quality of the sources of historical information. While the earliest events are reported in sources written outside the Yemen, richer indigenous chronicles, in the form of annals rather than biographical histories, provide a particularly valuable source of data from the start of the Rasulid period (early thirteenth century) onwards. The importance of the provenance of these chronicles, and the historical context in which they were produced, is demonstrated by the variations in the geographical distribution of reported seismicity.

In the thirteenth century, only earthquakes round San'a are mentioned, reflecting the continuing importance of the northern part of the country in the political struggles of the period. Thereafter, however, all macroseismic data for the fourteenth and fifteenth centuries

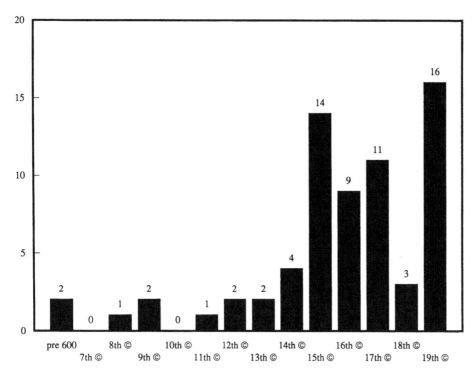

Figure 4.3. Time distribution of earthquakes in the Yemen to the end of the nineteenth century (from Table 2.1, all events).

concern only the southern part of the country, and particularly the area between Zabid, capital of the Rasulids (to 1454), and Aden, capital of the Tahirids (to 1526). Following the interlude of the Ottoman period in the Yemen (to 1635), the focus of the historical sources returns to events in the north, where the Zaidi Imams had restored Yemen's independence. Almost all the earthquakes reported in the seventeenth century are located in the north of the country, particularly around the capital, San'a. Thereafter, no earthquakes are reported for about a century, reflecting a cultural decline that seems to have been general throughout the Middle East. It is beyond the scope of this chapter to examine the causes of this decline,[3] but for our immediate purposes we can state that the lack of macroseismic data is primarily due to a deficiency of sources rather than a genuine absence of earthquakes. Arabic chronicles lose much of their value in the eighteenth century and the loss is not yet made good by European sources of information, which become increasingly important in the nineteenth century. In fact, the indigenous annalistic tradition survived more vigorously in the Yemen than elsewhere, and Yemeni chronicles continue to report macroseismic data, particularly in the second half of the nineteenth century. It is only after 1800 that earthquakes

are reported evenly throughout the length of the country.

Thus, the record of historical seismicity in the Yemen approaches completeness for different areas of the country at different periods. Particularly in the fifteenth to seventeenth centuries, earthquakes are systematically reported, first in the area round Zabid and then in the area round San'a. Since even relatively small events, and tremors felt at low intensities, are quite regularly reported, it encourages us to believe that few if any larger earthquakes are missing from the record, even if some are incompletely identified. There remains a good possibility of discovering additional macroseismic data in time, since a large proportion of the rich historiography of the country remains in manuscript and recently edited texts are still not readily accessible.

Turning to Egypt, the situation is not dissimilar, though much larger fluctuations can be observed in the proportion of events contributed by Egypt to the total reported each century (between around 5% and 70%). Figure 4.4 shows that a relatively high level of reporting in the ninth and tenth centuries is not repeated until the fourteenth century. The loss of many of the historical sources of the Fatimid period (969–1171) partly explains the intervening dearth of data, compounded by the fact that Syria, rather than Egypt, remained the focus of events under the Ayyubids (to 1250) and during the last years of the Crusader states in the Levant. The

---

[3] The decline of the Islamic world in the eighteenth century is discussed, among others, by Hodgson (1974), III, 134 ff.

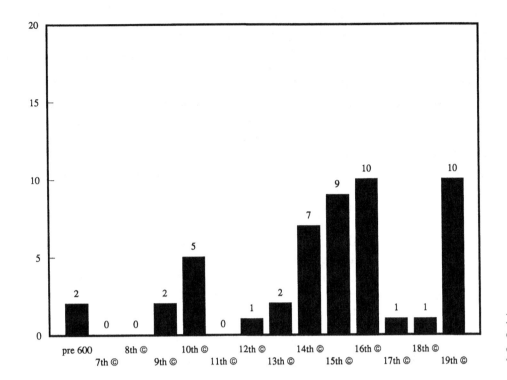

Figure 4.4. Time distribution of earthquakes in Egypt to the end of the nineteenth century (from Tabe 2.1, all events).

Mamluk period that followed (1260–1517) is the greatest age of Arabic historiography, with numerous chronicles providing at times almost a daily record of events, particularly in Cairo. This is amply reflected in the wealth of macroseismic data available in the fifteenth and sixteenth centuries (interestingly, a period completely blank in the catalogues of Lyons and Sieberg). In the seventeenth and eighteenth centuries, when Egypt was a province of the Ottoman empire, there is a dramatic decline in the number of reported earthquakes. (Future work in the Ottoman archives in Istanbul and Cairo might one day alter this picture.) Indigenous sources, though numerous, are less detailed than in the Mamluk period, and clearly less interested in reporting minor tremors. In the nineteenth century, European sources help to raise the amount of information available back to the levels reached in the sixteenth century.

The figures for Egypt would be much higher if reports were included of the effects of more distant earthquakes. Table 2.1 lists 40 events located in the Hellenic Arc or Eastern Mediterranean, all of which were reported in Egypt, and several other earthquakes in the northern Hejaz or the Dead Sea system, nearer at hand. At least half the earthquakes that affected Egypt, sometimes at damaging intensity, originated from epicentres elsewhere, and Egyptian data can help to identify their full extent.

From this, it is apparent that at certain times, particu-

larly during the Mamluk period, the record of events in Egypt approaches completeness even for events felt at low intensities, whether of local or more-distant origin. On the other hand, this low threshold of earthquake perception, or sensitivity to earthquakes, applies almost exclusively to Cairo, the only place stated or implied to have experienced the shocks reported in chronicles that were all written in the capital. The historical record is lacking in detail and often deficient in far-field information. In almost all cases, it is not possible to define precisely the epicentral area of the local events. The larger of these were probably associated with epicentral areas along the west coast of the Red Sea and Gulf of Suez.

Thus our data are relatively complete only for Cairo and, by extension, the immediately surrounding areas of Lower Egypt and the Nile Delta. Proper reporting of earthquakes in Upper Egypt only begins in the eighteenth century, with the advent of European sources of information. Even then, for reasons discussed more fully below (Section 4.2), information is restricted to the Nile Valley itself.

In conclusion, it is clear that as regards the number of earthquakes reported, the historical record is very incomplete. The general tendency is for information to become fuller with time, and more geographically homogeneous, so that coverage in the nineteenth century is relatively full, though still well below the twentieth-century macroseismic record. Even in areas with a relatively well-documented history, such as Egypt and the Yemen, there are periods when very few earthquakes are reported. Spatial variations in the record for the Yemen reflect political changes, notably in the relative importance of different capital cities; in Egypt, where Cairo was capital throughout, coverage is only adequate for the surrounding areas of the Delta and Lower Nile Valley. Elsewhere, such as in the Arabian Peninsula, the record is so sparse that it is not possible to detect significant variations in the level of reporting, nor is there any increase nearer the present time. It is more convenient to discuss these data gaps in terms of the geographical distribution of the earthquake record.

4.2　Regional distribution of seismicity

We have seen that in favourable circumstances, the surviving record of past earthquakes can be quite full. Figure 4.5 plots the distribution of earthquakes reported before 1900 (from Table 2.1, excluding dubious events); volcanic eruptions are distinguished by a separate symbol. It is clear from Figure 4.5 that seismicity is not evenly distributed throughout the region, but rather is concentrated in the Yemen, Ethiopia, Egypt and south of the Dead Sea. As has already been observed, these areas have a relatively well-documented history, and it is reasonable to suppose that cultural factors have influenced the apparent distribution of reported earthquakes.

Comparison of Figure 4.5 with a plot of the density and distribution of population in the region (Figure 4.6) reveals to what extent demographic patterns might have distorted the pattern of reported seismicity. The uplands of the Yemen, the central plateau of Ethiopia and the Nile Valley in Egypt are the most densely populated areas. Central Arabia and the Egyptian and Libyan deserts are, by contrast, barely inhabited. Naturally, the reporting of earthquakes in history depends on the presence of concentrations of people to experience and record them. The *location* of events is not, however, totally dependent on this factor, but depends also, for example, on communications.

The most obvious cultural bias exerted on the location of past earthquakes is seen in the case of Egypt, where the epicentres assigned tend to follow the course of the Nile. On the other hand, it has been possible to distinguish a few events in Sinai and the Gulf of Suez, which are areas of very low population density.

The absence of seismicity in the interior of the Arabian Peninsula also closely mirrors the sparse population of

the desert and its low literary output. A few data are available from Oman and Hadramaut, both regions with a surviving documentary record, though still very imperfectly known and little exploited. Oman, in particular, was important in maritime trade and some information could be expected from coastal regions (as for coastal Iran north of the Persian Gulf). Leaving aside the Yemen, the only region in Arabia for which macroseismic data exist in any number is the Hejaz, in the northwest. This is due not so much to population density as to the pilgrim routes passing through the area down to Mecca. On the other hand, similar routes crossed the northern Arabian desert from the Hejaz to the head of the Persian Gulf, and no data survive from this area. This suggests that the record for the northern Hejaz is genuinely representative of seismic activity there.

While the comparison between the distribution of earthquakes and the density of population does, therefore, suggest strong correlations in areas such as the Yemen, the Nile Valley and central Arabia, our sources are also adequate to distinguish the occurrence of earthquakes in barren regions such as the Gulf of Suez and the northern Hejaz. This is partly due to the communication networks crossing the area, and partly due to the size of earthquakes themselves.

Comparison of the historical record with that of the twentieth century provides another measure of the completeness of the long-term pattern of seismicity. Figure 4.7 plots all the events listed in Table 3.1. The greatest concentrations of earthquakes are near the rift systems at the north end of the Red Sea, in the Red Sea itself, the Gulf of Aden and northern Ethiopia. In the Red Sea and Gulf of Aden, the earthquakes are mainly close to the median rift, but the accuracy of location is not good enough to confirm whether shocks actually originated in the central Axial Trough.

Apart from the greater number of events recorded in the twentieth century, and ignoring for the present the question of size, there are some immediately striking differences in the pattern of epicentres between the two periods covered.

First, the historical record is almost devoid of earthquakes located in the Red Sea and the Gulf of Aden. Apart from the poorly known earthquake of 1121, no shocks are located in the northern Red Sea, while events located in the southern end are mainly volcanic eruptions and associated tremors. On the other hand, the historical record for the Yemen, and other mainland areas of the Arabian Peninsula, shows evidence of activity that has not yet been repeated during the present century.

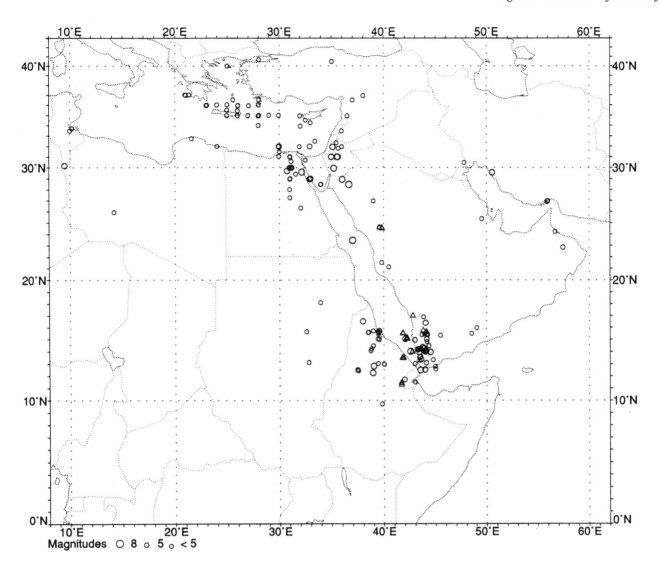

Figure 4.5. Plot of earthquakes and volcanic eruptions for pre-instrumental period, up to 1899 (see Table 2.1). Earthquakes are shown as circles and volcanic eruptions as triangles. Events designated X (probably of non-seismic origin) are omitted.

Size of symbols. The relationship between size of symbols and earthquake magnitude is shown beneath the maps. Earthquakes of magnitude 5 are plotted with the middle symbol; symbols for larger events increase linearly with magnitude, and an event of magnitude 8 would have the size of the largest symbol. All events of magnitude less than 5, or with no magnitude specified, are plotted using the smallest symbol. For those events for which no magnitude can be determined (particularly in the historical period), it must not be assumed that these events are necessarily small.

Randomisation. The events in our lists have their position specified with a precision of 0.1°, but many events, particularly those located macroseismically, are located less accurately and several events in a particular region may be allocated the same position. In order to display these better, in some plots a system of randomisation has been adopted, in which the plotted positions are shifted by random amounts of up to 0.05° (about 5 km) in latitude and longitude.

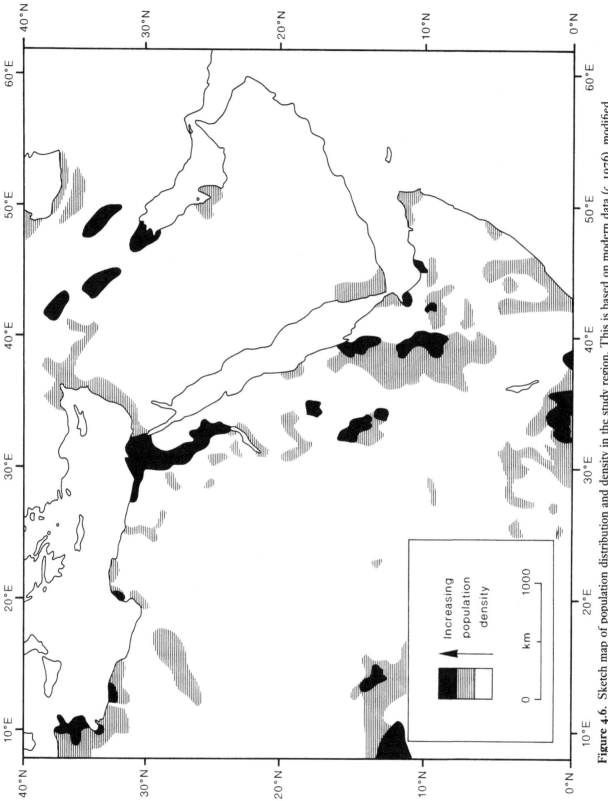

Figure 4.6. Sketch map of population distribution and density in the study region. This is based on modern data (*c.* 1976), modified slightly to more nearly approximate the pattern in earlier centuries, for which no statistical evidence is available. The concentration of population in the Nile Valley, the Ethiopian uplands and inland Yemen was a constant feature in the past, even if it is not possible to quantify it.

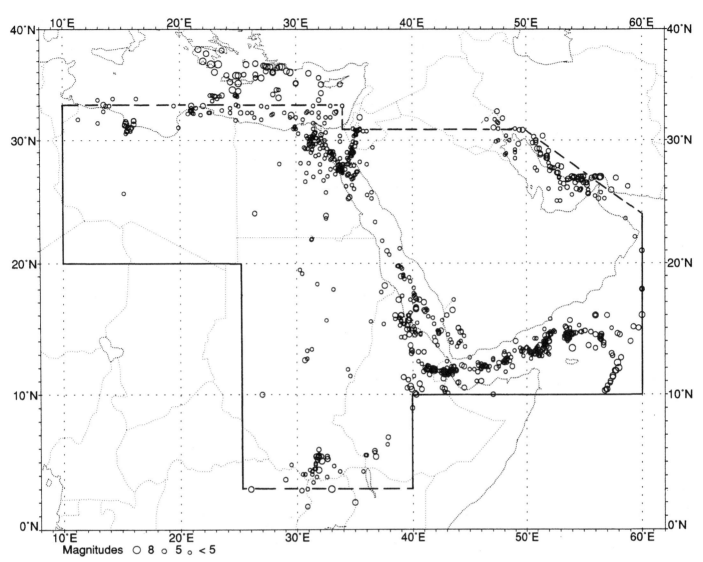

Figure 4.7. Plot of earthquakes located for the instrumental period (1899–1992), including some located macroseismically (Table 3.1). The boundaries of the study area are shown: note that they are slightly different from those fixed for the historical period (see Figure 1.1, and p. 119). Earthquakes outside the solid boundary are not considered; beyond a broken boundary some earthquakes are shown that could have been felt within our area. For the size of symbols, see caption to Figure 4.5.

In the case of the Yemen itself, we can be reasonably confident that the majority of earthquakes reported (see Figure 4.8) did indeed originate on land rather than in the southern Red Sea, though the true location of many shocks reported to have affected Zabid is not known. There are very few cases (1504 August 30 is one) when reports are available from both the Ethiopian and Arabian side of the water. Shocks located round Aden, in Hadramaut and coastal Oman, however, might easily have originated offshore. Macroseismic evidence alone is incapable of distinguishing offshore epicentres along such underpopulated coastlines. Figure 4.9 demonstrates how scarcely a handful of the many earthquakes instrumentally located in the Gulf of Aden are reported

to have been felt. On the whole, Figure 4.9 is similar to Figure 4.5, though Ethiopia shows up more strongly than the Yemen, raising the possibility that some historical events reported in southern Yemen originated in the Afar Triangle.

Secondly, although numerous twentieth-century events have been located along the Gulf of Aqaba and the Dead Sea system, there is as yet no evidence of large earthquakes in the northern Hejaz. In this respect, the long-term historical record supplements and fills in the pattern observed this century.

Thirdly, the true extent to which the concentration of population in the Nile Valley has distorted the location of historical earthquakes is clear from Figure

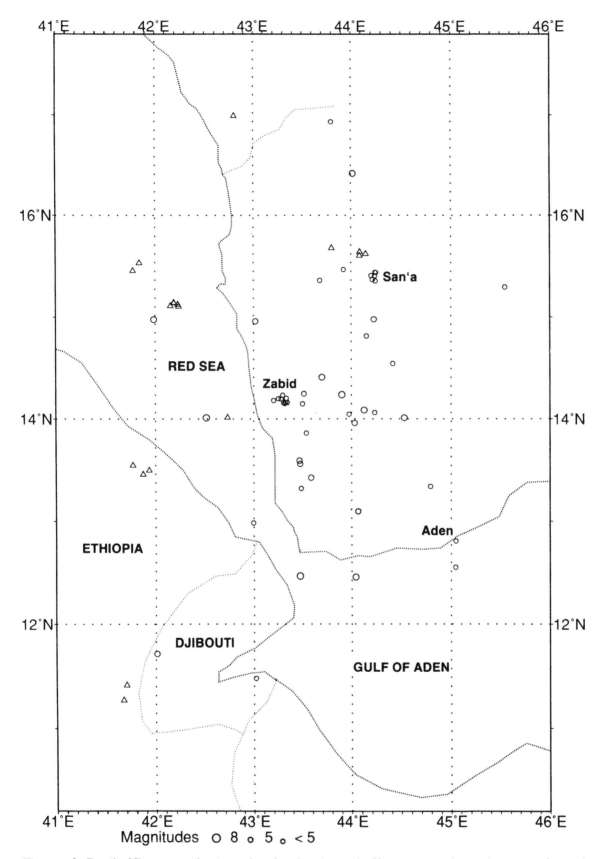

Figure 4.8. Detail of Figure 4.5, showing a plot of earthquakes in the Yemen area in the pre-instrumental period up to 1899. Sites of volcanic eruptions are shown by triangles. Positions have been randomised and symbol size is as explained in the caption to Figure 4.5.

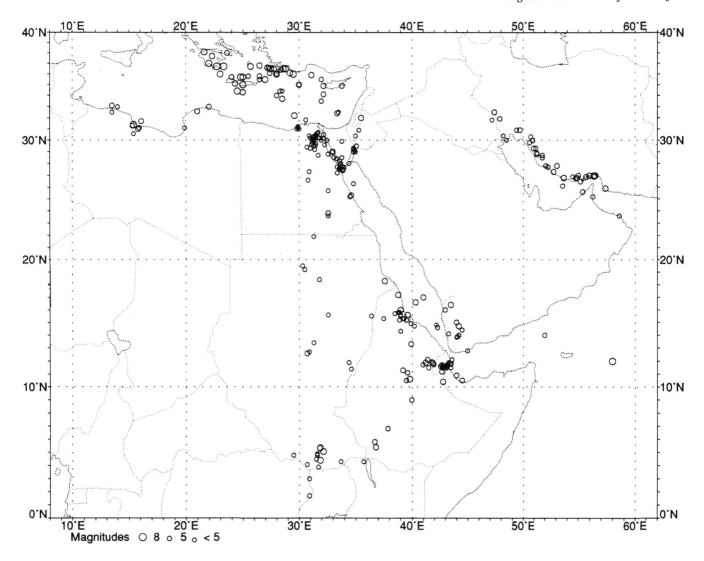

Figure 4.9. Earthquakes during the instrumental period (1899–1992) which have been reported as felt or causing damage. Some early events have been located by felt information alone, but most have been located instrumentally, with confirmatory felt reports. The large earthquake located in the Arabian Sea (12° N, 58° E) was felt on board ship, but not reported on land (see catalogue in Chapter 2 under 1904 October 3).

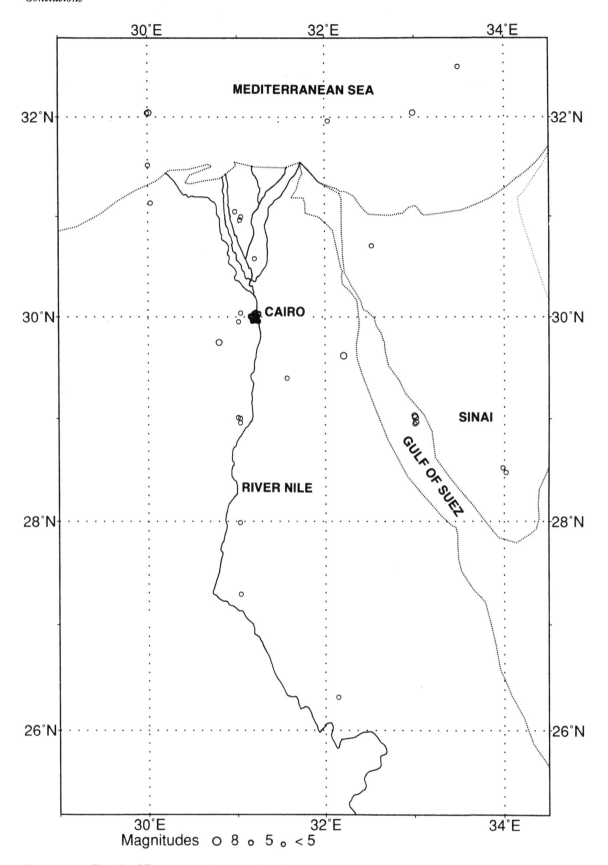

Figure 4.10. Details of Figure 4.5, showing a plot of earthquakes in Egypt in the pre-instrumental period up to 1899. Positions have been randomised and symbol size is as explained in the caption to Figure 4.5.

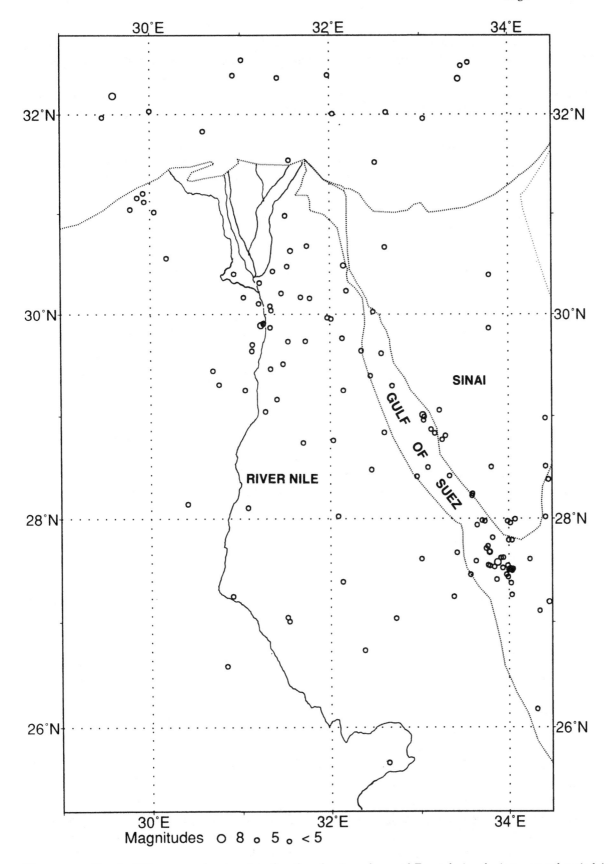

Figure 4.11. Detail of Figure 4.7, showing plot of earthquakes in and around Egypt during the instrumental period (1899–1992). Positions have been randomised and symbol size is as explained in the caption to Figure 4.5.

4.7. While the seismicity of the Gulf of Suez was correctly, if incompletely, depicted in the historical sources, there is no obvious zonation along the Nile during the twentieth century. Furthermore, no modern events appear to have originated in the Delta, and historical earthquakes located in this area should probably be associated with offshore epicentres. These differences can be seen more clearly in the detailed maps of Egypt provided in Figures 4.10 and 4.11.

Overall, the historical record complements and confirms the picture obtained from twentieth-century data. The absence of past seismicity in the Libyan desert, much of Egypt and the Sudan, as well as central Arabia, seems to be confirmed by the short-term instrumental record, suggesting it is not purely a function of a dearth of sources of information. Historical data confirm the long-term seismicity of the Yemen and Ethiopia, and give an incomplete, though suggestive picture of heightened activity in the northern Red Sea and the Dead Sea system.

4.3 Seismicity of Egypt, Arabia and the Red Sea

Finally, the validity of the long-term record is strengthened when one considers the levels of seismicity reported, rather than simply the number and location of individual events. Despite all the gaps in the macroseismic information available, the documentary sources are adequate to distinguish regions of high and low seismicity, even though precise details may be incomplete. Figure 4.12 combines historical data with twentieth-century earthquakes of magnitude 5 or greater, to produce a long-term picture of the largest earthquakes to affect the study region. For the whole area and period investigated, it is unlikely that all moderate and many large-magnitude shocks have been identified. It is very probable, however, that any major or large earthquake in the region has been noted in some form, although not necessarily accorded its true size.

In the historical period, unless there is definite evidence to the contrary, most of the earthquakes reported must have been moderately large to have been considered worth mentioning. With both sides of the Red Sea very sparsely populated, for example, only a few of the largest offshore shocks can be identified from their effects on land. The historical record is not representative of what is known of twentieth-century seismicity in the Red Sea and the Gulf of Aden, but earthquakes in these offshore regions have not constituted a hazard.

The activity in the southern Red Sea and Gulf of Aden extends through northern Ethiopia and the Ethiopian Rift into the southern province of Sudan, Equatoria. Several significant events have occurred here, including the earthquakes of 21 May 1915 (M_s 6.3) and 20 May 1990 (M_s 7.2), one of the largest events known to have occurred in Africa.

From Figure 4.12 it appears that the only area in which the twentieth-century record does not reflect true long-term seismicity is the northern Hejaz and the southern end of the Dead Sea system, where the return period of large events is longer than the period of instrumental monitoring. In such sparsely populated areas, only the largest events in the past would have attracted attention. Evidence of infrequent large earthquakes is apparent in the shocks of 1068, 1212, 1293, 1458 and 1588. As this region is traversed by the much-frequented pilgrim route to Mecca, the silence of historical sources about earthquakes along the southern part of this route is likely to reflect a genuine absence of such large events, both on land and in the central section of the Red Sea.

In the case of Lower Egypt, which has a long and well-documented history, it is clear that many of the earthquakes reported were relatively small, the largest of them probably occurring in the Gulf of Suez region. Inland areas of Egypt are relatively stable, but can be subject to infrequent events of significant size. An example is the earthquake of 9 December 1978 (M_s 5.0) in southwestern Egypt, about 650 km from the Nile Valley in an area previously considered stable.

Figure 4.12 also shows activity extending offshore from the Egyptian coast westwards to Cyrenaica. The largest of these events, on 21 February 1963, although only of magnitude 5.4, caused extensive damage and many casualties at Barce. There is also an area of repeated activity on the northern coast of Libya to the east of Tripoli, near 15°–16° E, not reflected in historical sources. The twentieth-century record for Libya highlights the defective nature of the early information available, and the apparently low earthquake hazard of a number of regions shown on seismic maps of Libya (e.g. Mallik and Morghem, 1977) may not be entirely genuine, reflecting rather the lack of long-term observations.

Inland areas of Libya and Sudan appear generally stable, though shocks were reported in the Fezzan in 1853, and central Sudan has also experienced isolated earthquakes, such as that on 9 October 1966 (M_s 5.6), located to the southwest of Khartoum.

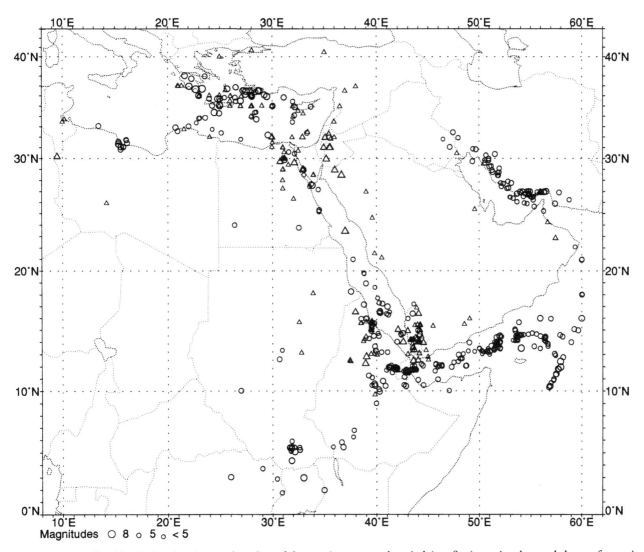

Figure 4.12. Combined plot showing earthquakes of the pre-instrumental period (to 1899) as triangles, and those of magnitude 5 or greater in the instrumental period (1899–1992) as circles. Positions have been randomised and symbol size is as explained in the caption to Figure 4.5.

The Arabian Peninsula is one of the most stable areas known. The central part of Arabia appears quite devoid of earthquakes and it is only in the Yemen, in the southwest corner, that significant earthquakes have occurred. The long-term record for the Yemen is quite full, and indicates relatively high seismicity, with occasional medium-sized earthquakes, often followed by long aftershock sequences, and numerous swarm events, characteristic of regions of active volcanism. The most recently observed eruption occurred near Dhamar in 1937.

The nine or ten centuries covered reasonably adequately in the sources used for the present study may still not be a long enough period to provide a complete picture of the seismicity of the region. Nevertheless, whatever the imperfections of the record for the best-documented areas in Egypt and the Yemen over the past millennium, there is no evidence of the occurrence of a major earthquake in either region, though in both cases relatively heavy losses have been sustained in earthquakes of only moderate magnitude.

References

Abbreviations and conventions

References are given in the main text, in the notes and in figure captions either by author and date, in the case of modern works, or by author alone in the case of historical texts in Arabic and other languages. The bibliography follows these two main divisions. There is not always a clear distinction between a 'historical' source and a 'modern' work. On the one hand, there are some twentieth-century Arabic chronicles that can be considered primary sources; on the other hand, European travellers are also primary sources, though 'modern' publications. The general tendency is for modern Arabic works to be included in the section entitled 'Printed editions of historical texts', and European travellers to be included in the section entitled 'Printed modern works'. In all cases, the method of citation indicates in which section of the bibliography a work will be found.

A further distinction is made between printed editions and works consulted in manuscript (Ms). When the latter are cited in the course of this book, reference is given to a folio number (fol.), recto (ro) or verso (vo), rather than a page number (p.). The manuscript sources used are listed separately below.

When an author has written more than one work, the correct reference in a given case is distinguished by the date of publication for modern works. In the case of historical texts, the *first* work listed is intended unless otherwise indicated. For example, all citations from Abu Shama refer to his work, *Kitāb al-rauḍatain*, unless Abu Shama, *Dhail*, is specified. It should also be noted that when a translation is available of an Arabic work, the translation is cited after the Arabic text: thus al-Suyuti, p. 34/35 refers to al-Suyuti's *Kashf al-ṣalṣala*, p. 34/ French translation by Nejjar (1974), p. 35, and Taher (1979), p. 106/45 refers to the Arabic part of Taher's thesis, followed (when appropriate) by the page in the French part. Generally, translations are given together with the original work, unless they are referred to separately in the text, in which case they are listed as modern works.

The following abbreviations are used to denote archives and other library collections.

AE = Archives Etrangères (in A.N.)
A.N. = Archives Nationales, Paris
BBA = Başbakanlık Arşivi, Istanbul
BL = British Library (India Office), London
Bodleian = Bodleian Library, Oxford
Cairo = Dar al-Kutub, National Library, Cairo
CUL = University Library, Cambridge
Dublin = Chester Beatty Library, Dublin
F.O. = Foreign Office archives, Public Record Office, London
IMV = Irade Meclis-i Vala (in BBA)
I.O.R. = India Office Records, London
Leiden = Bibliotheek der Rijksuniversiteit te Leiden
MD = Mühimme Defteri (in BBA)
MMD = Maliyeden Müdevver Defterler (in BBA)
Paris = Bibliothèque Nationale, Paris
SP = State Papers (in F.O.)

Other abbreviations found in the references and bibliography are as follows.

b. = ibn (son of)
l.c. = loc. cit.
B.E.O. = Bulletin d'Études Orientales
B.S.O.A.S. = Bulletin of the School of Oriental and African Studies, London
B.S.S.A. = Bulletin of the Seismological Society of America
C.S.C.O. = Corpus Scriptores Christianorum Orientalium
C.S.H.B. = Corpus Scriptorum Historiae Byzantinae
E.I. = Encyclopedia of Islam, 1st edn. 4 vols. (1913–42), 2nd edn. (1954 in prog.) Leiden
G.M.S. = E.J.W. Gibb Memorial Series, London

References

IFAO = Institut Français d'Archéologie Orientale, Cairo
J.E.S.H.O. = *Journal of the Economic and Social History of the Orient*
J.R.A.S. = *Journal of the Royal Asiatic Society*, London
MIFAO = *Mémoires de l'Institut Français d'Archéologie Orientale*
P.O. = *Patrologia Orientalis*
Ss Arab. = *Scriptores Arabici* (in *C.S.C.O.*)

Manuscript sources
Arabic historical texts

'Abd al-Bāsiṭ, *Nail al-amal fī dhail al-duwal*. Bodleian Ms. Huntington 610.
al-Ahdal, *Tuḥfat al-zaman*. BL Or. 1345.
al-'Ainī, *'Iqd al-jumān fī tārīkh ahl al-zamān*. Paris Ms. Arabe 1543 (621–79 H) and 1544 (799–832 H).
al-'Ainī, *Tārīkh al-Badr*. BL Or. Add. 22,360.
Anon., Fragment of a Damascus chronicle. BL Or. Add. 23, 278.
Anon., *Jawāhir al-sulūk fi'l-khulafā wa 'l-mulūk*. BL Or. 6854.
Anon., *Manah al-rabāniyya*. Paris Ms. Ar. 1536.
Anon., *Zubdat al-ikhtiṣār*. BL Or. Add. 9972.
al-Dhahabī, *Tārīkh al-Islām*. BL Or. 49 and 50; Paris Ms. Ar. 1581.
Ḥasan b. Yaḥyā, *Tuḥfat al-zaman fī akhbār mulūk al-Yaman*, BL Or. 3330.
Ibn 'Abd al-Majīd, *Bahjat al-zaman fī tārīkh al-Yaman*, Paris Ms. Ar. 5977.
Ibn Duqmāq, *Nuzhat al-anām fī tārīkh al-Islām*. Paris Ms. Ar. 1597.
Ibn Duqmāq (attrib.) Continuation of Ibn Duqmāq. Paris Ms. Ar. 5762.
Ibn al-Ḥimṣī, *Ḥawādith al-zaman wa wafiyāt al-shuyūkh wa 'l-aqrān*. CUL Ms. Dd.11.2.
Ibn al-Jauzī, *Shudhūr al-'uqūd fī tārīkh al-'uhūd*. CUL Or. 1476.
Ibn Shākir al-Kutubī, *'Uyūn al-tawārīkh*. CUL Add. 2923; also BL Or. 3005; and Cairo Ms. 1497 *tārīkh*.
'Imād al-Dīn, *al-Rauḍ al-nāḍir*. Bodleian Ms. Huntington 172.
al-Jazarī, *Jawāhir al-sulūk*. Paris Ms. Ar. 6739. [See also abr. trans. J. Sauvaget, *La chronique de Damas d'al-Jazari*. Paris 1949.]
al-Khazrajī, *al-'Asjad al-masbūk*, Ms. in private collection of Qadi Isma'il b. 'Ali al-Akwa', Yemen; also a facsimile edition, printed San'a 1981.
Muḥammad b. 'Ali, *Tārīkh*, Paris Ms. Ar. 1507.
Muḥsin b. al-Ḥasan Abū Ṭālib, *Ṭib al-kisā*, Cairo Ms. 1347 *tārīkh*.
al-Muqrī, *al-Nathr al-jumān*. Dublin Ms. 4113.
al-Nuwairī, Shihāb al-Dīn, *Nihāyat al-arab fī funūn al-adab*. Paris Ms. Ar. 1578 (658–701 H); Leiden Ms. Or. 2-0 (701–20 H).
al-Qūṣī, Aḥmad b. al-'Allāma, *al-Barākīn wa 'l-zalāzil* (1907). Cairo Ms. *ṭabi'iyat* 114 (Taimūr).
al-Ṣafadī, *Nuzhat al-mālik wa 'l-mamlūk fī mukhtaṣar sīrat man waliya min al-mulūk*. BL Ms Or. Add. 23,326.
al-Sakhāwī, *Dhail duwal al-Islām*. Bodleian Ms. Marshall 611.
Sibṭ b. al-Jauzī, *Mir'āt al-zamān*. Paris Ms. Ar. 1506; also BL Or. Add. 23,277.

al-'Umarī, *al-Āthār al-jāliyya fī ḥawādith al-arḍiyya*. Ms Baghdad Academy; also BL Or. 6300.
al-'Umarī, *Masālik al-abṣār fī mamālik al-amṣār*. Cairo Ms. 559, *ma'ārif 'āmma*, cited in Taher (1979).
Yaḥyā b. al-Ḥusain, *Anbā al-zaman*. Cairo Mss. 17075 *ḥadīth* and 1347 *tārīkh*.

European languages

Anonymous (1592) in: *Libro ordini e terminazioni*, Lib. 3, fol. 135–6. Archives of Zakinthos, Zante.
Anonymous (1609) in: BBA. *MD.*, 78/698 (10 Muḥarram 1018).
Anonymous Pilgrim, in: Queen's College Oxford, Ms. 357, fol. 33ro.; also: BL Ms. Harleian 2333, fol. 30ro.
Chronicle of the Augustins of Freiburg, Breisgau, Badisches Landesarchiv, Ms. 11221.
Congreve (1741) Letter to his brother, Stafford County Record Office, Ms. D1057/M/9/4/15.
Donati, V. (1759) *Giornale del viaggio fatto in Levante nell'anno 1759*. Ms. Biblioteca Reale di Torino, Varie 291, vol. 1, fols. 67–72.
Peresc (1633) *Incendies souterrains, tremblements de terre etc. observés en Arabie et en Ethiopie*; récits des PP Gilles de Loches et Cesarée de Rosgo, recueillis par Peresc. 25 juillet 1633, In: *Divers mémoires très singuliers servant à l'histoire de France*. Ms. 1864, Bibliothèque Inguimbertine, Carpentras.

Printed editions of historical texts
Arabic sources (including Persian and Turkish)

'Abd al-Laṭīf, *Kitāb al-ifāda*, facs. ed. and trans. K.H. Zand, J.A. and I.E. Videan, *The Eastern Key*. London 1965; also trans. Silvestre de Sacy, *Relation de l'Egypte*. Paris 1810.
'Abd al-Qādir, *Waqā'i'-i manāzil-i Rūm*, ed. Mohibbul Hasan. London 1968 (in Persian).
Abu 'l-Faraj, *Chronography* (see Bar Hebraeus, under *Other languages*).
Abu 'l-Fidā, *al-Mukhtaṣar fī akhbār al-bashar*. 4 vols., Cairo 1907.
Abū Makhrama, *Qilādat al-naḥr fī wafāyāt a'yān al-dahr*, partial ed. and trans. L.O. Schurmann, *Political history of the Yemen at the beginning of the 16th century*. Amsterdam 1960.
Abū Makhrama, *Tārīkh thughr 'Adan*, ed. O. Lofgren. 2 vols., Leiden and Uppsala 1936–50.
Abū Shāma, *Kitāb al-rauḍatain*, ed. Muhammad Hilmi Muhammad Ahmad. 2 vols., Cairo 1956–62.
Abū Shāma, *Dhail 'alā al-rauḍatain*, ed. M. Zahid al-Kauthari. Cairo 1947.
Agapius, *Kitāb al-'unwān*, ed. A. Vasiliev, *P.O.*, **V–XI**. Paris 1909–15.
al-'Aidarūsī, *al-Nūr al-sāfir 'an akhbār al-qarn al-'āshir*, ed. M. al-Safar. Baghdad 1934.
al-'Ainī, *'Iqd al-jumān fī tārīkh ahl al-zamān*, ed. M.M. Amin. 4 vols., Cairo 1987–92.
al-'Alawī, Ṣāliḥ al-Ḥāmid, *Tārīkh Ḥaḍramaut*. 2 vols., Jidda 1388/1968.
Anon., *Bustān al-jāmi'*, ed. C. Cahen, *B.E.O.* 7–8 (1937–8), 113–58.
Anon., *Kitāb al-'uyūn wa 'l-ḥadā'iq fī akhbār al-ḥaqā'iq*, vol. **IV**, ed. O. Saidi. 2 vols., Damascus 1972–3.
Anon., *Tārīkh al-daulat al-Rasūliyya fī 'l-Yaman*, ed. A.M. al-

Hibshi. Damascus 1405/1984; also ed. H. Yajima, *A chronicle of the Rasulid dynasty of the Yemen*. Tokyo 1976.

Anon., Ms. fragment, ed. K.V. Zetterstéen, *Beiträge zur Geschichte der Mamlūkensultane*. Leiden 1919.

al-'Aẓīmī, *Tārīkh al-mukhtaṣar*, ed. C. Cahen, *Journal Asiatique* **230** (1938), 353–448.

al-Azraqī, *Kitāb akhbār Makka*, ed. F. Wüstenfeld, *Die Chroniken der Stadt Mekka*, I. Leipzig 1858.

al-Bakrī, *Kitāb muʿjam mā istaʿjam*, ed. F. Wüstenfeld. Göttingen and Paris 1876.

al-Bilādī, 'Atīq b. Ghaith, *Mu'jam qabā'il al-Hijāz*. Mecca n.d.

al-Bīrūnī, *al-Āthār al-baqiyya 'an al-qurūn al-khaliyya*, ed. C.E. Sachau. Leipzig 1876.

al-Dā'udī, continuator of al-Suyūtī, in *Kashf al-ṣalṣala*, pp. 62–4; also ed. al-Ḥāfiẓ (1982) [see *Printed Modern Works*].

al-Dhahabī, *Tārīkh al-Islām*, ed. H.D. al-Qudsi, V. Cairo n.d. (?1369/1950).

al-Dhahabī, *Kitāb al-'ibar fī khabar man ghabara*, ed. S. Munajjid. 5 vols., Kuwait 1960–6.

al-Dhahabī, *Kitāb duwal al-Islām*, ed. Hyderabad 1337/1919.

al-Dwaihī, *Tārīkh al-azminat*, ed. F. Taoutel, *al-Machriq*, année 44. Beirut 1951.

al-Fāriqī, Ibn al-Azraq, *Tārīkh-i Mayyāfāriqīn*, ed. Cairo 1959.

al-Fāsī, *Shafā al-gharām bi-akhbār al-balad al-ḥarām*, ed. F. Wüstenfeld, *Die Chroniken der Stadt Mekka*, II. Leipzig 1859.

Faṣīḥ Khwāfī, *Mujmil-Faṣīḥī*, ed. M. Farrukh. 3 vols., Mashhad 1340/1961 (in Persian).

al-Ghuzzī, Continuator of al-Suyuti, in al-Ḥāfiẓ (1982).

Ḥājjī Halīfe (Katip Çelebi), *Takvīm al-tavārīh*, ed. with continuation to 1146/1733 by I. Müterferrika. Istanbul (in Turkish).

al-Hamadhānī, *Takmilat tārīkh al-Ṭabarī*, ed. Muhammad Abu'l-Fadl Ibrahim, *Dhuyūl tārīkh al-Ṭabarī*, vol. **XI**, 187–489. Cairo 1979.

al-Hamdānī, *Ṣifat jazīrat al-'arab*, ed. M.A. al-Akwa'. Riyad 1944/1974.

al-Hamdānī, *al-Iklīl*, vol. **VIII**, ed. and trans. N.A. Faris. Princeton 1938, 1940.

al-Ḥarāzī, *Riyāḍ al-rayyāḥīn wa anbā al-awwālīn wa ahl al-bait al-ṭāhirīn*, ed. Husain b. 'Abdallah al-'Umari. Damascus 1406/1986.

al-Ḥubaishī, *Tārīkh Waṣṣāb, al-I'tibār fī 'l-tawārīkh wa 'l-āthār*, ed. 'Abdallah Muhammad al-Hibshi. San'a 1979.

Ibn 'Abd al-Ghanī, *Audaḥ al-ishārāt fī man tawallā miṣr wa 'l-qāhira*, ed. 'Abd al-Rahim. Cairo 1978.

Ibn Abi'l-Surūr, *Kitāb al-Kawākib al-sā'ira fī akhbār miṣr wa'l-qāhira*, trans. S. de Sacy, *Notices et extraits des manuscrits de la Bibliothèque du Roi*, I (Paris 1788), 165–280.

Ibn Abī Zar', *al-Anīs al-muṭrib bi-rauḍ al-qirṭās*, ed. C.J. Tornberg. Uppsala 1843; trans. A. Beaumier. Paris 1860.

Ibn al-Athīr, *al-Kāmil fī 'l-tārīkh*. 12 vols. in 6, Bulaq, Cairo 1303/1885.

Ibn al-Athīr, *al-Tārīkh al-bāhir fī 'l-daulat al-Atābikiyya*, ed. A. Talimat. Cairo 1963.

Ibn al-Athīr, *'Usd al-ghāba fī ma'rifat al-ṣaḥāba*. 5 vols., Cairo 1285–7/1868–70.

Ibn al-Bannā, see Makdisi (1956).

Ibn Baṭṭūṭa, *Travels*, trans. H.A.R. Gibb, I. Hakluyt Soc., London 1958.

Ibn al-Daiba', *Bughyat al-mustafīd fī tārīkh madīnat Zabīd*, ed.

'A.M. al-Hibshi. San'a 1979; Latin trans. C.T. Johannsen, *Historia Iemenai*. Bonn 1828.

Ibn al-Daiba', *al-Faḍl al-mazīd 'alā bughyat al-mustafīd*, ed. I. Salihiyya. Kuwait 1982; also ed. J. Chelhod. San'a 1982.

Ibn al-Daiba', *Kitāb qurrat al-'uyūn bi-akhbār al-Yaman al-maimūn*, ed. al-Akwa'. 2 vols., Cairo 1971, 1977.

Ibn al-Dawādārī, *Kanz al-durar wa jāmi' al-ghurar*, vols. **VI–IX**. Cairo 1960–72.

Ibn al-Furāt, *Tārīkh al-duwal wa 'l-mulūk*, vol. **IV/1–V/1**, ed. Hassan al-Shamma. Basra 1967–70; vol. **VIII**, ed. C.K. Zurayk and N. Izzedin. Beirut 1939.

Ibn al-Fuwaṭī (attrib.), *al-Ḥawādith al-jāmi'a wa 'l-tajārib al-nāfi'a fi 'l-mā'at al-sābi'a*, ed. M. Jawad. Baghdad 1932.

Ibn Ḥabīb, *Tadhkirat al-nabīh*, ed. M.M. Amin. 3 vols., Cairo 1976–86.

Ibn Ḥabīb, *Durrat al-aslāk fī daulat al-atrāk*, ed. P. Leander, *Le Monde Oriental*, Uppsala, vol. 7 (1907), 1–81.

Ibn Ḥajar, *Inbā al-ghumr bi-abnā al-'umr*, ed. H. Habashi. 3 vols., Cairo 1969–72.

Ibn Ḥātim, *Kitāb al-simṭ al-ghālī al-thaman fī akhbār al-mulūk min al-ghuzz bi 'l-Yaman*, ed. G.R. Smith. G.M.S., London 1974.

Ibn Hishām, *Kitāb sīrat Rasūl Allāh*, 2 vols., Cairo 1955; trans. A. Guillaume, *The Life of Muhammad*. London 1955.

Ibn 'Idharī, *al-Bayān al-mughrib*, ed. G.S. Colin and E.Lévi-Provençal. 2 vols., Leiden 1948, 1951.

Ibn al-'Imad, *Shadharāt al-dhahab*. 8 vols., ed. Cairo 1350–1/1931–2.

Ibn Iyās, *Badā'i' al-zuhūr fī waqā'i' al-duhūr*, ed. P. Kahle and M. Mostafa. 5 vols., Cairo and Wiesbaden 1960–75. Partly trans. G. Wiet, *Histoire des Mamlouks Circassiens*, II. Cairo 1945; and G. Wiet, *Journal d'un bourgeois du Caire*. 2 vols., Paris 1955, 1960.

Ibn al-Jauzī, *Kitāb al-muntazam fī tārīkh al-mulūk wa 'l-umam*, ed. Hyderabad 1359/1940.

Ibn Jubair, *The travels of Ibn Jubayr*, ed. W. Wright; 2nd edn, J.M. de Goeje. G.M.S., London 1907.

Ibn Kathīr, *al-Bidāya wa 'l-nihāya fī 'l-tārīkh*. 13 vols., ed. Cairo 1351–8/1932–9.

Ibn Majīd, *Kitāb al-fawā'id fī uṣūl al-baḥr wa 'l-qawā'id*, trans. G. Tibbetts, *Arab Navigation in the Indian Ocean before the coming of the Portuguese*. Oriental Trans. Fund, vol. 44. London 1971.

Ibn al-Muyassir, *Tārīkh Miṣr*, ed. A.F. Sayyid. IFAO, Cairo 1981.

Ibn Qāḍī Shuhba, *al-Kawākib al-duriyya fī 'l-sīrat al-Nūriyya*, ed. M. Zayidh. Beirut 1971.

Ibn al-Qalānisī, *Dhail*, ed. H. Amedroz, *History of Damascus, 363–555 a.h.* Leiden 1908.

Ibn Shaddād, *al-A'lāq al-khaṭīra fī dhikr umarā al-Shām wa 'l-Jazīra*, pt. II, ed. S. Dahhan, *Topographie Historique d'Ibn Šaddād*: Liban, Jordanie, Palestine. Damascus 1963.

Ibn al-Shiḥna, *Rauḍat al-munāẓir*, on the margins of Ibn al-Athīr, ed. Bulaq, Cairo 1874, vols. **VII–IX**.

Ibn Taghribirdī, *al-Nujūm al-zāhira*, ed. F.M. Shaltut et al. 16 vols., Cairo 1929–72; also ed. and partly trans. W. Popper, *The History of Egypt 1382–1469 A.D.* Berkeley and Los Angeles 1915–60.

Ibn Taghribirdī, *Ḥawādith al-duhūr fī madā 'l-ayyām wa 'l-shuhūr*, ed. W. Popper. Publs in Semitic Philol., vol. **VIII**, Berkeley and Los Angeles 1930–42; part. trans. W. Popper, New Haven 1967.

Ibn Ṭūlūn, *Mufākahat al-khillān fī ḥawādith al-zamān*, ed. M. Mostafa. 2 vols., Cairo 1962, 1964.

Ibn al-Wardī, *Tatimmat al-mukhtaṣar fī akhbār al-bashar*, ed. Beirut 1970.

Ibn Wāṣil, *Mufarrij al-kurūb fī akhbār banī Ayyūb*, vol. III, ed. M. Shayyal. Cairo 1962.

Ibn al-Wazīr, *Tārīkh al-Yaman* (*Ṭabaq al-ḥalwā*), ed. M. 'Abd al-Rahim Jazim. San'a 1405/1985; also Chester Beatty Library, Dublin, Ms. 4094.

Ibn Ẓāfir, *Akhbār al-duwal al-munqaṭi'a*, ed. A. Ferré. IFAO, Cairo 1972.

'Īsā b. Luṭf-Allāh, *Rauḥ al-rūḥ fī-mā jarā ba'd al-mā'a al-tāsi'a min al-fitan wa' l-futūḥ*. Facs. edn. San'a 1401/1981.

al-Isḥāqī al-Manūfī, *Laṭā'if akhbār al-uwal fī-mā taṣarrafa fī miṣr min arbāb al-duwal*, ed. Cairo 1310/1892; trans. Digeon (1781).

al-Jabartī, *'Ajā'ib al-āthār fī'l-tarājim wa'l-akhbār*, ed. H.M. Jauhar *et al.* 7 vols., Cairo 1958–67; trans. Chefik Mansur Bey *et al.* 9 vols., Cairo 1888–96.

al-Jauharī, *Nuzhat al-nufūs wa 'l-abdān fī tawārīkh al-zamān*, ed. H. Habashi. 3 vols., Cairo 1973.

al-Jauharī, *Inbā al-haṣr bi-abnā al-'aṣr*, ed. H. Hibshi. Cairo 1970.

al-Jazzār, *Taḥsīn al-manāzil min haul al-zalāzil*, ed. M. Taher, *Annales Islamologiques*, 12 (1974), 131–59.

al-Khazrajī, *al-'Uqud al-lu'lu'iyya fī tārīkh al-daulat al-Rasūliyya*, ed. Cairo 1329/1911; ed. Muhammad 'Asal. G.M.S., 2 vols., London 1913, 1918. See also trans. and commentary, Redhouse (1906–8).

al-Kutubī, see Ibn Shākir.

al-Makīn b. al-'Amīd, *Kitāb al-majmū' al-mubārak*, ed. and Latin trans. Th. Erpenius, *Historia Saracenica*. Leiden 1625; also French trans. P. Vattier. Paris 1657.

al-Makīn, ed. C. Cahen, La chronique des Ayyoubides d'al-Makīn ibn al-'Amīd. *B.E.O.*, 15 (1955–7), 109–84.

al-Maqrīzī, *Kitāb al-sulūk li-ma'rifat duwal al-mulūk*, ed. M. Mustafa Ziada and Sa'id A.F. 'Ashur. 4 vols. in 8, Cairo 1934–72; partial trans. in Quatremère (1837).

al-Maqrīzī, *Itti'āz al-ḥunafā*, ed. Jamal al-Din al-Shayyal. 3 vols., Cairo 1967–73.

al-Maqrīzī, *Kitāb al-Mawā'iz wa 'l-i'tibār fī dhikr al-khiṭaṭ wa'l-āthār* [*Khiṭaṭ*]. 2 vols., ed. Bulaq, Cairo 1270/1853–4; partial ed. G. Wiet, *MIFAO*, XXX–LIII. Cairo 1911–25.

al-Mas'ūdī, *Murūj al-dhahab*, ed. and trans. C. Barbier de Meynard and Pavet de Courteille. 9 vols., Société Asiatique, Paris 1861–77.

al-Mas'ūdī, *al-Tanbīh wa'l-ishrāf*, ed. J.M. de Goeje, Bibl. Geograph. Arab. vol. III. Leiden 1894.

Mufaḍḍal b. Abi 'l-Faḍā'il, *al-Nahj al-sadīd wa 'l-durr al-farīd fī-mā ba'd tārīkh Ibn al-'Amīd*, ed. and trans. E. Blochet, *Histoire des sultans mamlouks*. *P.O.*, XX. Paris 1929; and ed. S. Kortantamer, *Agypten und Syrien zwischen 1317 und 1341*. Freiburg 1973.

Mujīr al-Dīn, *Dhail*, ed. L.A. Mayer, A sequel to Mujir ad-Din's chronicle. *Jnl Palest. Orient. Soc.*, 11 (1931), 85–97.

al-Nahrawālī, *Kitāb al-i'lām bi-a'lām bait-allāh al-ḥarām*, ed. F. Wüstenfeld, *Die Chroniken der Stadt Mekka*, III. Leipzig 1857.

al-Nāṣirī, Aḥmad b. Khālid, *Kitāb al-istiqṣā li-akhbār duwal al-maghrib al-aqṣā*, ed. Casablanca 1954.

Nāṣir-i Khusrau, *Safar-nāma*, ed. and trans. C. Schefer. Paris 1881 (in Persian).

al-Nuwairī, Muḥammad b. Qāsim, *Kitāb al-Ilmān*, vol. IV, ed. A.S. Atiya. Hyderabad 1390/1970.

Oruç, trans. R.F. Kreutel, *Der Fromme Sultan Bayezid*. Graz 1978.

al-Qalqashandī, *Ma'āthir al-ināfa fī mu'ālim al-khilāfa*, ed. A.A. Farraj. 2 vols., Kuwait 1964.

al-Rāzī, *Tārīkh madīnat Ṣan'a*, ed. H. al-'Amri and 'A.J. Zakkar. San'a 1974.

Rif'at Pāshā, Ibrāhīm, *Mir'āt al-ḥaramain*. 2 vols., Cairo 1344/1925.

Sa'īd b. Biṭrīq (Eutychius), *Kitāb al-tārīkh*, ed. L. Cheikho *et al.*, *C.S.C.O.*, *Ss Arab.* ser. 3, vols. VI–VII. Paris 1906, 1909.

al-Sakhāwī, *al-Ḍau al-lāmi' li-ahl al-qarn al-tāsi'*. 12 vols., Cairo 1934–6.

al-Sālimī, *Tuhfat al-a'yān bi-sīrat ahl 'Umān*, ed. A.I. Atfish. 2 vols., Cairo 1380/1961.

al-Samhūdī, *al-Wafā bi-mā yajib li-ḥaḍrat al-muṣṭafā*. 4 vols., ed. Beirut 1981.

Ṣanī' al-Daula, *Muntazam-i Nāṣirī*. 3 vols., Lith. Tehran 1880–2 (in Persian).

Sāwīrus b. al-Muqaffa', *History of the Patriarchs of the Egyptian Church*, ed. and trans. A.S. Atiya *et al.* 4 vols., Publ. Soc. Archeol. Copte, Cairo 1943–74; also ed. and trans. B. Evetts, *P.O.*, V. Paris 1910.

al-Shādhilī, Continuator of al-Suyūṭī, in *Kashf al-ṣalṣala*, pp. 62–4.

Sibṭ b. al-'Ajami, *Kunūz al-dhahab fī tārīkh Ḥalab*, trans. J. Sauvaget, *Matériaux pour servir à l'histoire de la ville d'Alep*. Beirut 1950.

Sibṭ b. al-Jauzī, *Mir'āt al-zamān*, vol. VIII, printed ed. Hyderabad 1951.

al-Subkī, *Ṭabaqāt al-Shāfi'iyyat al-kubrā*, ed. 'Abd al-Fattah Muhammad al-Hulw and M.M. al-Tannahi. 8 vols., Cairo 1964–71.

Süheyl Efendi, *Tārīh-i Miṣr el-Kadīm*. Istanbul, 1142/1730 (in Turkish).

al-Suyūṭī, *Kashf al-ṣalṣala 'an waṣf al-zalzala*, ed. A. Sa'adani, Fez 1971; French trans. Nejjar (1974).

al-Suyūṭī, *Ḥusn al-muḥāḍara fī akhbār Miṣr wa'l-Qāhira*, ed. Cairo 1882.

al-Suyūṭī, *Tārīkh al-khulafā*, ed. 'Abd al-Hamid. Cairo 1964.

Synaxarium Alexandrinum, ed. I. Forget, *C.S.C.O.*, *Ss Arab.* ser. 3, vols. XVIII–XIX. Paris 1922, 1926.

al-Ṭabarī, *Tārīkh al-rusul wa 'l-mulūk*, ed. J.M. de Goeje. 3 vols. in 15, Leiden 1879–1901.

Takvimler 1153 AH /AD 1741. Istanbul (in Turkish).

Tarihī Takvimler, ed. O. Turan. Ankara 1954.

al-'Ulaimī, *al-Uns al-jalīl bi-tārīkh al-Quds wa 'l-Khalīl*, ed. Cairo 1283/1866; abr. trans. J. Sauvaire, *Histoire de Jerusalem et d'Hebron*. Paris 1876; also 2 vols., ed. Najaf 1388/1968.

'Umāra b. 'Ali, *al-Mufīd fī akhbār Ṣan'a wa Zabīd*, 3rd edn, al-Akwa' al-Hawali. San'a 1985; also trans. Kay (1892).

al-Wāsi'ī, *Tārīkh al-Yaman*. Cairo 1346/1927–8.

al-Yāfi'ī, *Mir'āt al-janān wa 'ibrat al-yaqzān*. 4 vols., Hyderabad 1337–39/1918–20.

Yaḥyā b. Ḥusain, *Ghāyat al-amānī fī akhbār al-qaṭr al-Yamānī*, ed. 'Abd al-Fattah 'Ashur. 2 vols., Cairo 1968.

Yaḥyā b. Sa'īd al-Anṭākī, *Dhail tārīkh Sa'īd b. Biṭrīq*, ed. L. Cheikho *et al.*, *C.S.C.O.*, *Ss Arab.* ser.3, vol. VII. Paris 1909; ed. and trans. I.I. Kratchovsky and A. Vasiliev, *P.O.*, XVIII/5 and XXIII/3, Paris 1924, 1932.

al-Ya'qūbī, *Tārīkh*, ed. T. Houtsma. 2 vols., Leiden 1883.

Yāqūt al-Ḥamawī, *Muʻjam al-buldān*, ed. F. Wüstenfeld. 4 vols., Leipzig 1866–73.

al-Yūnīnī, *Dhail mirʼāt al-zamān*. 4 vols., ed. Hyderabad 1954–61.

Zabāra, Muḥammad, *Aʼimmat al-Yaman biʼl-qarn al-rābiʻ ʻashar li ʼl-hijra*. Cairo 1376/1956.

Zabāra, Muḥammad, *Nashr al-ʻarf li-nubalā al-Yaman baʻd al-alf*. 2 vols., Cairo 1386/1966.

Other languages (Latin, Greek, Syriac and Hebrew etc.)

Agathias, *Agathiae Myrinei Historiarum*, ed. B.G. Niebuhr. *C.S.H.B.*, vol. 3. Bonn 1828.

Amadi, Francesco, *Chroniques d'Amadi et de Strambaldi*, vol. 1, *Chronique d'Amadi*, ed. Mas Latrie. Coll. de doc. inédit. sur l'hist. de France, Paris 1891.

Ammianus Marcellinus, *Ammiani Marcellini Rerum Gestarum*, trans. J.C. Rolfe. Loeb Classical Lib. London 1956.

Bar Hebraeus (Abu ʼl-Faraj), *Chronography*, trans. E.A.W. Budge. London 1932.

Benjamin of Tudela, *The itinerary of Benjamin of Tudela*, ed. and trans. M.N. Adler. London 1907.

Bustron, Florio, *Chronique de l'Île de Chypre*, ed. R. de Mas Latrie. Doc. inédits sur l'hist. de France, Paris 1886.

Cedrenus, *Compendium Historium*, ed. I. Bekker. *C.S.H.B.*, vol. 24, 2 vols., Bonn 1838–9.

Chronicon Pseudo-Dionysius, Incerti auctoris Chronicon pseudo-Dionysianum vulgo dictum, ed. J.B. Chabot. *C.S.C.O.*, vol. 121, *Scriptores Syri*, no. 66. Louvain 1949.

Dietrich von Schachten, in: R. Röhricht and K. Meisner (eds) *Deutsche Pilgerreisen nach dem Heiligen Lande*. Berlin 1880; new edn. Innsbruck 1900.

Elias Nisibinus, *Eliae Metropolitae Nisibeni opus chronologicum*, ed. E.W. Brooks. *C.S.C.O.*, vol. 63, *Scriptores Syri*, no. 23. Louvain 1910 (reprinted 1954).

Ethiopic Synaxarium (Maṣḥafa Sĕnkĕsâr), in: E.A.W. Budge, *The Book of the Saints of the Ethiopian Church*, 4 vols., Cambridge 1928.

Eusebius, S., *Hieronymi interpretatio chronicae Eusebii Pamphili*, ed. J.-P. Migne. *Patrologiae Cursus Completus*, Series Latina, vol. 27 (1846).

George the Monk, *George Monachos Hamartolos, Chronicle*, ed. C. de Boor. 2 vols., Leipzig 1904.

Gestes des Chiprois, ed. G. Raynaud. Publ. de la Soc. de l'Orient Latin, vol. 5. Paris 1887.

Glycas, M., *Michaelis Glycae annales*, ed. I. Bekker. *C.S.H.B.*, vol. 16, Bonn 1836.

John Cassian, *Collationes*, ed. M. Petschenig. *Corpus Scriptorum Ecclesiast. Latinorum*, vol. 13 (1866).

John Nikiu, *The Chronicle of John, Bishop of Nikiu*, trans. R.H. Charles. London 1916.

Malalas, J., *Chronographia Joannis Malalae. C.S.H.B.*, vol. 8. Bonn 1831.

Marchisius, *Marchisii Scribae Annales*, ed. R. Rohricht. Publs. de la Soc. de l'Orient Latin, ser. historiques, vol. 3. Geneva 1882.

Michael the Syrian, *Syrian Chronicle*, ed. and trans. J.B. Chabot. 4 vols., Brussels 1963.

Oliverus, *Oliveri Scolastici historia Damiatina*, ed. J.F. Gleditschi,

in: J.G. Eckhart, *Corpus historicum medii aevi*, vol. 2, part vii. Leipzig 1723.

Pachymeris, George, *De Michaele et Andronico Palaeologis*, ed. I. Bekker. *C.S.H.B.*, 2 vols., Bonn 1835.

Philostratus, *The Life of Apollonius of Tyana*, trans. F.C. Conybeare. Loeb Classical Lib. London 1960.

Severus of Antioch, *The sixth book of the selected letters of Severus Patriarch of Antiochia*, ed. and trans. E.W. Brooks. 4 vols., London 1902–4.

Solomon ben Yehuda, see Mann (1920).

Sozomenes, *Ecclesiastical history, History of the Church (324–440)*, trans. E. Walford. H.G. Bohn's Eccl. Lib., London 1855.

Strabo, *The Geography of Strabo*, ed. and trans. H.L. Jones. Loeb Classical Lib. London 1969.

Theophanes, *Theophanis Chronographia. C.S.H.B.*, vol. 26/i. Bonn 1839.

Trebellius Pollio, *Gallieni duo Trebellii Pollionis. Scriptores Hist. Augustae*, vol. 3, trans. D. Magie. Loeb Classical Lib. London 1968.

Printed modern works

Abir, M. (1980) *Ethiopia and the Red Sea. The rise and decline of the Solomonic dynasty and Muslim-European rivalry in the region*. London.

Abu-Lughod, J.L. (1971) *Cairo, 1001 years of The City Victorious*. Princeton.

Adams, R.D. (1983a) Incident at the Aswan Dam. *Nature*, **301**, 14.

Adams, R.D. (1983b) Seismograph networks in the Arab region: requirements, existing facilities and plans, in Cidlinský and Rouhban, pp. 127–57. Paris.

Adams, R.D. and Barazangi, M. (1984) Seismotectonics and seismology in the Arab region: A brief summary and future plans. *B.S.S.A.*, **74**, 1011–30.

Agamennone, G. (1896) Liste des tremblements de terre en Orient 1894–1896. *Bull. Météorol. et sismique de l'Observ. de Constantinople*. Constantinople [Istanbul].

Agamennone, G. (1900) Liste des tremblements de terre observés en Orient et en particulier dans l'Empire Ottomane pendant l'année 1896. *Beiträge zur Geophys.*, **4**, 118–99.

Agamennone, G. (1904) Le tremblement de terre dans l'île de Chypre du 29 juin 1896. *Beiträge zur Geophys.*, **6**, 108–37.

Agamennone, C. and Issel, A. (1894) Intorno ai fenomeni sismici osservati nell'Isola di Zante durante il 1893. *Ann. Uff. Centr. de Meteor. e Geod.*, vol. **15**. Rome.

Alexandre, P. (1990) *Les séismes en Europe occidentale de 394 à 1259. Nouveau catalogue critique*. Observatoire Royal de Belgique, Brussels.

Alsinawi, S. and Ghalib, H. (1975) Historical seismicity of Iraq. *B.S.S.A.*, **65**, 541–7.

Amari, M. (1854) *Storia dei musulmani di Sicilia*. 3 vols. Also in *Bibliotheca Arabo-Sicula*, 1857 and 1881.

Ambraseys, N.N. (1961) On the seismicity of South-West Asia: data from a XV-century Arabic manuscript. *Rev. Étude des Calamités*, **37**, 18–30.

Ambraseys, N.N. (1962) A note on the chronology of Willis's list of earthquakes. *B.S.S.A.*, **52**, 77–80.

Ambraseys, N.N. (1965) The seismic history of Cyprus. *Rev. de l'Union Internat. de Secours*, **3**, 25–48.

Ambraseys, N.N. (1971) Value of historical records of earthquakes. *Nature*, **232**, no.5310, 375–9.

Ambraseys, N.N. (1975) Studies in historical seismicity and tectonics. *Geodynamics Today*, pp. 7–16. Royal Society, London.

Ambraseys, N.N. (1983) Seismicity in the Arab region: retrieval and evaluation of macroseismic data, in Cidlinský and Rouhban, pp. 25–41. Paris.

Ambraseys, N.N. (1984) Material for the investigation of the seismicity of Tripolitania (Libya), in Brambati, A. and Slejko, D. (eds) *The O.G.S. Silver Anniversary Volume*, pp. 143–53. Trieste.

Ambraseys, N.N. (1988) Engineering seismology. *Earthq. eng. struct. dyn.*, **17**, 1–105.

Ambraseys, N.N. (1991) Seismicity of Egypt, in Cosgrove, J. and Jones, M. (eds), *Neotectonics and Resources*, pp. 148–57. London and New York.

Ambraseys, N.N. and Adams, R.D. (1986a) Seismicity of the Sudan. *B.S.S.A.*, **76**, 483–93.

Ambraseys, N.N. and Adams, R.D. (1986b) Seismicity of West Africa. *Annales Geophysicae*, **4** B.6, 679–702.

Ambraseys, N.N. and Adams, R.D. (1993) Seismicity of the Cyprus region. *Terra Nova*, **5**, 85–94.

Ambraseys, N.N. and Finkel, C.F. (1987) Seismicity of Turkey and neighbouring regions 1899–1915. *Annales Geophys.*, **5** B, 701–26.

Ambraseys, N.N. and Finkel, C. (1988) The Anatolian earthquake of 17 August 1668, in Lee *et al.*, pp. 173–80. San Diego.

Ambraseys, N.N. and Finkel, C. (1990) The Marmara Sea earthquake of 1509. *Terra Nova*, **2**, 167–74.

Ambraseys, N.N. and Jackson, J.A. (1990) Seismicity and associated strain of central Greece between 1890 and 1988. *Geophy. J. Int.*, **101**, 663–708.

Ambraseys, N.N. and Karcz, I. (1992) The earthquake of 1546 in the Holy Land. *Terra Nova*, **4**, 253–62.

Ambraseys, N.N. and Melville, C.P. (1982) *A history of Persian earthquakes*. Cambridge University Press.

Ambraseys, N.N. and Melville, C.P. (1983) Seismicity of Yemen. *Nature*, **303**, 321–3.

Ambraseys, N.N. and Melville, C.P. (1988) An analysis of the Eastern Mediterranean earthquake of 20 May 1202, in Lee *et al.*, pp. 181–200. San Diego.

Ambraseys, N.N. and Melville, C.P. (1989) Evidence for intraplate earthquakes in northwest Arabia. *B.S.S.A.*, **79**, 1279–81.

Ambraseys, N. *et al.* (1983) Notes on historical seismicity. *B.S.S.A.*, **73**, 1917–20.

Amiran (1951) see Kallner-Amiran.

Anonymous of Wittenberg (1546) *Zeittung von einen grossen und erschrecklichen Erdtbiden, etc.* Wittenberg.

Anonymous (1580) *Les espouventables tremblemans de terre . . . etc.* Lyon.

Anonymous (1687) *Histoire abrégée de l'Europe pour le mois de mars 1687.* Paris.

Anonymous (1693) *Unglücks-Chronica vieler grausahmer und erschrecklicher Erdbeben.* Hamburg.

Anonymous (1817) *Description of the Holy Mount Sina.* N. Glyka, Venice (in Greek).

Arieh, E. (1969) Red Sea earthquake. *B.S.S.A.*, **59**, 2117–18.

Arieh, E. *et al.* (1983) *Revised and updated catalogue of earthquakes in Israel and adjacent areas.* Publ. Z6/1216/83(3), Inst. Petrol. Research and Geophys. Holon.

Arvanitakis, G.L. (1903) Essai sur le climat de Jerusalem. *Bull. Inst. Egypt.*, ser. 4, 178–83. Alexandria.

Arvanitakis, G.L. (1911) Erdbeben in der Umgebung von Kairo am 22 August 1911. *Petermanns Geograph. Mitteil.*, **57**, 205.

Arya, A.S., Srivastava, L. and Gupta, S. (1985) Survey of damage during the Dhamar earthquake of 13 December 1982. *B.S.S.A.*, **75**, 597–610.

Aubin, J. (1973) Le royaume d'Ormuz au début du XVIe siècle. *Mare Luso-Indicum*, **2**, 77–179.

Auchterlonie, J.P. and Safadi, Y. (1977) *Union catalogue of Arabic serials and newspapers in British libraries.* London.

Ayalon, D. (1985) Regarding population estimates in the countries of medieval Islam. *J.E.S.H.O.*, **27**, 1–19.

Ayhan, E. and Alsan, E. (1988) *Turkiye ve dolaylari deprem katalogu.* Publ. Bogazici Univ., Istanbul.

BAAS (1899–) Seismological Investigation Reports. British Association for the Advancement of Science.

Baethgen, F. (1884) Fragmente Syrischer und Arabischer Historiker. *Abhandlungen f. die Kunde des Morgenlandes*, **VIII**/3. Leipzig.

Baratta, M. (1901) *I terremoti d'Italia.* Turin.

Barazangi, M. (1983) A summary of the seismotectonics of the Arab region, in Cidlinský and Rouhban (eds), pp. 43–58. Paris.

Barbiani, D.-G. and Babiani, B.-A. (1863) Mémoire sur les tremblements de terre dans l'île de Zante. *Mémoires de l'Acad. de Dijon*, **11**, 1–112.

Basset, R. (1881) Etudes sur l'histoire d'Éthiopie. *Journal Asiatique*, vol. **17**, 315–434, vol. **18**, 93–183, 285–389. Paris.

Batman, S. (1581) *The doome warning all men to the Judgemente.* London.

Baumgarten, M. (1594) *Peregrinationae.* Nurenberg. Also in *Churchill's Collection of Voyages*, The travels of Martin Baumgarten, vol. **I** (1704).

Bayle St John (1849) *Adventures in the Libyan desert.* London.

Bayle St John (1850) *Five views in the oasis of Siwa.* London.

Bel, J.M. (1988) *Architecture et peuple du Yemen.* Paris.

Belzoni, G. (1822) *Narrative of the operations and recent discoveries in Egypt and Libya.* 2 vols., London.

Ben-Menahem, A. (1979) Earthquake catalogue for the Middle East (92 B.C. to 1980 A.D.). *Bol. Geof. Teor. ed Applic.*, vol. **21**.

Ben-Menahem, A., Nur, A. and Vered, M. (1976) Tectonics, seismicity and structure of the Afro-Eurasian junction: the breaking of an incoherent plate. *Phys. of the Earth and Planet. Interiors*, **12**, 1–50.

Bentor, Y.K. (1989) Geological events in the Bible. *Terra Nova*, 1/iv, 326–38.

Berchem, M. van (1891) Notes d'archéologie arabe. Monuments et inscriptions Fatimites. *Journal Asiatique*, vol. **17**, 411–95, vol. **18**, 46–86.

Berloty, R. (1927) Sur le tremblement de terre de Palestine 11 juillet 1927. *Annales Obs. Ksara, Section Seism.*, pp. 62–93. Zahle.

Berryat, J. (1761) *Liste chronologique des éruptions de volcans, etc.*, *Coll. Acad.* vol. 6, 488–676. Paris.

Beuther, T. (1601) *Compendium terrae motuum.* Strassburg.

Bevis, R. (1973) *Bibliotheca Cisorientalia. An annotated checklist of early English travel books on the Near and Middle East.* Boston, Mass.

Blanchard-Lemée, M. (1984) Cuicul le 21 juillet 365: critiques archéologique et historique de l'idée de séisme, in *Tremblements de terre, histoire et archéologie*, pp. 207–19. Valbonne.

Bonito, N. (1691) *Terra tremante, etc.* Naples (repr. 1980).

Braslavskii, J. (1938) The earthquake that blocked the Jordan in 1546. *Zion*, 3, 323–36 (in Hebrew).

Brehm, A.E. (1862) *Reiseskizze aus Nord-Ost Afrika.* Jena.

Breitung, E. (1905) Seebeben im Arabischen Meer. *Annalen der Hydrographie & Maritimen Meteorologie*, Hamburg.

Brennan, P.F. (1963) *Earthquake damage: the Libyan earthquake of February 21 1963*. Internal Rept., Libyan-American Oil Co., Benghazi.

BRGM (1979) *Séismes de 1941 et 1955 à la frontière Arabo-Yemenite. Enquête macrosismique.* Bureau de Recherches Géologiques et Minières, Marseille.

Brice, W.C. (1981) *An historical atlas of Islam.* Leiden.

Brown, R.H. (1892) *The Fayum and Lake Moeris.* London.

Buist, G. (1857) The recent earthquake in the Levant. *Trans. Bombay Geogr. Soc.*, 13, xxxix–xli.

Burckhardt, J.L. (1829) *Travels in Arabia.* London.

Burgoyne, M.H. and Richards, D.S. (1987) *Mamluk Jerusalem: an architectural study.* London.

Burton, Sir R.F. (1893) *Personal narrative of a pilgrimage to al-Madinah and Meccah.* 2 vols., repr. New York 1964.

Bustronius, G. (1989) *Georgiou Voustroniou diigisis* (Greek cont. and trans.). Nicosia.

Caetani, L. (1913–23) *Chronographia Islamica ossia riassunto chronologico della storia di tutti i popoli musulmani.* 5 vols., Paris.

Cahen, C. (1965) *Jean Sauvaget's Introduction to the history of the Middle East. A bibliographical guide.* Berkeley and Los Angeles.

Cailliaud, F. (1826) *Voyage à Meroe, au fleuve Blanc, etc.* 4 vols., Paris.

Camp, V.E. *et al.* (1987) The Madinah eruption, Saudi Arabia: Magma mixing and simultaneous extrusion of three basaltic chemical types. *Bull. Volcan.*, 49, 489–508.

Campbell, A.S. (1968) The Barce (al Marj) earthquake of 1963, in Barr, F.T. (ed.), *Geology and archaeology of Northern Cyrenaica*, Libya, pp. 183–95. Petrol. Explor. Soc. of Libya, 10th Annual Field Conference, Amsterdam.

Carré, J.-M. (1956) *Voyageurs et écrivains français en Égypte.* 2 vols., IFAO, Cairo.

Chaine, M. (1925) *La chronologie des temps Chrétiens de l'Égypte et de l'Éthiopie.* Paris.

al-Chalabi, D. (1974) *Yasin bin Khayr Allah al-'Umari's Athar al-jaliya fi al-hawadith al-ardiya.* Najaf.

Chaplin, Th. (1883) Observations on the climate of Jerusalem. *Palest. Explor. Fund, Qtly Stat.*, pp. 11–32, Table 8. Jerusalem.

Cheikho, L. (1907) Les Archevêques du Sinai. *Mélanges de la Fac. Orient. de l'Université St-Joseph*, Beirut, 2, 408–21.

Choy, G. and Kind, R. (1987) Rupture complexity of a moderate-sized earthquake: the north Yemen earthquake of 13 December 1982. *B.S.S.A.*, 77, 28–46.

Christophorides, M. (1969) Oi seismoi kai oi seismikai doniseis en Kypro. *Kyriakos Logos* (1969), 227–9, 267–70, 323–5; (1970), 28–30, 89–91 (in Greek).

Cidlinský, K. and Rouhban, B.M. (1983) (eds) *Assessment and mitigation of earthquake risk in the Arab region.* Arab Fund for Economic and Social Development, and Islamic Development Bank. UNESCO, Paris.

Clédat, J. (1920, 1923) Notes sur l'Isthme de Suez. *Bull. I.F.A.O.*, 17, 103–19 and 21, 55–106.

Clément, J.-F. (1984) Jalâl al-Dîn al-Suyût'î, séismosophe, in *Tremblements de terre, histoire et archéologie*, pp. 253–87. Valbonne.

Combe, E., Sauvaget, J. and Wiet, G. (1944) *Répertoire Chronologique d'Épigraphie Arabe*, vol. XIII. Cairo.

Corbett, E.K. (1890) The history of the mosque of Amr at Old Cairo. *J.R.A.S.*, XXII, 759–800.

Cornu, G. (1983–5) *Atlas du monde Arabo-Islamique à l'époque classique, IXe–Xe siècles.* Leiden.

Coronelli, P. (1693) De'tremuoti accaduti dal diluvio, etc., in *Epitome Cosmografica*, pp. 286–324. Venice.

Coronelli, P. and Parisotti, A. (1688) *Isola di Rodi, geografica-storica antica e moderna.* Libraria della Geografia, Venice.

Coumbary, A. (1870a) Le tremblement de terre du 24 juin 1870. *Nouvelles Météorologiques*, 3, 200–1. Paris.

Coumbary, A. (1870b) Les tremblements de terre en Turquie et en Grèce. *Nouvelles Météorologiques*, 3, 100. Paris.

Craveri, M. (1870) Note sur un tremblement de terre ressenti à Alexandrie. *Bull. Soc. Géogr.*, 5th ser., 21, 233–4.

Creswell, K.A.C. (1932–40) *Early Muslim architecture.* 2 vols., Oxford.

Creswell, K.A.C. (1969) *Early Muslim architecture.* 2nd edn, vol. I (in two pts), Oxford.

Critikos, N. (1928) Le tremblement de terre de la mer de Crete du 26 juin 1926. *Annales Obs. Natl. Athens*, 10, 39–53.

Cyprianos, P. (1788) *Historia chronologiki tis nisou Kyprou eranisthisa ek diaforon historikon*, ed. N. Glyca, Venice. Also ed. Nicosia 1933 (in Greek).

Daressy, G. (1929) Ménélaïs et l'embouchure de la branche Canopique. *Rev. de l'Ancienne Égypte*, 2 no. 3, 20–32.

Daressy, G. (1934) Les branches du Nil sous la XVIIe dynastie. *Bull. Soc. R. Géogr. Égypte*, 18, 45–52.

Darrouzés, J. (1954, 1957, 1958) Notes pour servir à l'histoire de Chypre. *Kypriakai Spoudai*, vol. 17, 83–102; vol. 20, 33–63; vol. 22, 224–50. Nicosia.

Degg, M. (1993) The 1992 'Cairo Earthquake': Cause, effect and response. *Disasters*, 17/iii, 226–38.

Degrandpré, L. (1801) *Voyages dans l'Inde et au Bengale fait dans les années 1789 et 1790.* 2 vols., Paris.

Despeyroux, J. and Rouhban, B.M. (1983) *Le Séisme de Dhamar du 13 Decembre 1982.* UNESCO Field Mission rept., Paris.

Digeon, J.N. (1781) *Nouveaux contes turcs et arabes*, vol. 1, *Abrégé chronologique de l'histoire de la maison ottomane.* Paris.

Dinome, A. (1862) Voyage de M. de Beurmann en Afrique. *Nouvelles Annales de Voyages*, no. 4.

Dols, M.W. (1977) *The Black Death in the Middle East.* Princeton.

Dositheos, B. (1715) *Istoria peri ton en Ierosolymois patriarchisanton*, Lib. xi, cpt. 7, par. 3. Bucharest.

Dressdnische Gelehrte Anzeigen (1756) Nachrichten von Erdbeben, no. 2–40. Dresden.

Drioton, E. and Vandier, J. (1962) *Les peuples de l'Orient méditerranean, II: L'Égypte.* 4th edn, Paris.

Duveyrier, H. (1883) Tremblement de terre à Ghadames. *Comptes Rend. Séances Soc. géogr.* (Paris), pp. 454–5.

EERI (1992) *Cairo, Egypt, earthquake of October 12, 1992.* Special earthquake report, Earthquake Engineering Research Institute, Berkeley.

Eckenstein, L. (1921) *History of Sinai.* London.

Enlart, C. (1896) Notes sur le voyage de Nicolas de Martoni en Chypre. *Rev. l'Orient Latin*, 4, 623–32.

Eustratiades, S.A. (1932) Oi en Byzantio megaloi kai katastreptikoi seismoi. *Romanos o Melodos*, **1** (Paris), 121–6.

Fahmi, K.J., Ayar, B.S. and Al Salim, M.A. (1987) The Iraqi seismological network: current status and future trends. *Eos*, **68/x**, 137–43.

Fairhead, J.D. (1968) *The seismicity of the East African Rift system, 1955–1968*. M.Sc. Dissertation, Univ. of Newcastle.

Faulkner, R.O. (1969) *The Ancient Egyptian Pyramid Texts*. Oxford.

Fenech, A. and Froud, W. (1857) The Levant earthquake of 12th October 1856. *Trans. Bombay Geogr. Jl*, **13**, Appendix C, 9–12.

Ferrari, G. and Marmo, C. (1985) Il 'quando' del terremoto. *Quaderni storici*, NS **60/iii**, 691–715.

Fisher, W.B. (1978) *The Middle East*. 7th ed. London.

Forbes, R.H. (1921) Siwa oasis. *The Cairo Scientific Jl*, **10**, 1–8.

Freeman-Grenville, G.S.P. (1963) *The Muslim and Christian calendars*. London.

Fuchs, K. (1876) *Les volcans et les tremblements de terre*. Paris.

Fuchs, C.W.C. (1886) Statistik der Erdbeben von 1865–1885. *Sitzungsber. Kaiserlich. Akad. d. Wissenschaft.*, Bd. XCII, Heft. **3**, 215–625. Vienna.

Galanopoulos, A. (1941) Das Riesenbeben der messennischen Küste vom 27 August 1886. *Praktika Akademias Athenon*, **16**, 120–7.

Galanopoulos, A. (1953) Katalog der Erdbeben in Griechenland für die Zeit von 1879 bis 1892. *Annales Géol. Pays Helléniques*, **5**, 114–229. Athens.

Garçin, J.-C. (1976) *Un centre musulman de la Haute-Égypte médiévale: Qūṣ*. IFAO, Cairo.

Gassendi, P. (1657) *Vita Perescii*, trans. W. Rand, *The Mirrour of true Nobility ... the Life of N.C. Fabricius, Lord of Peiresk*. London.

Geelan, J.P.M. (1977) Arabic geographical names, in Grimwood-Jones *et al.*, pp. 129–33.

Ghawanmeh, Y. (1989) Earthquakes effects on Belad al-Sham settlements. Paper presented at IVe Congrés sur l'Histoire et l'Archéologie de Jordanie, 30 May–4 June, Lyon.

Gil, M. (1992) *A History of Palestine, 634–1099*. Cambridge.

Girardi, F. (1664) *Il Mercurio del decimosttimo secolo 1601–1650*. Naples.

Goby, J.-E. (1955) Phénomènes sismiques observés dans l'Isthme de Suez. *Bull. Soc. Étud. Histor. Géogr. Isthme Suez*, **6**, 33–6.

Goodchild, R. (1968) Earthquakes in Ancient Cyrenaica, in Barr, F.T. (ed.), *Geology and Archaeology of Northern Cyrenaica, Libya*, pp. 41–4. Amsterdam.

Gordon, D. and Engdahl, E. (1963) An instrumental study of the Libyan earthquake of Feb.21, 1963. *Earthquake Notes*, **34**, 50–6.

Gouin, P. (1979) *Earthquake history of Ethiopia and the Horn of Africa*. Publ. Internat. Develop. Research Centre, Ottowa.

Griffiths, J. (1805) *Travels in Europe, Asia Minor and Arabia*. London.

Grigoriadis, P. (1875) *The Holy monastery Tor Sina. Topographic and historical and administrative*. Jerusalem.

Grimwood-Jones, D., Hopwood, D. and Pearson, J.D. (1977) (eds) *Arab Islamic bibliography*. Hassocks.

Grumel, V. (1958) *Traité d'études Byzantines: La Chronologie*. Paris.

Gubbins, R.E. (1944) *Incidence of earthquakes in Iraq*. Unpubl. report no. 714/3226, Director of Central Irrigation, Baghdad.

Guest, A.R. (1902) A list of writers, books, and other authorities mentioned by El Maqrizi in his *Khiṭaṭ*. *J.R.A.S.*, **1902**, 103–25.

Guidoboni, E. (1989) (ed.) *I terremoti prima del Mille in Italia e nell'area mediterranea*. SGA, Bologna.

Gutenberg, B. and Richter, C. (1948) Deep-focus earthquakes in the Mediterranean region. *Geofis. Pura ed Applic.*, **12/iii–iv**, 1–4.

Gutenberg, B. and Richter, C. (1956) Magnitude and energy of earthquakes. *Annali di Geofis.*, **9**, 1–15.

al-Ḥāfiẓ, M.M. (1982) Nuṣūṣ ghair manshūra 'an al-zalāzil. *B.E.O.*, **32–3** (for 1980–1), 256–64 (in Arabic).

Al Hakeem, K. (1988) Studying of historical earthquakes activity in Syria (*sic*), in Margottini, C. and Serva, L. (eds), *Workshop on historical seismicity of central-eastern mediterranean region, Proceedings*, pp. 19–32. ENEA, Rome.

Hammer-Purgstall, J. von (1827–35), *Geschichte des Osmanischen Reiches*. 2nd edn, 10 vols., trans. K. Krokidas, *Historia tes Othomanikes Autokratorias*. Athens (1870).

Hasselquist, Fr. (1762) *Reise nach Palastina in den Jahren 1749–1752*, ed. C. Linnaeus, Rostock. Also in C.D. Cobham, *Excerpta Cypria*. Cambridge (1908).

Heuglin, M. von (1868) *Reise nach Abessinien in der Jahren 1861 und 1862*. Jena.

al-Ḥibshī, 'Abd-Allāh (1972) *Marāji' tārīkh al-Yaman*. Damascus (in Arabic).

al-Ḥibshī, A. (1982) article on earthquakes in *al-Thaura* (San'a), 19 December 1982 (in Arabic).

Hodgson, M.G.S. (1974) *The Venture of Islam*. 3 vols., Chicago and London.

Hoff, K. von (1840) Chronik der Erdbeben und Vulkan-Ausbrüche etc., *Gesch. Ueberlief. nachgew. natürl. Veraender. Erdoberfläche*, vols. **2–4**, Gotha.

Holt, P.M. (1968) Ottoman Egypt (1517–1798): an account of Arabic historical sources, in Holt, P.M. (ed.) *Political and social change in modern Egypt*, pp. 3–12. London.

Holt, P.M. (1973) *Studies in the history of the Near East*. London.

Holt, P.M. (1983) *Memoirs of a Syrian prince. Abu'l-Fidā', Sultan of Ḥamāh (672–723/1273–1331)*. Wiesbaden.

Holt, P.M. and Daly, M.W. (1988) *A history of the Sudan*. 4th edn, London.

Hopwood, D. (1977) Press and periodicals, in Grimwood-Jones *et al.*, pp. 101–22.

Hughes, R.E. (1988) The geotechnical study of soils used as structural materials in historic monuments, in Marinon, P.G. and Koukis, G.C. (eds), *The Engineering geology of ancient works, monuments and historical sites*, pp. 1041–8. Rotterdam.

Hume, W.F. (1928) The Corinth earthquake of 22 April 1928. *Bull. Soc. R. Géograph. d'Égypte*, **16**, 263–8.

Huntingford, G.W.B. (1989) *The Historical geography of Ethiopia from the first century AD to 1704*, ed. R. Pankhurst. Oxford.

Ibrahim, E.M. (1985) Seismic activity in different tectonic provinces of Egypt. *Bull. Internal. Inst. Seism. and Earthq. Eng.* (Tokyo), **21**, 139–76.

Irwin, R. (1986) *The Middle East in the Middle Ages. The early Mamluk Sultanate 1250–1382*. London.

Itier, J. (1848) *Journal d'un voyage en Chine en 1843–6*. 3 vols., Paris.

Jackson, J.A. and McKenzie, D.P. (1984) Active tectonics of the

Alpine–Himalayan belt between western Turkey and Pakistan. *Geophys. Jl.R. Astron. Soc.*, **77**, 185–264.

Jacques, F. and Bousquet, B. (1984) Le cataclysme du 21 juillet 365: phénomène régional ou catastrophe cosmique?, in *Tremblements de terre, histoire et archéologie*, pp. 183–98. Valbonne.

Jalil, W. (1986) Proposition de carte d'agressions sismiques de l'Arabie Saoudite. *Proc. 1er Colloque Natl. Génie Parasismique*, pp. 45–53. St-Rémy.

Jomard, E.F. (1848) Climat de l'Egypte. *Bull. Soc. Géograph.*, 3rd ser. vol. **9**, 276–8. Paris.

Kabābī, M.A. (1963) *Bandar 'Abbās va khalīj-i Fārs.* Tehran 1342 sh. (in Persian).

Kahhāla, U.R. (1968) *Mu'jam qabā'il al-'Arab al-qadīma wa'l-ḥadītha.* 3 vols., Beirut 1388 H (in Arabic).

Kaiser, A. (1922) Die Sinaiwüste. *Mitteilungen Thurgauischen Naturforschenden Gesellschaft*, no. 24.

Kallner-Amiran, D. (1951) A revised earthquake catalogue of Palestine. *Israel Explor. Jl*, vol. **1**, 223–46; vol. **2** (1952), 48–65.

Kárník, V. (1968, 1971) *Seismicity of the European area.* 2 vols., Prague and Dordrecht.

Kay, H.C. (1892) *Yaman, its early mediaeval history.* London.

Kebeasy, R.M. (1980) Seismicity and seismotectonics of Libya, in *The Geology of Libya*, vol. **III**, 954–62. Academic Press, New York.

Kebeasy, R.M., Maamoun, M. and Ibrahim, E. (1981) Aswan Lake induces earthquakes. *Bull. Internat. Inst. Seism.* (Tokyo), **19**, 155–60.

Keeling, B.F. (1906) Earthquakes in Egypt. *Cairo Sci. Jl*, **1**, 182–3.

Khatjikyan, L.S. (1955) *XV dari hayeren dzer'agreri hishatakaranner.* Erivan.

Klirides, N. (1936–7) Palaeographika apo tin Potamia. *Kypriaka Grammata*, **III** (in Greek).

Knolles, R. (1603) *The generall historie of the Turkeiis*, ed. E.A. Slip. London.

Kopp, H. (1983) Das Erdbeben vom Dezember 1982. Eine erste Bilanz. *Jemen-Report*, **14**/ii, 5–11.

Kriaris, P. (1930–7) *History of Crete.* Athens (in Greek).

Lamare, P. (1924) L'Arabie Heureuse: le Yemen. *La Géographie*, **1**, 1–23.

Lamare, P. (1930) Les manifestations volcaniques post-crétacées de la Mer Rouge et des pays limitrophes. *Mém. Soc. Géol. France*, vol. **14**/iii, 21–48.

Lampros, S. (1910) Enthymiseon syllogi proti. *Neos Hellinimnimon*, vol. **7**. Athens.

Landry, J. (1983) *Le séisme de Dhamar du 13.12.1982.* Rapp. BRGM, no.83. AEG-001, Orléans.

Lane, E.W. (1896) *Cairo fifty years ago.* London.

Lane-Poole, S. (1901) *History of Egypt.* London.

Langer, C.J., Bollinger, G.A. and Merghelani, H.M. (1987) Aftershocks of the December 13 1982, North Yemen earthquake: conjugate normal faulting in an extensional setting. *B.S.S.A.*, **77**, 2038–55.

Lee, W.H.K., Meyers, H. and Shimazaki, K. (1988) (eds) *Historical seismograms and earthquakes of the world.* Academic Press, San Diego.

Legrain, G. (1900) Rapports sur l'écroulements de onze colonnes dans la salle hypostyle du grand Temple d'Amon à Karnak le 3 Octobre 1899. *Annales du Service des Antiquités de l'Égypte*, **1**, 120–40. IFAO, Cairo.

Lepelley, Cl. (1984) L'Afrique du Nord et le séisme du 21 juillet 365: remarques méthodologiques et critiques, in *Tremblements de terre, histoire et archéologie*, pp. 199–206. Valbonne.

Leslie, J. (1986) A building education programme in North Yemen. *Disasters*, **10**/iii, 163–71.

Le Strange, G. (1887) Description of the Noble Sanctuary at Jerusalem in 1470 A.D. by Kamâl (or Shams) ad Dîn as Suyûtî. *J.R.A.S.*, **19**, 247–305.

Le Strange, G. (1890) *Palestine under the Moslems.* London.

Letronne, M. (1833) *La statue vocale de Memnon.* Paris.

Lipparini, T. (1940) Tettonica e geomorfologia della Tripolitania. *Boll. Soc. Geol. Italiana*, **59**, 251.

Little, D.P. (1970) *An introduction to Mamlūk historiography.* Frieburger Islamstudien, Wiesbaden.

Lorimer, J.G. (1915) *Gazetteer of the Persian Gulf, Oman and Central Arabia.* 4 Parts, Calcutta.

Luttrell, N. (1857) *A brief historical relation of State Affairs, from September 1678 to April 1714.* 6 vols., Oxford.

Lyons, H.G. (1907) Earthquakes in Egypt. *Survey Notes*, Cairo, **1**, no.10, 277–86.

Maamoun, M. (1979) Macroseismic observations of principal earthquakes in Egypt. *Helwan Observatory Bull.* no.183. Helwan.

Maamoun, M. and Ibrahim, E.M. (1978) Tectonic activity in Egypt as indicated by earthquakes. *Helwan Observatory Bull.* no.170. Helwan.

Maamoun, M. and El-Khashab, H. (1978) Seismic studies of the Shedwan, Red Sea, earthquake. *Bull. Helwan Inst. Astron. & Geophys.* no.171.

Maamoun, M., Allam, A. *et al.*, (1981) Neotectonics of Eastern Arabian region – regional studies. *Bull. Helwan Inst. Astron. & Geophys.*, **1**, ser. B, 173–235.

Maamoun, M., Meghahed, A. and Allam, A. (1980) The Gilf Kebir, Egypt, earthquake of December 9 1978. *Bull. Internat. Inst. Seism. Earthq. Eng.* (Tokyo), **18**, 1–9.

Maamoun, M., Meghahed, A. and Allam, A. (1984) Seismicity of Egypt. *Helwan Inst. Astron. and Geophys.*, Publ. **4**.

Maillet, B. de (1735) *Description de l'Égypte.* Paris.

Makdisi, G. (1956) Autograph diary of an eleventh-century historian of Baghdad, part II. *B.S.O.A.S.*, **18**, 239–60.

Mallet, R. (1853–5) Third report on the facts of earthquake phaenomena. *British Ass. Adv. Sci.*, pt 1 (1853), 1–176; pt 2 (1854), pp. 117–212; pt 3 (1855), pp. 1–326. London.

Mallik, D. and Morghem, F. (1977) Earthquake zoning in Libya. *Proc. 6th World Conf. Earthq. Eng.*, **II**, 487–8. Rome.

Mann, J. (1920–2) *Jews in Egypt and Palestine under the Fatimid Caliphs.* 2 vols., Oxford University Press.

Maravelakis, M.I. (1939) Beitrag zur kenntnis der Erdbebengeschichte in Griechenland und den nachbarländern aus den Erinnerungen. *Epistimoniki Epeteris*, **5**, 67–148. Aristotelian University of Salonica.

Maspero, G. (1914) *Manual of Egyptian archaeology.* New York.

Maspero, G. and Wiet, G. (1919) Liste de provinces, villes et villages d'Égypte cités dans les tomes I et II des *Khiṭāṭ* de Maqrizi. *MIFAO*, vol. **36**. Cairo.

Mayer, E. (1856a) Über das Erdbeben in Egypten am 12 Oct. 1856. *Zeitschr. für allg. Erdk.*, N.F., vol. **1**, no.6, 551–3.

Mayer, E. (1856b) Sur le tremblement de terre ressenti au Caire et à Boulak, 11 au 12 octobre 1856. *Jl de l'Union des Deux Mers*, **10**, 159–60.

Mazzarelli, G. (1947) Nota preliminare sul grande terremoto del Mediterraneo Orientale del 12 ottobre 1856. *Boll. Soc. dei natural.*, **55**, 120–2. Naples.

Meinardus, O. (1962) Atlas of Christian sites in Egypt. *Publs de la Soc. d'Archéol. Copte*, Cairo.

Melly, G. (1851) *Khartum and the Blue and White Niles*. 2 vols., London.

Melville, C. (1984a) Sismicité historique de la mer rouge septentrionale, in *Tremblements de terre, histoire et archéologie*, pp. 95–107. APDCA, Valbonne.

Melville, C.P. (1984b) The use of historical sources for seismic assessment, in Brambati, A. and Slejko, D. (eds) *The O.G.S. Silver Anniversary Volume*, pp. 109–19. Trieste.

Melville, C.P. (1985) The geography and intensity of earthquakes in Britain – the eighteenth century, in *Earthquake Engineering in Britain*, pp. 7–23. Inst. Civil Engineers, London.

Melville, C. (1989) Earthquake Hazard Atlas, 1. Israel. Book review in *Disasters*, **13**/iv, 371–4.

Melville, C. and Muir Wood, R. (1987) Robert Mallet, first modern seismologist, in Guidoboni, E. and Ferari, G. (eds), *Mallets's macroseismic survey of the Neapolitan earthquake of 16th December, 1857*. Vol.1, 17–48. SGA, Bologna.

Merghelani, H.M. and Gallanthine, S.K. (1980) Microearthquakes in the Tihamat-Asir region of Saudi Arabia. *B.S.S.A.*, **70**, 2291–3.

Milne, J. (1911) A catalogue of destructive earthquakes A.D. 7 to A.D. 1899. *Rept. B.A.A.S.*, 649–740. London.

Minami, K. (1965) *Relocation and reconstruction of the town of Barce, Libya*. UNESCO Rept. no.WS/0865.76–AVS. Paris.

Montessus de Ballore (1906) *La géographie séismologique*. Paris.

Muratori, L.A. (1723–51) *Rerum Italicarum Scriptores*. 25 vols., Milan.

Naval Intelligence Division (1946) *Western Arabia and the Red Sea*. Geographical Handbook Series, B.R. 527 (restricted). London.

Neale, F.A. (1852) *Eight years in Syria, Arabia and Asia Minor, from 1842 to 1850*. 2 vols., London.

Nectarios Kretes, Patriarch of Jerusalem (1677) *Epitomi tis ierokosmikis historias*, 1st edn; also ed. S. Milias, Venice (1758).

Neimann, von (1856) Bericht uber das Erdbeben zu Kairo 12 Oktober 1856. *Petermanns Mitteil.*, pp. 488–9.

Nejjar, S. (1974) (trans.) *Jalal ad-Din as-Suyut'i, Kashf aç-Çalçala 'an waçf az-Zalzala*. Rabat.

Netton, I.R. (1983) *Middle East materials in United Kingdom and Irish libraries: a directory*. London.

Neumann van Padang, M. (1963) *Catalogue of the active volcanoes of Arabia and the Indian Ocean*. Catalogue of the Active volcanoes of the World, part 16. Publ. Internat. Assoc. Volcanology, Rome.

Neuville, R. (1948) Heurs et malheurs des consuls de France à Jerusalem au XVII – XIX siècles. *Publ. M.E. Soc. of Jerusalem*, privately printed. Jerusalem.

Nowroozi, A.A. (1971) Seismo-tectonics of the Persian plateau, Eastern Turkey, Caucasus and Hindu-Kush regions. *B.S.S.A.*, **61**, 317–41.

Oppolzer, Th. Ritter (1962) *Canon of Eclipses*, trans. Owen Gingerich. Dover Publs, New York.

Palazzo, L. (1915) Cronistoria dei terremoti Etiopici anteriori all'anno 1913. *Bolletino della Soc. Sismolog. Italiana*, **19**, 293–350.

Palgrave, W.G. (1869) *Central and Eastern Arabia 1862–3*. London.

Panzac, D. (1985) *La Peste dans l'empire Ottoman*. Louvain.

Papamichalopoulos, C.N. (1912) *I moni tou orous Sina*. Athens and Cairo.

Papazachos, B. and Comninakis, P. (1982) *A catalogue of earthquakes in Greece and surrounding area*. Publ. no.5, Geophys. Lab. Univ. Thessaloniki, Thessaloniki.

Parker, R.B. and Sabin, R. (1974) *A practical guide to Islamic monuments in Cairo*. Cairo.

Pellat, C. (1986) *Cinq Calendriers Égyptiens*. Textes Arabes et Études Islamiques, vol. **XXVI**, IFAO, Cairo.

Pelly, L. (1863) Remarks on the tribes, trade and resources around the shore line of the Persian Gulf. *Trans. Bombay Geogr. Soc.*, **17**, 32–112.

Perrey, A. (1848) Note sur les tremblements de terre en 1847. *Bull. Acad. Sci. Bruxelles*, **15**, 442–54.

Perrey, A. (1850a) Sur les tremblements de terre ressentis dans la Péninsule Turco-Hellénique et en Syrie. *Mémoires Couronn. Acad. R. Sci. de Belgique*, vol. **23**. Brussels.

Perrey, A. (1850b) Note sur les tremblements de terre ressentis en 1849. *Bull. Acad. Sci. Bruxelles*, **17**, 216–35.

Perrey, A. (1852) Tremblements de terre ressentis en 1851. *Bull. Acad. R. Sci. Bruxelles*, **19** pt.1, 353–96.

Perrey, A. (1855) Note sur les tremblements de terre en 1854, avec suppléments pour les années antérieures 1852–1853. *Bull. Acad. R. Sci. Bruxelles*, **22** pt.1, 526–72.

Perrey, A. (1856) Note sur les tremblements de terre en 1855, avec suppléments pour les années antérieures 1843–1854. *Bull. Acad. R. Bruxelles*, **23**, 23–68.

Perrey, A. (1860) Note sur les tremblements de terre en 1857, avec suppléments pour les années antérieures. *Mém. Couron. Bruxelles*, **10** pt. 4, 1–114.

Perrey, A. (1862a) Note sur les tremblements de terre en 1858, avec suppléments pour les années antérieures 1843–. *Mém. Couron. Bruxelles*, **12** pt. 4, 3–68.

Perrey, A. (1862b) Note sur les tremblements de terre en 1859, avec suppléments pour les années antérieures 1843–. *Mém. Couron. Bruxelles*, **13** pt. 4, 3–78.

Perrey, A. (1862c) Note sur les tremblements de terre ressenti en 1860, avec suppléments pour les années antérieures. *Mém. Couron. Bruxelles*, **14** pt. 3, 1–74.

Perrey, A. (1864) Note sur les tremblements de terre en 1861, avec suppléments pour les années antérieures, 1843–. *Mém. Couron. Bruxelles*, **16** pt. 5, 3–112.

Perrey, A. (1865) Note sur les tremblements de terre en 1863, avec suppléments pour les années 1843–1862. *Mém. Couron. Bruxelles*, **17** pt. 5, 1–213.

Perrey, A. (1867) Note sur les tremblements de terre en 1865, avec suppléments pour les années antérieures de 1843–1861. *Mém. Couron. Bruxelles*, **19** pt. 3, 1–125.

Perrey, A. (1870) Note sur les tremblements de terre en 1866 et 1867, avec suppléments pour les années antérieures de 1843–1865. *Mém. Couron. Bruxelles*, **21** pt. 5, 1–223.

Perrey, A. (1872a) Note sur les tremblements de terre en 1868, avec suppléments pour les années antérieures de 1843–1867. *Mém. Couron. Bruxelles*, **22** pt. 3, 1–116.

Perrey, A. (1872b) Note sur les tremblements de terre en 1869, avec suppléments pour les années 1843–1868. *Mém. Couron. Bruxelles*, **22** pt. 4, 1–116.

Perrey, A. (1873) Note sur les tremblements de terre en 1870, avec supplément pour 1869. *Mém. Couron. Bruxelles*, **24** pt. 3, 1–146.

Petry, C.F. (1981) *The Civilian elite of Cairo in the later Middle Ages*. Princeton.

Pirazzoli, P.A. (1986) The early Byzantine tectonic paroxysm. *Zeitschr. Geomorphol. N. Folg. Suppl.*, **62**, 31–49.

Pirenne, J. (1958) *À la découverte de l'Arabie*. Paris.

Plafker, G., Agar, R., Asker, A. and Hanif, M. (1987) Surface effects and tectonic setting of the 13 December 1982 North Yemen earthquake. *B.S.S.A.*, **77**, 2018–37.

Plassard, J. (1960) Catalogue des séismes 1903–56 du Proche-Orient. *Observ. de Ksara*, unpubl. report.

Platakis, E. (1950) Earthquakes in Crete. *Kritika Chronika*, vol. **4**, no. 3. Heraklion (in Greek).

Playfair, R.L. (1861) On the outburst of a volcano near Edd, on the African coast of the Red Sea. *Proc. Geol. Soc. London*, **67**, 552–3.

Playfair, R. (1863) Eruption of a volcano near Edd. *Trans. Bombay Geogr. Soc.*, **16**, 41–5.

Poirier, J. and Taher, M. (1980) Historical seismicity of the Near and Middle East, etc. *B.S.S.A.*, **70**, 2185–201.

Popper, W. (1951) *The Cairo Nilometer*. Univ. of California Publ. in Semitic Philology, vol. **13**, Berkeley.

Preisigke, F. (1915–26) *Sammelbuch Griechischer Urkunden aus Aegypten*. 4 vols., Strasbourg, Berlin and Leipzig.

Quatremère, M. (1837–45) *Histoire des sultans Mamlouks de l'Égypte*. Oriental Trans. Fund, 4 vols. in 2, London.

Quittmeyer, R.C. and Jacob, K.H. (1979) Historical and modern seismicity of Pakistan, Afghanistan, northwestern India, and southeastern Iran. *B.S.S.A.*, **69**, 733–823.

Qureshi, I. (1968) The Jebel Dumbeir earthquake of 1966. *Sudan Notes and Records*, **49**, 128–35.

Qureshi, I. and Sadig, A. (1967) Earthquake and associate faulting in Central Sudan. *Nature*, **215**, 263–5.

Rabino, M.H.L. (1937) Le monastère de Sainte Catherine (Mont-Sinai); souvenirs épigraphiques des anciens pèlerins. *Bull. Soc. R. de Géogr. d'Égypte*, **19**, 21–126.

Raşīt, A. (1875) *Tārih-i Yemen ve San'ā*. 2 vols., Istanbul (in Turkish).

Raulin, V. (1869) *Description physique de l'isle de Créte*, Paris.

Redhouse, J.W. (1906–8) *The Pearl-strings; a history of the Resuliyy dynasty of Yemen*. Translation and Commentary, G.M.S., 3 vols., London.

Richard, Père (1657) *Relation de ce qui s'est passé de plus remarquable à Saint-Erini*. Paris.

Rothé, J.P. (1969) *The Seismicity of the Earth 1953–1965*. UNESCO, Paris.

Roux, G. (1934) Notes sur les tremblements de terre ressentis au Maroc avant 1933, *Mém. Soc. Sci. Natur. Maroc*, **29**, 41–71.

Rudolph, E. (1905) Katalog der im Jahre 1903 bekannt gewordenen Erdbeben. *Beitr. z. Geogphysik*, Ergänzungsband (Supplement) no. 3.

Runciman, Sir S. (1971) *A history of the Crusades*. 3 vols., Harmondsworth.

Russell, K.W. (1980) The earthquake of May 19, A.D. 363. *Bull. Amer. Schools of Oriental Res.*, **238**, 47–64.

Russell, K.W. (1981) *The earthquake chronology of ancient Palestine and Arabia from the 2nd to the 8th century A.D.* Publ. Dept. Anthropology, Univ. of Utah.

Russell, K.W. (1985) The earthquake chronology of Palestine and Northwest Arabia from the 2nd century through the mid-8th century A.D. *Bull. Amer. Schools of Oriental Res.*, **260**, 37–59.

Saartain, E.M. (1975) *Jalāl al-Dīn al-Suyūṭī, I. Biography and background*. Cambridge.

Sacy, Silvestre de (1787–98/9) *Notice et extraits des manuscrits de la Bibliothèque du Roi*. 4 vols., Paris.

St Ours, P. de (1976) *Seismic and volcanic risks in the Yemen Arab Republic*. Report no.19.77.GE-11130. UNDRO, Geneva.

Sāmī, Amīn Pāshā (1928) *Taqwīm al-Nīl*. Cairo (in Arabic).

Sathas, C.N. (1873) Chronigraphoi vasileiou Kyprou. *Messaioniki Vivliothiki*, vol. **II**. Athens.

Sayyid, A.F. (1974) *Sources de l'histoire du Yemen à l'époque musulmane*. IFAO, Cairo (in Arabic).

Schmidt, J. (1879) *Studien über Erdbeben*. Leipzig.

Schmidt, J. (1880) Vulkan-eruptionen und Erdbeben im Oriente. *Archiv fur mittel – und neugriechische Philologie hrsg. von Michael Deffner*, **I**, no. 1–2.

Schöne, W. (1940) *Die Relation des Jahres 1609*. Leipzig.

Schreiner, P. (1975, 1977, 1979) *Die byzantinischen Kleinchroniken*. 3 vols., Oesterr. Akad. Wissen., Vienna.

Serjeant, R.B. (1950) Materials for South Arabian history. *B.S.O.A.S.*, **13**/ii, 281–307; **13**/iii, 581–601.

Serjeant, R.B. (1962) Historians and historiography of Ḥaḍramawt. *B.S.O.A.S.*, **25**/ii, 239–61.

Serjeant, R.B. and Lewcock, R. (1983) *San'a', an Arabian Islamic city*. London.

Seyfart, J. (1756) *Algemeine Geschichte der Erdbeben*. Frankfurt.

Shahid, I. (1970) Pre-Islamic Arabia. *Cambridge History of Islam*, **I**, 3–29.

Shalem, N. (1928) Il terremoto in Palestina, Julio 1927. *Boll. Soc. Sism. Ital.*, **27**, 169–83.

Shalem, N. (1955) *The tremor of the 13th September 1954*. Rept. Geol. Survey Israel.

Sieberg, A. (1929) Das Korinther Erdbeben vom 22. April 1928. *Jenaische Zeitschrift f. Natürwiss.*, **64**, 1–20.

Sieberg, A. (1932a) Erdbebengeographie, in Gutenberg, B. *Handbuch der Geophysik*, vol. **IV**, Berlin.

Sieberg, A. (1932b) Erdbeben und Bruchschollenbau im Östlichen Mittelmeergebiet. *Denk. d. Medizin.-Natürwiss. Ges. zu Jena*, **18**, no. 2, Jena.

Simkin, T. *et al.* (1981) *Volcanoes of the World. A Regional Directory, Gazetteer, and Chronology of Volcanism during the last 10,000 years*. Smithsonian Institute, Stroudsburg, Pennsylvania.

Smith, G.R. (1978) *The Ayyubids and early Rasulids in the Yemen (567–694/1173–1295)*, II. G.M.S., London.

Smith, G.R. (1984) The Ṭāhirid sultans of the Yemen (858–923/1454–1517) and their historian Ibn al-Dayba'. *Journal of Semitic Studies*, **29**, 141–54.

Sonnini, C.S. (1799) *Travels in Upper and Lower Egypt*. 3 vols., London.

Sopwith, T. (1857) *Notes of a visit to Egypt*. London.

Sprenger, A. (1843) As-Soyuti's work on earthquakes. *J.R. Asiatic Soc. Bengal*, **12**, 741–9.

Stark, F. (1939) Some pre-Islamic inscriptions on the frankincense route in South Arabia. *J.R.A.S.*, 479–98.

Steindorff, G. (1904) Durch die Libysche Wüste zur Umonsoase. *Lande und Leute, Monographien zur Erdkunde*, no. 19. Bielefeld and Leipzig.

Striem, H.L. (1986) *Macroseismic effects in Israel due to earthquakes in the 1950s*. Rept. Inst. Petrol. Res. & Geophys. no. Z1/567/79 (43), Holon.

Stucchi, S. (1965) *L'Agora di Cirene. I, Collazione*. Monog. Arch. Libica, no. 7. Rome.

Stucchi, S. (1975) *Architettura Cirenaica*. Monog. Arch. Libica, no. 9. Rome.

Sykes, L. and Landisman, M. (1964) The seismicity of East Africa, the Gulf of Aden and the Arabian and Red Sea. *B.S.S.A.*, **54**, 1927–40.

Taher, M.A. (1974) Traité de la fortification des demeures contre l'horreur des séismes. *Annales Islamologiques*, **12**, 131–59.

Taher, M.A. (1979) *Corpus des textes arabes relatifs aux tremblements de terre de la conquête arabe au xii H./xviii J.C.* Doctoral thesis, Sorbonne, 2 vols., Paris.

Tamari, S. (1977) Two further inscriptions from Qal'at al-Jundi, in Rosen-Ayalon, M. (ed.) *Studies in memory of Gaston Wiet*, pp. 261–5. Hebrew Univ. Jerusalem.

Thevet, A. (1554–6) *Cosmographie du Levant*. Lyon.

Tholozan, J. (1879) Sur les tremblements de terre qui ont eu lieu en Orient, etc. *Comp. Rend. Acad. Sci.*, **88**, 1063–6.

Tibbetts, G.R. (1971) see Ibn Majīd [*Printed editions*].

Tibbetts, G.R. (1978) *Arabia in Early Maps*. Cambridge.

Tsafrir, Y. and Foerster, G. (1992) The dating of the 'Earthquake of the Sabbatical year' of 749 C.E. in Palestine. *B.S.O.A.S.*, **55**/ii, 231–5.

Tucker, W.F. (1981) Natural disasters and the peasantry in Mamluk Egypt. *J.E.S.H.O.*, **24**/ii, 215–24.

Tully, R. (1817) *Narrative of ten years' residence in Tripoli in Africa*. London.

Turner, W. (1820) *Journal of a tour in the Levant*. 3 vols., London.

United Arab Republic (1963) *A Bibliographical list of works about the Arabian Peninsula*. Cairo.

Vaněk, J., Kárník, V., Zátopek, A., Kondorskaya, N.V. *et al.*

(1962) Standardization of magnitude scales. *Izvest. Acad. Sci. U.S.S.R., Geophys. ser.*, **2**, 153–8.

Vincent, H. and Abel, F.M. (1922) *Jerusalem, II. Jerusalem nouvelle*. Paris.

Vogel, E. (1862) Edouard Vogel et son exploration de l'Afrique Centrale (by Ch. Grad). *Bull. Soc. de Géogr.*, 5e sér., **4**, 77–184.

Volkoff, O.V. (1971) *Le Caire, 969–1969*. IFAO, Cairo.

Volkoff, O.V. (1973) *Voyageurs russes en Égypte*. IFAO, Cairo.

Wagstaff, J.M. (1985) *The evolution of Middle Eastern landscapes. An outline to A.D. 1840*. London.

al-Waisī, Ḥusain b. Muḥammad (1962) *Al-Yaman al-kubrā*. Cairo (in Arabic).

Walther, B.S. (1805) *Die Erdbeben und Vulkane, physisch und historisch betrachtet*. Leipzig.

Willis, B. (1928) Earthquakes in the Holy Land. *B.S.S.A.*, **18**, 73–103.

Willis, B. (1933) Earthquakes in the Holy Land: a correction. *Science*, **77**, p. 351.

Wilson, R.T.O. (1989) *Gazetteer of historical north-west Yemen*. Hildesheim.

Winden, Johann (1613) *Neue Reysbeschreibung eines gefangenen Christen etc.* Nürnberg.

Wirth, P. (1966) Zur Byzantinischen Erdbebenlist. *Byzant. Forschungen* (Amsterdam), **1**, 393–9.

Witkam, J.J. (1989), Qur'ān fragments from Ḍawrān. Manuscripts and manuscripts, 6. *Manuscripts of the Middle East*, **4**, 155–74.

al-Wohaibi, Abdullah (1973) *The Northern Hejaz in the writings of the Arab geographers, 800–1150*. Beirut.

Wood, F. (1966) Seismicity of Alaska, in *The Prince William Sound, Alaska earthquake, 1946*. Washington.

Wulzinger, K. (1932) Die Apostelkirche und die Mehmedije zu Konstantinopel. *Byzantion*, **7**, 7–39.

Wüstenfeld, F. (1857–61) *Die Chroniken der Stadt Mekka*. 4 vols., Leipzig.

Zeki, A. (1908) in: *Les Pyramides*, 27 January 1908. Cairo.

Index

This index does not aim to be exhaustive. Summary information about the location and main characteristics of each earthquake in the catalogue can be found in Tables 2.1 and 3.1. A page number followed by 'n' indicates a reference to the footnotes on the page in question.